The Lure of Texas

The Lure of Texas

By

Robert D. Morritt

CAMBRIDGE SCHOLARS

PUBLISHING

The Lure of Texas, by Robert D. Morritt

This book first published 2011

Cambridge Scholars Publishing

12 Back Chapman Street, Newcastle upon Tyne, NE6 2XX, UK

British Library Cataloguing in Publication Data
A catalogue record for this book is available from the British Library

ISBN (10): 1-4438-2713-4, ISBN (13): 978-1-4438-2713-3

TABLE OF CONTENTS

FOREWORD

Texas has always fascinated me. In my youth I recall my first impressions of 'The West' on a Greyhound 'Scenicruiser' Bus for over three days in the days of Route 66 'America's Highway'.(Sadly no longer an Interstate Route.) Many people have vivid memories of the Highway.

I recall travelling through Missouri, Kansas, Oklahoma down through the Texas Panhandle to Amarillo then West through Texas.

I have always been interested in history. I later took Prehistory and Early Agriculture at University. From my own interest in this area I have tried to combine a history of the State together with earlier accounts of the Spanish influence in Texas and the harrowing tales of the early settlers of the frontiersmen and the real story of the creation of the State of Texas.

Whilst the book pertains to the present area known as Texas there are areas that *overlap as there were no 'State line' the French and Spanish conrol over the area is also portrayed within.* I consider that to omit the geographical extent of early explorers in the Southwest would detract offering the reader a more comprehensive overview of an incredible journey culminating eventually into the Lone Star State. The book concludes the pre-Statehood period. It pertains to go beyond the material found within modern travel guidebooks to give the reader an opportunity to 'be there' a chance to travel back in time to eyewitness the birth of modern Texas, won out of adversity the many who risked their lives to create by their own personal valor.

PREFACE

The book describes archaice hunters and food gathereers, the Mogollam and Mississipi mound-builder influence. Flint quarries discovered from the Panhandle to as far away as West Texas. The Paleo-Indian era evident from a Clovis campsite in Denton County circa 8000 BCE.

Cabeza de Vaca who reported rumours of gold in the north who as a member of a Spanish expedition was *shipwrecked on the Texas coast around 1528* swam to shore and lived as a slave of an Indian group and later escaped into the interior and became both a trader and a medicine man.Francisco Vazquez de Coronado, his search for the mythical "Seven Cities of Gold". De Niza who set out on his famous expedition in 1539, giving the earliest descriptions of people in the Southwest.

The Spanish Mission c.1655 built near Corpus Christi de la Isleta near present day Ysleta. Father Massanet who in 1690 founded San Francisco de los Tejas, the first Mission in East Texas. The French who in 1690 built a Fort at Biloxi Bay. The Comanche raids on Missions in 1758. All French territory west of the Mississippi River ceded by France to Spain after the French-Indian Wars (1754-1763).

The Anglo-American settlers in the 1820's and Moses and Stephen Austin, their colonization attempts espcially with many people from the new Louisiana Territory. Extensive coverage from the mid 1830's

Little known facts of the real first attempt to eastablish a Texan Republic, which predates the Battle of the Alamo. and takes a look at the efforts of Lt.Fannin and his troops with contempoary accounts of The Goliad massacre by the few that escaped.

Stephen Austin, Jim Bowie, The Battle of the Alamo and recent discovery of contemporary combatant's diary giving an accurate description of the death of Crockett at the Alamo.

PRE-COLUMBIAN HISTORY

Texas lies at the juncture of two major cultural spheres of Pre-Columbian North America, the Southwestern and the Plains areals. The area now covered by Texas comprised three major indigenous cultures which had reached their developmental peak prior to the arrival of European explorers and are known from archaeology. These are The Pueblo from the upper Rio Grande region, centered west of Texas; the Mound Builder culture of the Mississippi Valley region, centered east of Texas, ancestral to the Caddo nation; the civilizations of Mesoamerica, centered south of Texas. Influence of Teotihuacan in northern Mexico peaked around AD 500 and declined over the 8th to 10th centuries.

The Paleo-Indians that lived in Texas between 9200 – 6000 B.C. may have links to Clovis and Folsom cultures; these nomadic people hunted mammoths and bison latifrons using atlatls. They extracted Alibates flint from quarries in the panhandle region. Beginning during the 3rd millennium BC, the population of Texas increased despite experiencing a changing climate and the extinction of giant mammals. Many pictograms drawn on the walls of caves or on rocks are visible in the state, including at Hueco Tanks and Seminole Canyon.

Native Americans in East Texas began to settle in villages shortly after 500 BC, farming and building the first burial mounds. They were influenced by the Mound Builder civilizations that lived in the Mississippi basin. In the Trans-Pecos area, populations were influenced by Mogollon culture.

From the eighth century, the bow and arrow appeared in the region, manufacture of pottery developed and Native Americans increasingly depended on bison for survival. Obsidian objects found in various Texan sites attest of trade with cultures in present day Mexico and the Rocky Mountains.

No one culture was dominant in the present-day Texas region and many different peoples inhabited the area. Native American tribes that lived inside the boundaries of present-day Texas include the Alabama, Apache,

Atakapan, Bidai, Caddo, Coahuiltecan, Comanche, Cherokee, Choctaw, Coushatta, Hasinai, Jumano, Karankawa, Kickapoo, Kiowa, Tonkawa, and Wichita. The name ***Texas*** derives from *táysha?*, a word in the Caddoan language of the Hasinai, which means "friends" or "allies."

Native Americans determined the fate of European explorers and settlers depending on whether a tribe was friendly or warlike Friendly tribes taught newcomers how to grow indigenous crops, prepare foods, and hunting methods for wild game. Warlike tribes made life unpleasant, difficult and dangerous for explorers and settlers through their attacks and resistance to European conquest.

A remnant of the Choctaw tribe in East Texas still lives in the Mt. Tabor Community near Amberly, Texas. Currently, there are three federally-recognized Native American tribes which reside in Texas: the Alabama-Coushatta Tribes of Texas, the Kickapoo Traditional Tribe of Texas, and the Ysleta Del Sur Pueblo of Texas.

Texas prehistory extends back at least 11,200 years and is witnessed by a variety of Indian cultural remains.

This abbreviated list of site types suggests the diverse ways in which ancient Indians utilized the terrain and took advantage of its plant and animal resources. At many sites, the only surviving cultural remains are stone tools. If we are to understand what was done at the site, it is vital that we know the context of the stone tools. For instance, at kill-sites, proper excavation will usually discover projectile points and cutting or butchering tools in association with animal bones. The most common type of kill-site in Texas is the bison-kill of Paleo-Indian times, from 9200 to 6000 B.C. For example, at Bonfire Shelter, near Langtry in Val Verde County, excavations revealed a mass of bison remains associated with Folsom and Plainview points, accompanied by flakes and bifaces used for processing the slaughtered animals.

At quarries or lithic processing areas, controlled surface collection will often yield great numbers of large, crudely chipped bifaces, rocks in the early stage of tool-making known as quarry blanks. Rarely are projectile points or other finished tools found, since this is a locality where the basic levels of stone-working took place-securing good chipping materials, using a hammerstone to remove the rough exterior from the cobbles, and roughly shaping the blanks for further reduction elsewhere. Though they have long been ignored, lithic processing areas are important sites for archeological study, as they shed a great deal of light on a fundamental activity of prehistoric cultures. One quarry site that is open to the public is the famed Alibates Flint Quarries, on the Canadian River in the Panhandle.

Campsites are found throughout the state along streams or other water sources; most are "open occupation" sites, though caves and rockshelters were also often used for habitation. Many represent the villages of hunters and gatherers, whose foraging was the main way of life throughout Texas until later times, when farming was introduced in East Texas and in parts of the Panhandle and far West Texas. Campsites, the locales of daily life, were perhaps occupied for a few weeks or months before the group moved on to exploit the plant and animal foods of another area. These are the most common sites and contain great quantities of stone tools, flakes, and other debris

Context is particularly important in these sites. Even the surface collecting, by hobbyists, of an eroded campsite can ruin fragile patterns of tool distribution which, under controlled conditions, might tell the archeologist a great deal about site function and the ways in which different parts of the site were used. Excavation presents an even larger challenge.

Test pits can plumb the depths of the site, sometimes giving us information on the sequence of occupations by recovering stone tool types from different levels. However, to understand the behavior of the ancient inhabitants and the activities they carried out, a large block or open-area excavation is necessary. In it, we can plot in place the projectile points, scrapers, choppers, flakes, animal bones, snail shells, and other items and study the patterns of their horizontal distribution.

The distribution often shows the archeologist where tool-making took place, where animals were skinned and butchered, where bone tools were made or wooden spear shafts fashioned. The relationships of the tools to the areas of the site and to other stone tools provide, then, contextual information critical to archeological interpretation.

Much of what is found in Texas prehistoric sites is artifacts of chipped stone (such as projectile points), pottery, antler, bone, and shell. If excavated with systematic methods, the context of these artifacts-taken as a whole-provides a picture of ancient life at certain periods of time. Critical to our interpretations is the dating of these materials and contexts.

The most widely used method for absolute dates is radiocarbon analysis, in which associated organic remains (such as wood charcoal from a hearth) can be assayed to yield a date for cultural remains at the same level of the hearth.

Excavations can also provide relative dates by determining which styles of artifacts are earlier or later than others. Once a chronology is established at several regional sites, types of known dates can then be "cross-dated" by distinctive artifacts to other sites. Gradually, a framework of prehistoric cultures can be built up in a sequential fashion. In Texas, research has shown that in most regions, distinctive changes occurred in the shapes of projectile points through time

These artifacts, called "arrowheads" by non-archàologists, occur in two forms: as dart points-large, heavy points ("arrowheads") used on the tips of spears thrown with the spearthrower or atlatl, common in the Paleo-Indian and Archaic periods; and as arrow points ("bird points" to collectors)-tiny, thin points that tipped arrow shafts, often made of cane, when the bow and arrow was introduced to ancient Texas cultures around A.D. 700. This weapon appears to have wholly replaced the spearthrower, as it was more accurate and more effective at longer distances (tiny arrow tips could penetrate a bison, a man, or a smaller creature).

Many of the dart and arrow points can be sorted into "types" of distinctive shapes that are restricted in distribution in both time and space. This makes the points "time-sensitive"; they are often valuable chronological aids for archeological research.

Archeological work has continued in parts of Texas for more than eighty years. Some areas, such as Central Texas, have been intensively studied, and detailed archeological sequences of them have been established. In other regions, such as South Texas, research intensified only in the 1970s, and much remains to be learned about them. Cultural change proceeded at somewhat different rates over the vast area of Texas; in some regions, hunting and gathering cultures persisted throughout prehistory; in others, cultures with farming and settled village life appeared.

Research has divided the Texas archeological record into four general periods:

- Paleo-Indian (9200–6000 B.C.)
- 6000 B.C. to around the beginning of the Christian era
- Late Prehistoric (roughly A.D. 700–1600
- and Historic.

Paleo-Indian (9200–6000 B.C.) Archaic, Late Prehistoric (roughly A.D. 700–1600), and Historic.

Paleo-Indian

Although some claims have been made for greater antiquity, the earliest known inhabitants of the state, during the late Pleistocene (Ice Age), can be linked to the Clovis Complex around 9200 B.C. The distinctive Clovis fluted point is widespread and was used at least in some cases in mammoth hunting. A mammoth kill-site, Miami, is found in Roberts County in the Panhandle. *The Gault Site* in Central Texas has a Clovis occupation that includes incised pebbles, a blade core, and several Clovis points, including one made of Alibates material from the Canadian River quarries.

At a deeply buried Clovis campsite at the *Aubrey Site* in Denton County, a Clovis point, Clovis blades, and thousands of flakes were found. The *Lewisville Site near Denton* is another Clovis campsite. The Folsom Complex, around 8800–8200 B.C., is distinguished by Folsom fluted points and is known from sites where now-extinct forms of bison were killed and butchered (Bonfire) or from campsites (Adair-Steadman) where the points are found along with other stone tools.

The Clovis and Folsom materials might be considered to fall within the early part of this period. Although fluting ceases to be an important trait of Paleo-Indian points after Clovis and Folsom, later Paleo-Indian points maintain an overall lanceolate, parallel-sided form, often with careful parallel flaking and with the basal edges dulled to facilitate hafting. One unfluted type that may well be "early Paleo-Indian" is Midland, known from excavations at the Scharbauer Site, near Midland, in the early 1950s. A portion of a human cranium found at that site may be linked to this early cultural pattern.

Dalton and San Patrice points may date around 8000 B.C. in East Texas; Plainview points found from the Panhandle into South Texas date from around 8200–8000 B.C. and are associated with kills of Pleistocene bison at Plainview and Bonfire. By around 8000 B.C., the end of the Pleistocene, remnants of the animals of that era-mammoth, bison, camel, horse, sloth-disappeared. Climates became more like those of modern times, yet in some regions, group mobility and stone toolmaking continue to follow the patterns of earlier times. There is a great diversification of point types, several of which we still cannot precisely date, in post-Pleistocene, late Paleo-Indian times.

Early Archaic

Excavations done in the 1980s and 1990s at the Wilson-Leonard Site in Williamson County, Central Texas, may help to resolve some of these issues, as well as provide archeologists with a broader view of the cultural patterns associated with distinctive Paleo-Indian points.

The Scottsbluff points in East Texas are from around 6500 B.C.; in the lower Pecos and South Texas, hunters and gatherers used Golondrina points, radiocarbon dated at 7000 B.C. Excavations at Baker Cave, a dry rockshelter on the Devils River drainage, has yielded a wide array of information on the climate, which was essentially modern though probably drier, and the diet of peoples there 9000 years ago (the Golondrina Complex). A well-preserved cooking pit yielded the remains of small game, especially rabbits, rodents, and several species of snakes; the cave also yielded charred walnut and pecan hulls as well as other organic remains. The Angostura projectile point marks the end of the Paleo-Indian period; radiocarbon dates from the Wilson-Leonard Site and the Richard Beene Site near San Antonio date it at around 6800 B.C. The peoples who made these points, like the peoples of the Golondrina complex, were hunters and gatherers who used resources quite similar to those of the modern era.

Much of Texas prehistory is subsumed within a long time span of hunting and gathering cultural patterns known collectively as the Archaic. The period begins around 6000 B.C. and is notable for changes in the style of projectile points and tools, the distribution of site types, and the introduction of grinding implements and ground-stone ornaments, all reflecting a gradually increasing population that utilized abundant plant and animal resources of environments similar to those of modern times.

As noted earlier, the primary weapon during the Archaic was the spearthrower or atlatl, and the bow and arrow had not yet been introduced. Climatic patterns surely vacillated during the Archaic, though we have little detailed knowledge of them; a dry, warm episode known as the Altithermal (about 5000–3000 B.C.) was clearly present, but we are uncertain about its effects on local populations.

The details of the Archaic sequence vary from region to region within the state. In general, the span can be divided into Early, Middle, Late, and Transitional eras. Each period is represented by changes in cultural patterns, often including specific artifact forms, hunting patterns, types of site utilized, and other elements. In some regions we have enough information to subdivide these periods into "phases" or "intervals."

The Early Archaic (6000–2500 B.C.) is poorly known in its earliest phases, though a number of point and tool types can be linked to that era. In general, settlement appears more scattered than in later times, and populations were still rather small and quite mobile. There are broader relationships among several regions, as indicated by the widespread occurrence of distinctive points, such as the Martindale, Uvalde, Early Triangular, Andice, and Bell (the latter two part of a cultural pattern known as Calf Creek, which encompasses Oklahoma and parts of Arkansas).

The Middle Archaic *(2500 B.C.-1000 B.C.)* marks a time throughout the state of significant population increase, large numbers of sites, and abundant artifacts, especially projectile points of various forms. This appears to have been a time when Indian cultures became more specialized on a regional basin. For example, most regions appeared to be typified in the Middle Archaic by one or two distinctive points: Gary and Kent points in East Texas, for example, Pedernales in Central Texas, Langtry in the lower Pecos, and Tortugas in South Texas.

In some regions, specific types of site are present, especially the burned-rock middens of Central Texas (apparently used for cooking wild plants of various sorts, especially the bulbs of sotol) and shell middens on the Texas coast. Additionally, cemeteries with large numbers of interments begin to appear late in the period, perhaps reflecting territoriality on the part of some hunting and gathering societies. Similarly, trade connections are established and artifacts of stone and shell are brought from distant areas, especially Arkansas.

The Late Archaic (1000 B.C.-300 B.C.) sees the continuation of hunting and gathering in most of Texas, again distinguished by certain types of projectile points and stone tools. In East Texas, pre-Caddo sites mark the beginning of settled village life shortly after 500 B.C. Cemeteries are more notable in some regions, such as Southeast Texas. Bison appear to be an important game resource in Central Texas and in the lower Pecos, where another bison-kill occurs at Bonfire Shelter. Other bison-kills are known in the Panhandle and South Plains at this time.

The Transitional Archaic (300 B.C.-A.D. 700) marks an interval which in some ways is little more than a continuation of the Late Archaic. Still, it features distinctive point styles, such as Ensor, Darl, Frio, and Fairland. Although this period is important in the Archaic sequences of Central and lower Pecos Texas, it is not part of the East Texas archeological record, where village sites such as the George C. Davis Site of the Gibson Aspect make their initial appearance and fully develop only during the subsequent Late Prehistoric period.

These sites often have large mounds, flat-topped ones sometimes used to support structures and conical ones for burials. Such sites mark the introduction of, and reliance upon, agriculture which leads to this population growth and the emergence of social and political systems Many Indian rock art sites in Texas, especially in the lower Pecos, date from the Archaic. The Archaic pictographs in the lower Pecos can be recognized by the presence of spear-thrower motifs in the panels of polychrome Pecos River Style art. Studies using a specialized type of radiocarbon dating, known as AMS (accelerator mass spectrometry), suggest that this style may date as early as 4000 B.C.

Late Prehistoric

This period (A.D. 700 to historic times) is particularly noticeable in the archeological record throughout the state. The bow and arrow is introduced, along with other distinctive types of stone tools. Pottery is also present, even among hunters and gatherers in Central, South, and coastal Texas. Bison hunting appears to be very important in most regions. The occurrence of tiny arrow points marks the spread of the bow and arrow throughout the state.

Many local types develop: Livermore in the Trans-Pecos, Friley and Catahoula on the Texas-Louisiana border, Lott and Garza on the Llano

Estacado, and McGloin and Bulbar Stemmed on the coast. In some areas we can discern distinct shifts in arrow point styles through time, especially with Scallorn (Austin Phase) and, later, Perdiz (Toyah Phase) in Central Texas. The Toyah Phase is of particular interest because it represents a widespread bison-hunting tradition in Central and South Texas from around A.D. 1300–1600; in addition to Perdiz points, its material culture includes end scrapers for hide-working, beveled knives for bison butchering, and a distinctive bone-tempered ceramic.

On the central Gulf Coast, the Rockport Complex represents a population that may be ancestral to the historic Karankawas; these peoples hunted and fished along the bayshores and oftentimes moved inland to hunt bison. An asphalt-lined, thin-walled pottery called Rockport Ware is diagnostic of this complex. In the Rio Grande Delta, the Brownsville Complex is unique for its trade with frontier Mesoamerican cultures (e.g., the Huastecs of Veracruz), which began around A.D. 1300–1400. Representatives of the Brownsville Complex made shell beads and other ornaments in large numbers and traded these to the Huastecs in return for pottery vessels, jadeite ornaments, and obsidian, all found in Brownsville Complex sites in the lower Rio Grande valley.

Although a hunting and gathering continues in the Late Prehistoric as in the Archaic, the material culture, hunting patterns, settlement types and other facets of the era mark a fairly distinctive break with the past. In East Texas, agriculture provides the base for the Gibson Aspect, which marks the earliest Caddoan culture; mound building, specific types of pottery and arrow points, sedentary villages, ceremonial centers, and an established social hierarchy are salient features. Around A.D. 1200, Gibson gives way to the Fulton Aspect, which continues into the Historic era and is clearly linked with the Caddos. In the Panhandle and Llano Estacado, settled villages (also engaged in bison hunting) are found in the Antelope Creek Phase on the Canadian River around A.D. 1400 and in Andrews County. Village sites with links to southeast New Mexico appear around the same time.

In the Trans-Pecos, a sequence of settled horticulturists with strong ties to the Southwest Mogollón culture begins in the early centuries A.D. and develops more fully around A.D. 600. It is marked especially by pithouse dwellings. Down the Rio Grande, near Presidio, another center of agriculturally based villages, the Bravo Valley Aspect, dates to around A.D. 1200–1400.

One distinctive aspect of the Late Prehistoric was widespread, long-distance trade, best reflected in the distribution of obsidian artifacts in parts of Texas. Artifact-quality obsidian (volcanic glass, usually black to gray in color) does not occur in Texas. Yet at sites in deep South Texas, across Central Texas, and into the Panhandle, obsidian artifacts are often reported. The geologic origin of the obsidian can be traced using methods of nuclear chemistry, such as X-ray fluorescence or neutron activation analysis of a chemical "fingerprint" of trace minerals in a specimen.

The specimen can then be linked to a specific obsidian quarry. In the Panhandle, most of the obsidian comes from sources in the Jemez Mountains of northern Mexico, and was part of Plains-Pueblo trade in Late Prehistoric times. However, some of the obsidians found in South and Central Texas can be definitively traced to sources in southern Idaho (Malad), Wyoming (Obsidian Cliff), and central Mexico. These facts reflect long-distance trade networks, especially in the case of the Idaho and Wyoming obsidian, which were part of a north-south trade system through the Great Plains that continued into Historic times.

The transition from Late Prehistoric to Historic is difficult to discern in many parts of the state. The initial European expeditions had little, if any, effect on the native cultures, which were largely unchanged for another 100–150 years. Texas archeologists refer to this brief span as the "Protohistoric." Perhaps it is best exemplified by sites of the sixteenth and seventeenth centuries on Galveston Island and in South Texas, where no tangible evidence of contact (e.g., glass beads) is found. However, by the early eighteenth century most peoples of these areas were affected by the Spanish missions, and their cultures began to unravel.

Historic

The Historic era (after ca. A.D. 1600) marks the beginning of the end for the Indian cultures of the state. The Spanish and French brought change to both agriculturalists and hunter-gatherers, though the latter were decimated by the introduction of the Spanish mission system and the intrusion of Apache, and later, Comanche groups. Archeologically, we can recognize certain sites as Historic Caddo on the basis of their pottery and arrow points. Similarly, some arrow point types such as Harrell and Washita are found with historic hunter-gatherers and village farmers in north central Texas and the Panhandle.

Rock art sites incorporate such historic motifs as churches and horse-borne Indian warriors or Spaniards. With the advent of the Spanish mission system, the Indians who adopted mission life continued for a while to make stone tools, and a distinctive point type, Guerrero, is often found in missions, ranchos, and Indian campsites of that era. However, by the late eighteenth century, stone tools gave way to glass, and brass and iron points replaced those chipped from stone, thus signaling the end of an 11,000-year tradition.

Bibliography

Lawrence E. Aten, *Indians of the Upper Texas Coast*, (New York: Academic Press, 1983).

Thomas N. Campbell, *The Indians of Southern Texas and Northeastern Mexico: Selected Writings of Thomas Nolan Campbell* (Austin: Texas Archeological Research Laboratory, 1988).

David S. Dibble and Dessamae Lorrain, *Bonfire Shelter: A Stratified Bison Kill Site, Val Verde County, Texas*, Texas Memorial Museum Miscellaneous Papers [Austin: University of Texas, 1968]).

Thomas Hester, *Digging into South Texas Prehistory: A Guide for Amateur Archaeologists* (San Antonio: Corona Press, 1980). T. R. Hester et al., *From the Gulf to the Rio Grande: Human Adaptation in Central, South and Lower Pecos Texas* (Arkansas Archeological Survey Research Series 33 ,Fayetteville: University of Arkansas, 1989]). T. R. Hester et al., *Field Methods in Archaeology* (Palo Alto, California: Mayfield, 1975). J. L. Hofman et al., *From Clovis to Comanchero: Archeological Overview of the Southern Great Plains* (Arkansas Archeological Survey Research Series 35(Fayetteville: University of Arkansas, 1989). Jack T. Hughes, "Prehistoric Cultural Developments on the Texas High Plains," *Bulletin of the Texas Archeological Society* 60 (1989). Alex D.Krieger, *Culture Complexes and Chronology in Northern Texas with Extension of Puebloan Datings to the Mississippi Valle y* (University of Texas Publication 4640, Austin, 1946).

H. P. Newell and A. D. Krieger, *The George C. Davis Site, Cherokee County, Texas* (Society for American Archaeology Memoir 5, Menasha, Wisconsin, 1949). Parker Nunley, *A Field Guide to Archeological Sites in Texas* (Austin: Texas Monthly Press, 1989). R. A. Ricklis, *Aboriginal Life and Culture on the Upper Texas Coast: Archaeology at the Mitchell Ridge Site, 41GV66, Galveston Island* (Corpus Christi: Coastal Archaeological Research, 1994). Martin

Salinas, *The Indians of the Rio Grande Delta* (Austin: University of Texas Press, 1990). Harry Shafer and Jim Zintgraff, *Ancient Texans: Rock Art and Lifeways along the Lower Pecos* (Austin: Texas Monthly Press, 1986).

Alan Simmons, *Human Adaptations and Cultural Change in the Greater Southwest* (Arkansas Archeological Survey Research Series 32 [Fayetteville: University of Arkansas, 1989]). Dee Ann Storyl, *The Archeology and Bioarcheology of the Gulf Coastal Plain* (Arkansas Archeological Survey Research Series 38 [Fayetteville: University of Arkansas, 1990]). Dee Ann Suhm ., "An Introductory Handbook of Texas Archeology," *Bulletin of the Texas Archeological Society* 25 (1954). A. J. Taylor and C. L. Highley, *Archeological Investigations at the Loma Sandia Site (41LK28), a Prehistoric Cemetery and Campsite in Live Oak County, Texas* (Texas Archeological Research Laboratory Studies in Archeology 20, University of Texas at Austin, 1995). Ellen Sue Turner and Thomas R. Hester, *A Field Guide to Stone Artifacts of Texas Indians* (Austin: Texas Monthly Press, 1985).

Thomas R. Hester and Ellen Sue Turner

SOURCE *Handbook of Texas Online, s.v:*

SPANISH TEXAS

By a twist of fate, the next remarkable experience in Texas involved Englishmen. In 1568 John Hawkins placed several dozen of his countrymen ashore near Tampico after he suffered a nearly disastrous defeat by the Spanish fleet in Veracruz harbor.

One of the men, David Ingram, along with two others, walked from Pánuco along the Gulf and Atlantic coasts to near Cape Breton, Nova Scotia, and lived to tell about the trek. Ingram's experiences, like those of Álvarez de Pineda, Cabeza de Vaca, Coronado, and Moscoso did little to encourage immediate settlement in Texas. Legends, however, died hard for Spaniards.

Despite the reports of Coronado and Moscoso about the dearth of readily exploitable wealth in the north country, Tierra Nueva (as it was then called) continued to attract the attention of gold-hungry men in New Spain. Within five years after Coronado's return, the presumed wealth of Gran Quivira was again a topic of interest. Future explorers looked for the pearls of the Jumanos and the Great Kingdom of the Tejas.

In the second half of the sixteenth century, the discovery of rich silver deposits in northern Mexico drew Spaniards into the area like a magnet attracting nails. The first of the mining boomtowns was Zacatecas, where a mountain of silver ore was discovered in 1546. By the 1570s additional strikes brought about the founding of settlements in southern Chihuahua near the headwaters of the Río Conchos. The town of Santa Bárbara in that locale became the principal staging area for entradas into New Mexico and Texas.

From Handbook of Texas on line
Handbook of Texas Online, s.v.:
http://www.tshaonline.org/handbook/online/articles

Earliest Spanish Expeditions

Development of the mining frontier, however, had spawned the Chichimeca Wars (1550–1590s) and Spain's institutional response to them. The latter comprised missions and presidios, frontier agencies designed to convert Indians and pacify rebellious ones. From their inception, missions and presidios served as interrelated agencies of church and state and validated Spain's claim to frontier regions.

The primary function of missions was the propagation of the Catholic faith. But missions also served the state by Hispanicizing the Indian population, thereby making Indians in theory into tractable and tax-paying citizens. Presidios, as the nuclei of military presence on the frontier, were clearly agencies of the state, but they also served as necessary adjuncts to the security of the missions and the discipline of the neophytes within-a lesson painfully learned by the friars in the early days of Spanish Texas.

As the frontier of New Spain advanced northward, the northeastern field of missionary work, which encompassed Coahuila, Nuevo León, New Mexico, and Texas, became the primary responsibility of the Franciscan order. In the American Southwest the Friars Minor, the Little Brothers of St. Francis, established their first missions in New Mexico. Forty years after the return of the Coronado expedition, three Franciscans journeyed to Pueblo country with Francisco Sánchez Chamuscado (1581), and within a year all of the friars had suffered martyrdom.

A subsequent expedition to New Mexico, led by Antonio de Espejo and Diego Pérez de Luxán (1582), brought the first Europeans into extreme Southwest Texas as they returned to Chihuahua. In the following year, the Spanish crown authorized the pacification of New Mexico by a private individual, but no formal agreement was reached for a dozen years. Finally, Juan de Oñate received a contract (1595) that led to the occupation of New Mexico in 1598 and to the establishment of more than twenty missions by 1680.

In the eighty-two years of continuous Spanish presence in New Mexico, Texas along the Rio Grande from modern Presidio to El Paso bordered the path from the mines, missions, and ranches of northern Mexico to the land of the Pueblos. The interior of Texas, however, remained for the most part *tierra incognita*. It was penetrated by some, however. Between 1629 and 1654 expeditions from New Mexico entered Texas to search for Indians allegedly instructed in Christian doctrine by María de Agreda, the miraculously bilocating "Woman in Blue," and to establish trade with the Jumano Indians. The Jumanos carried aspects of Spanish material culture as far as the eastern Gulf Coast and East Texas.

Source; *Handbook of Texas on line* - , s.v.:
http://www.tshaonline.org/handbook/online/articles

FRAY MARCOS OF NIZA, 1539

Fray Marcos of Niza (Mark of Nice) and his companion Estevanico (a freed black African Moorish slave) Spanish authorities in Mexico City had sent him to verify earlier reports by Cabeza de Vaca and three other Spaniards of fertile valleys, prosperous Indian settlements and rumours of gold in the north. De Vaca was a member of a Spanish expedition that was *shipwrecked on the Texas coast around 1528*

A.D, De Vaca swam to shore and lived as a slave of an Indian group on the Texas coast for over a year when he escaped into the interior. He survived by becoming both a trader and a medicine man. After five years he became re-united with three other survivors of the same shipwreck. Together they traveled through western Texas and northern Mexico and were the first Europeans to set foot in the Southwest.

After several months of wandering they were reunited with their fellow countrymen in western and central Mexico. They were the first Europeans to enter Pimeria Alta which encompassed the region later known as Arizona. A land inhabited by the pueblo-dwelling Hopis. Indian farmers know as the Hohokam and nomadic Apaches.

In 1539 De Niza encountered, during his famous expedition to the north of the Gila. A tribe whom he designated as as the Pintados, from the fact that they painted their faces. These were probably the Papagoes, who are of the same nation as the Pimos and speak the same language.

Estevanico was killed by natives. De Niza then planted a cross in Cibola (Zuni) and took possession of the country in the name of Spain. De Niza journey was to observe the lifestyles of the native people in the Southwest, to estimate the size of their population, to describe the flora and fauna of the area. De Niza wrote, "I came to a pueblo, in green irrigated land, where many people came to meet me, both men and women.

They were clothed in cotton, some wearing skins of the cattle (buffalo), which in general they consider better material than cotton. In this pueblo they were all bedecked with turquoises, which hung from their noses and ears and which they called *cacona*."

His reports mentioned the fabled seven lost cities of Cibola in the unmapped region known as 'Arizuma' (as it the region was earlier referred to). He described how the natives,"Used gold, that to make into vessels and jewels for the ears, and into little blades with which they wipe away their sweat." This prompted the subsequent *visit* by Captain (later General) Francisco Vasquez de Coronado, between 1540 and 1542, with 300 Spaniards and 800 Indians, marched across Arizona, to the Moqui pueblos and beyond, along the route they had fought several battles with the natives. They also travelled as far as the southern plains of the territory later known as Kansas. The lives of the native people would never be the same.

In the year 1539 both traveled north from Culiacan and reached the Gila valley

Sources; De Neva and partial excerpts on Kino – *Ancient Peoples of the American Southwest*, Stephen Plog -2nd.Edition-1997, Thames & Hudson.

QUIVIRA

Quivira is a place first mentioned by Francisco Vazquez de Coronado in 1541, *who visited it* during his searches for the mythical "Seven Cities of Gold".

In 1539, the Spaniard Francisco Vasquez de Coronado led a large expedition north from Mexico to search for wealth instead of wealth, he found farming peoples living in the flat-roofed adobe towns in what are today Arizona and New Mexico. These were the Hopi, Zuni, and Rio Grande Pueblo Indians of today. Coronado was disappointed by the lack of wealth among the Pueblos, but he heard from an Indian the Spanish called "the Turk" of a wealthy civilization named *Quivira far to the east*, where the chief supposedly *drank from golden cups hanging from the trees.* Following this tale he led his army of more than one thousand Spaniards and Indian allies onto the Great Plains. The Turk was to guide him to Quivira.

Coronado traversed the panhandle of Texas in 1541. He found two groups of Indians, the Querechos and the Teyas. He was heading southeast when the Teyas told him that t*he Turk was taking him the wrong direction and that Quivira was to the north.* It appears the Turk was *luring the Spaniards away* from New Mexico with tales of wealth in Quivira, hoping perhaps that they would get lost in the vastness of the Plains. Coronado sent most of his slow-moving army back to New Mexico. With 30 mounted Spaniards, priests and Indian followers, the Turk, and Teya guides he forced into service, he set off northward to Quivira. After a march of more than thirty days, he found a large river, probably the Arkansas, and soon met several Indians hunting buffalo who led him to Quivira.

Apache, but they were not unified tribes in the modern sense. Early histories tended to call the different groups of Apaches and Navajos by various names that were not consistent from 1500s to the 1800s. The one consistent name was the name the *people called themselves which was Dine'.*[1]

Anthropologists and linguists since the late 19th century have accepted that the Navajo and Apache are close relatives of the northerne Dene peopless. There is a theory that a major volcanic eruption in the St. Elias range may have occurred between 310 and 525 AD as based on language differences between the Dene and the Apache a split between the two groups may have occurred due to this disaster. Traces of volcanic ash up to 30 centimetres deep have been found in the Yukon valley south of White River. There are versions told by the Hare, Mountain, and Slavey peoples who have partial memories of the catastrophe.

Father Leopold Osternman wrote in 1905, the Navajo called their northern relatives the "Dene nahodloni" meaning 'they who are also Navajos."The Navajo have a tradition that a party once set out to meet the Dene, but after living with them briefly, the Navajo returned home, unable to convince the Dene tp come south with them. Source; 'Drum Songs' – McGill-Queen's University Press, 1993- Kerry Abel.

Source; William B. Workman; Prehistory of the *Aishihik-Kluane Area, Southwest Yukon Territory, Mercury Series*, Archaeological Survey Paper no.74, Ottawa; National Museums of Canada, 1976.

[1] Dine': An indigenous reference to the Dine states the northern Tutchone, "*Tthi Ts'achan Huch'an*" recall their origins mention a series of eruptions 'Nelruna'.Scientific research dates these to 4,200 years ago. Other scientific data states that these same people lived along the edge of Ice-Age glaciers at a place now called Tatlmain Lake. Apparently the First Peoples came into this region soon after the ice sheets melted approximately 8-10,000 years ago.*(Source; -Jerry Alfred 'The Medicine Beat-1996).* The Navajos (Nabajos) appear to oriniated from the area of present day Yukon territory in Canada.

FRENCH AND SPANISH INTERESTS

When the Pueblo Revolt of 1680 forced prolonged abandonment of New Mexico, El Paso del Norte, where settlement west of the river had occurred as early as 1655, became the focal point of Spanish presence on the extreme northern frontier. Its sparse population was severely tested by the arrival of nearly 2,000 Spanish and Indian refugees from New Mexico. To accommodate the Indian exiles, Spaniards founded the first mission and pueblo within the present boundaries of Texas, Corpus Christi de la Isleta, at the site of modern Ysleta.

In the following years, efforts were made to found missions among the Jumano Indians at the junction of the Conchos and Rio Grande near the site of present-day Presidio. However, in the middle 1680s intelligence of French designs in the Gulf of Mexico downgraded the importance of that undertaking.

Concern centered on the extraordinary threat to Spanish realms posed by Frenchmen descending the Mississippi River from Canada. In 1682 René Robert Cavelier, Sieur de La Salle, explored the great river to its mouth and formally named the region "Louisiana" in honor of Louis XIV. La Salle established that the Mississippi emptied into the Gulf of Mexico between Spanish Florida and Pánuco

If Spain closed that gap and occupied the lower Mississippi valley, Canada would lose its access to the sea and be threatened from the south. On the other hand, a French colony placed on the lower part of the river would be close to the rich mines of New Spain. La Salle returned to France in 1683 to lay his colonization plans before the court at Versailles. After some delays, occasioned by a rival and international considerations, he received generous support for his plan to challenge the Spanish empire.

His expedition sailed from France in 1684, but because of misperceptions and the reading of *faulty maps* it *overshot the Mississippi by some 400 miles* and landed at Matagorda Bay in early 1685. By the time La Salle discovered that "his river" was not where he had landed, he was stranded on the Texas coast and had become the object of a resolute Spanish manhunt. The Spanish dispatched five sea and six land expeditions in search of La Salle's elusive colony at Fort St. Louis. In 1689

Alonso De León finally discovered its ruins on Garcitas Creek. By then La Salle, a victim of assassination, had been dead for two years. His colony had failed, as historian Robert S. Weddle has noted, not because of Spanish vigilance, but rather due to bad luck, hostile environment, fatal diseases, deadly Karankawa arrows, and the enmity of Frenchman toward Frenchman. But because La Salle overshot the Mississippi River and landed on the Texas coast, Spain reacted defensively. In reality, La Salle's expedition focused Spanish attention on East Texas, where success was hard to come by. The Spanish also underestimated the impending French threat to Louisiana.

La Salle

Failure to establish a colony – The First Mission in East Texas

Officials in Mexico City viewed the disastrous failure of La Salle's colony as evidence of God's "divine aid and favor." They were also impressed by Alonso De León's favorable account of the land and people of eastern Texas. Father Damián Massanet, who had accompanied the expedition of 1689, volunteered his services and those of his brethren of the College of Santa Cruz de Querétaro should missions be authorized among the Tejas. De Leon, however, wisely suggested the placement of presidios to bridge the gap between settlements in Coahuila and the proposed new mission field. If his suggestions had been accepted, the subsequent disasters in East Texas might well have been lessened, if not avoided. Nevertheless, officials in the capital concluded, with Massanet's endorsement, that military presence would impede spreading the Gospel.

In May 1690 Father Massanet founded San Francisco de los Tejas, *the first mission in East Texas*, perhaps near the site of modern Augusta in northeastern Houston County. Instead of leaving fifty soldiers at the mission, as recommended by Alonso De León, the expedition left only three and an equal number of friars when it returned to Coahuila. Before De León departed, Massanet secured a promise from the Indian governor that he would not mistreat the three padres. However, according to the late Franciscan historian Lino Gómez Canedo, San Francisco de los Tejas was abandoned in October 1693 because of floods, epidemics, the relaxation of the French challenge, and the contempt of the Indians.

The priests, expecting a rebellion, buried the cannons and bells, ignited the mission, and went back to Coahuila. But the mission effort in East Texas had familiarized Spaniards with the geography and Indians of Texas and convinced both church and government officials that future missions must be sustained by presidios and civilian settlements.

For one of the missionaries, Father Francisco Hidalgo, unfinished work among the Tejas Indians became a consuming passion. At the close of the seventeenth century, because of mining, ranching, and Hidalgo's unswerving commitment to missionary goals in East Texas, Spanish settlement in northern Mexico had advanced to the Rio Grande. Mission San Juan Bautista, which Weddle has appropriately called the "Gateway to Spanish Texas, was founded on January 1, 1700, at the site of present-day Guerrero, Coahuila. By then the French had resurrected La Salle's plan to settle the lower Mississippi valley.

Under the leadership of Pierre Le Moyne d'Iberville, a fort had been constructed at Biloxi Bay in 1699. In early 1700 the arrival of fresh supplies and reinforcements brought from France by Iberville strengthened French presence in Louisiana. Accompanying him was a relative by marriage, Louis Juchereau de St. Denis, *a Canadian-born adventurer who changed the course of Texas history*. Despite Iberville's best efforts, the crown colony of Louisiana suffered neglect due in part to resumption of the costly wars of Louis XIV (1702–13).

French Presence

In an effort to reduce royal expenses, Louisiana was assigned as a proprietary colony to a wealthy Frenchman, Antoine Crozat. Crozat's choice as governor of the colony fell on Antoine de La Mothe, Sieur de Cadillac, who arrived in Louisiana in May 1713.

Cadillac's duties were obvious. He had to run the colony in a businesslike manner and find a way to turn a profit. He concluded that the desired avenue of riches lay in establishing trade with New Spain, a commerce he knew to be forbidden by Spanish mercantile restrictions.

At that juncture, however, Cadillac received a letter, drafted two years earlier, from Father Hidalgo. Hidalgo had despaired of winning support from the Spanish crown for his plans to reestablish missions among the Hasinais. His letter asked the French governor to assist him in accomplishing that goal. For Cadillac it was an invitation to do God's work and his own. He called on St. Denis, who had become skilled in diplomacy and Indian languages and had already made explorations beyond the Red River, to make overtures to the Spanish.

St. Denis set out in late September and traveled to the site of Natchitoches, where he stored some of his merchandise. Twenty-two days after crossing the Sabine River, he reached the first Tejas villages and began trading for livestock, loosely managed by Indians, and buffalo hides. But the French adventurer had more ambitious goals. Justifying his actions on the grounds that he had not found Father Hidalgo living among the Indians, he planned to push on toward Spanish settlements .In July 1714,

St. Denis arrived at San Juan Bautista, bearing a French passport and news that the Tejas Indians earnestly desired the return of Spanish missionaries. His appearance alerted officials that the French were contacting Indians in East Texas and brought about the permanent Spanish occupation of Texas.

The presidio commander, Diego Ramón, placed St. Denis under house arrest while he awaited instructions from Mexico City. The Frenchman's genial incarceration provided an opportunity to court and win a promise of marriage from Ramón's beautiful granddaughter, Manuela Sánchez.

St. Denis was subsequently sent under guard to the capital, where he managed to charm the viceroy and obtain his freedom. Indeed, the chief executive of New Spain appointed St. Denis as commissary and guide for the expedition to reestablish Spanish missions in East Texas. St. Denis returned to San Juan Bautista, married his intended, and departed for Texas on April 27, 1716.

Spanish and French - Establishment of Missions

Capt. Domingo Ramón, a son of the presidio commander, led more than seventy priests, lay brothers, soldiers, and civilians along a path that became the Old San Antonio Road, or *camino real*.responsibility for founding new missions in Texas was divided equally between the missionary colleges of Querétaro and Zacatecas. The former supplied two of the most famous Franciscans in Texas history, Fathers Isidro Félix de Espinosa and Francisco Hidalgo, and from the College of Nuestra Señora de Guadalupe de Zacatecas came the renowned Father Antonio Margil de Jesús, whose cause for canonization is ongoing.

The religious contingent reestablished Mission San Francisco at a different site and renamed it San Francisco de los Neches. It also founded five new missions-Nuestra Señora de la Purísima Concepción de Acuña, Nuestra Señora de Guadalupe de los Nacogdoches, San José de los Nazonis, Nuestra Señora de los Dolores de los Ais, and San Miguel de Linares de los Adaes (near the site of present Robeline, Louisiana).

At the western edge of the mission field, Ramón built Nuestra Señora de los Dolores de los Tejas Presidio for his soldiers. The re-establishment of missions and a presidio in East Texas was very important historically, because it gave Spain *a valid claim to land north of the Rio Grande*, did much to determine that Texas would be Spanish, not French, and helped advance the eventual boundary between Texas and the United States to the Sabine River. For the enterprising St. Denis, the presence of Spaniards near Louisiana opened the door for the contraband trade that became a way of life on the Texas frontier.

Although the Spanish had received a friendly welcome from the Indians of East Texas, the latter were not willing to congregate in missions. To do so would require them to give up their idols and religious temples, as well as an established way of life. To try to force compliance with the wishes of the missionaries was viewed as foolish, for the strength of the military guard was inadequate. Unless Spanish presence could be augmented and a halfway station established between the Rio Grande and the eastward missions, the occupation of Texas seemed destined to go as it had gone in 1693.

Fortunately, the arrival in Mexico City of a new viceroy, the Marqués de Valero, signaled a positive trend for the struggling missions and presidio. The viceroy, at the instigation of Father Antonio de San Buenaventura y Olivares, made the suppression of illicit trade from Louisiana a primary objective. He also pledged support for the Franciscan missions in Texas. Father Olivares had earlier visited a site on the San Antonio River in 1709, and from that time forward he was determined to found a mission and civilian settlement there.

The viceroy gave formal approval for a halfway mission and presidio in late 1716, and assigned responsibility for their establishment to Martin de Alarcón, the governor of Coahuila and Texas.

A series of delays, however, occasioned in part by differences between Alarcón and Olivares, postponed definitive action until 1718. On May 1 on the San Antonio River the governor founded San Antonio de Valero Mission (later famous as the Alamo[qv]), and on May 5 established San Antonio de Béxar Presidio. As it turned out, these institutions came none too soon.

In 1719 a brief war in Europe between Spain and France affected Texas. From Natchitoches Philippe Blondel and six soldiers easily captured the poorly defended Adaes mission, but in the confusion a lay brother escaped. He spread fear of an impending French attack throughout the mission outposts all the way to Presidio Dolores. There Captain Ramón viewed the situation as untenable. He ordered immediate abandonment of the six missions and the military garrison, thus bringing an inglorious close to the second effort at establishing the Spanish in East Texas. Fortunately, the retreating Spaniards found refuge at the new settlements on the San Antonio River. Soon after his arrival there, Father Margil began building a second mission, San José y San Miguel de Aguayo, named for the governor of Coahuila and Texas who replaced Alarcón in 1719.

Spanish – Protection of Missions

The Marqués de San Miguel de Aguayo was a veteran soldier and a wealthy man, thanks to his wife, Ignacia Xaviera de Echeverez, who owned vast estates in Coahuila. At his own expense, he accepted the viceroy's request to deal with the troublesome situation in Texas and to undertake the establishment of missions in East Texas for a third time. In the early spring of 1721, Aguayo set out for Texas. He had recruited 500 men and collected 2,800 horses, 4,800 cattle, and 6,400 sheep and goats. Although livestock had accompanied previous entradas, Spanish ranching in Texas began with the arrival of these large herds in 1721. Aguayo reestablished all of the abandoned missions in East Texas; founded a new presidio, Nuestra Señora del Pilar, at Los Adaes; and made a lasting peace with St. Denis, who had become commandant of the French settlement at Natchitoches.

By the spring of 1722, Domingo Ramón, under orders of Aguayo, had begun construction of Nuestra Señora del Espíritu Santo de Zúñiga Mission and Nuestra Señora de Loreto Presidio, both commonly called La Bahía, at the site of La Salle's Fort St. Louis. When Aguayo departed from Texas in May 1722, he left a province that was soon separated from Coahuila.

It was secured by 268 soldiers at four presidios, two of which-at Los Adaes and La Bahía-were located where foreign aggression was most feared. Across the province were ten missions and hundreds of potential neophytes. Although there was a continuing problem of smuggling between French Louisiana and Spanish Texas, the threat of French dominance in Texas was ended.

Peace in Europe and its implications for America prompted the king of Spain to order reforms within New Spain in the interest of economy. To that end the viceroy sent Gen. Pedro de Rivera y Villalón to make a thorough inspection of frontier posts in Texas. Rivera's recommendations in 1727 led to the reduction of troop strength at Los Adaes and to the abandonment of Presidio Dolores. Without military support, the missions of San Francisco, Concepción, and San José were removed to the Colorado River and subsequently to San Antonio in 1731. Unfortunately for the settlement at San Antonio, the reduction of military strength in Texas left the settlement vulnerable to raids by the Apaches at the very time it was attempting to integrate an influx of very important settlers.

Historians have generally marked *the beginning of civilian settlement in San Antonio* with the arrival of fifty-five Canary Islanders on March 9, 1731, but it is well to remember that Alarcón's expedition of 1718 was not a purely military undertaking.

The presidio was to protect the missions in the area and serve as a way station between the Rio Grande and the East Texas missions. San Antonio was also to be the site of a Spanish villa (San Fernando de Béxar), and to this end Alarcón had recruited frontiersmen from Coahuila and Nuevo León. As Jesús F. de la Teja has demonstrated, "From its founding in 1718 to 1731, forty-seven couples married and 107 children were baptized at Mission Valero."

Thus, a first generation of native Bexareños was already living in San Antonio by 1731. The arrival of the Canary Island settlers temporarily disrupted the racially harmonious community, but the threat of Indian attacks and frontier isolation soon eroded the Islanders' aloofness. Indian attacks by the Apaches began in the 1720s and worsened in the 1760s with the appearance of the Comanches at San Antonio.

In the summer of 1768, Bexareños had to fight off a twenty-two-day siege without outside assistance. Again, as De la Teja has remarked, "Shared roles, kinship ties, and the frontier experience tied much of Bexar's population into a dynamic community." Oakah L. Jones, Jr., has similarly demonstrated that outside of San Antonio there was little by way of class rivalry among the Spanish population in Texas.

THE FRENCH AT RED ROVER

Spanish establishment of towns and missions

As San Antonio grew to become the most important and viable community in Spanish Texas, relations with the French remained generally peaceful. Arroyo Hondo, a small stream between Los Adaes and Natchitoches, became the accepted boundary between the two empires until the cession of Louisiana to Spain in 1762.

The French, however, were much more adept than the Spanish in establishing trade with Texas Indians. At the Red River, the French won favor with the Kadodachos (a Caddo group) and other Indians of the region. They were likewise successful in establishing trade with the Wichitas and Tawakonis in northern Texas.

To the south, French traders crossed the Sabine in the early 1730s and initiated contact with the Orcoquizas and Bidais along the lower San Jacinto and Trinity rivers. In response to that continuing threat, the Spanish in 1756 established San Agustín de Ahumada Presidio and Nuestra Señora de la Luz Mission in the general vicinity of Anahuac. Neither establishment enjoyed much success and both were abandoned within fifteen years.

More serious external threats stemmed from continued French incursions and the English colony of Georgia in 1733. Spaniards considered Georgia to be a part of Florida and viewed the presence of foreigners there as a threat to unsettled Gulf Coast regions. Concern deepened with the outbreak of war between England and Spain in 1739, a conflict that presaged the larger War of the Austrian Succession (1740–48). Spain regarded the Costa del Seno Mexicano, between Tampico and Matagorda Bay, as especially vulnerable to English and French designs.

To counter those threats, the viceroy of New Spain approved the appointment of José de Escandón as military commander and governor of the recently established province of Nuevo Santander. In a burst of energy, Escandón and his lieutenants founded twenty-four towns and fifteen missions between 1747 and 1755. Escandón also moved the mission and presidio at La Bahía from their second location on the Guadalupe River to the site of present Goliad.

The founding of Laredo completed Escandón's colonization. Less successful was a simultaneous effort to expand the mission system along the San Gabriel River, called San Xavier by the Spanish, to the northeast of San Antonio. In the area of modern Rockdale, Texas, three Franciscan missions and a presidio (collectively known as the San Xavier missions) were founded in the late 1740s and early 1750s. Trouble between the presidio commander and the friars, disease, drought, and hostile Apaches doomed this undertaking to eventual failure.

In 1757 the entire properties of the San Xavier missions were redirected toward a new effort on the San Saba River, the most disastrous mission-extension project in the history of Spanish Texas. In 1757 near the site of present Menard, Col. Diego Ortiz Parrilla and Franciscan missionaries headed by Father Alonso Giraldo de Terreros set up San Luis de las Amarillas Presidio and Santa Cruz de San Sabá Mission.

The settlements were established at the request of Lipan Apaches, a group generally hostile to the Spanish. The Lipans, because of increasing pressure from the Comanches and their allies, especially the Wichitas, had been forced to set aside their dislike for the Spanish. But in forging an alliance of convenience, they further infuriated their Indian adversaries.

MISSIONS ATTACKED BY COMANCHES AND ALLIES

In March 1758 the Comanches[1] and their allies, including Tejas, Tonkawa, and Bidai Indians, attacked the mission in great numbers. They pillaged and burned it and killed eight persons, including Father Terreros. Santa Cruz de San Sabá became the only mission in Texas to be destroyed by outright Indian attack.

In the following year, a major assault on the San Sabá livestock herd cost the lives of twenty soldiers and resulted in the loss of more than 700 horses, mules, and cattle. Several months later, Colonel Ortiz Parrilla led a sizable force of presidial soldiers and Apache allies on a punitive expedition to the Red River, only to suffer a humiliating defeat. The San Sabá mission was never rebuilt. Missionary efforts on behalf of the Apaches continued for a time from outposts on the upper Nueces River, but they, too, had to be abandoned within a few years

Presidio San Luis de las Amarillas was strengthened and maintained for another ten years, but it was abandoned in 1770. Thus, Spanish efforts toward expansion in Texas during the years 1731–62 were a failure, except at La Bahía and along the lower Rio Grande. Missions and presidios, although proven frontier institutions, had clearly failed north of San Antonio.

Elsewhere, events in Europe and the course of the French and Indian War in America (1754–63) had a profound effect on Texas. Before the Treaty

---

[1] The Comanche were of a wild disposition. Together with their southern neighbours the Yamparike (Spanish Llamparicas) they would raid Apache rancherias on the plains. They exchanged horses for French guns with the eartern Pawness and trade bison meat and skins for corn, vegetables and blankets in the middle of the 18th century they turned south to the Red River area and beyond. They wouls set up Tipis's often near larger streams or in the breaks of the arroyos where grass, water and kindling for their fires was available. The Horse represented wealth to the Comanches. Source; *Great Indian Tribes*, Daniel Jacobson, Hamond Incorporated. 1970.

of Paris, France had ceded to Spain *all French territory west of the Mississippi River.* Louisiana had been a total financial disaster for France, and it became no less burdensome for Spain. Lacking resources and manpower, Spain could not defend or develop this vast territory, and it brought assertive English rather than the tolerant French to its borders. Twenty years later the English were supplanted by even more aggressive citizens of the United States. For Spain, adjustments were necessary to meet changing circumstances that confronted its North American empire.

In the 1760s Charles III, an exceptionally able king of Spain, ordered a sweeping inspection of New Spain and its frontier provinces to determine what reforms were needed. The *visitador general* of importance for Texas was the Marqués de Rubí, who arrived in 1767. Rubí visited all major settlements in Texas and concluded that they could not be maintained without consolidation and new resources. His recommendations called for the strengthening of San Antonio by moving the settlers and remaining missions in East Texas to that site.

Second, he suggested a new approach aimed at developing friendly relations with the powerful Comanches and Wichitas, coupled with a war of extermination against their enemies, the Apaches. The forced abandonment of East Texas was a bitter blow to its inhabitants. Because they were no longer needed as a barrier against the French, their settlements were removed to shore up the defenses of San Antonio.

Closure of Missions in East Texas

Comanche Raids and flooding

In the summer of 1773, Baron Juan María de Ripperdá, the governor of Texas, reluctantly ordered the closure of missions in East Texas, removal of the entire Spanish population, and designation of San Antonio as capital in place of Los Adaes. The displaced settlers often complained that native Bexareños denied them arable and irrigable lands. They petitioned to return to their homes, but were denied that right by Hugo Oconór, then inspector general on the frontier.

Not dissuaded, the East Texans selected Antonio Gil Ibarvo as their spokesman and dispatched him to Mexico City for an audience with Viceroy Antonio María de Bucareli y Ursúa. Ibarvo was only partially successful. The viceroy did permit the resettlement of East Texas, but would not consent to dwellings within 100 leagues of Natchitoches, Louisiana. Still, the refugees in San Antonio viewed any concession as encouraging.

In August 1774 they founded the settlement of *Bucareli* on the Trinity River at a site in what is now Madison County. The town had attracted 347 inhabitants by 1777, but it was plagued by floods and Comanche raids. Without authorization, the population moved again in 1779 to Nacogdoches. That community, along with La Bahía, San Antonio, and Laredo, became another permanent settlement within the boundaries of the future state of Texas.

Also during the mid-1770s, as part of Charles III's sweeping administrative reforms, the *Provincias Internas* (1776) administration was instituted for the northern provinces of New Spain, including Texas. That political arrangement, however, was in a state of flux to the end of the colonial era. The Internal Provinces began as an independent jurisdiction responsible only to the king, then passed under the viceroy of New Spain, and were finally divided into shifting military districts.

One of the major military concerns of the period was the enactment of Rubí's recommendations for an Indian policy that would destroy the Apaches and make peace with the northern tribes. Spaniards wisely called on French agents, who had remained in Louisiana after its transfer to Spain, to help implement that policy. *Athanase de Mézières*, the former son-in-law of St. Denis, masterminded this arrangement. He was fluent in several Indian languages.

He enjoyed some success with the Wichitas-less with the Comanches-but was never able to piece together a general alliance of the northern tribes against the Apaches. Nevertheless, the talented frontiersman became highly respected by Spanish officials, and in 1779 they appointed him governor of Texas. He died, however, at San Antonio before assuming the post. Troubles with the Comanches continued until 1785, when a treaty with them brought peace that lasted, with a few interruptions, for nearly thirty years. Less successful was a final wave of missionary effort among the Karankawas and other coastal tribes.

This undertaking was not new. The original Espíritu Santo Mission had been established for Karankawas. In 1754 Nuestra Señora del Rosario de los Cujanes Mission had been founded a short distance west of the site of present Goliad by the Zacatecan Franciscans. Indians in the area, however, were difficult neophytes. After they fled the mission in 1781, it was closed. For a time, the nearby La Bahía Mission served Indians of the region. The Franciscans again directed their attention to Rosario when former mission Indians asked that it be reopened. Their request was granted in December 1789.

Four years later, the last of the Texas missions, Nuestra Señora del Refugio, was founded for Indians who had deserted the missions of La Bahía and Rosario. Because of unhealthful conditions, friars moved the mission in 1794 and finally reestablished it in January 1795 at the site of present Refugio. Success among the coastal tribes had proved difficult, and in the end the missions at La Bahía, Rosario, and Refugio, as elsewhere in Texas outside of San Antonio, disappointed their Franciscan founders.

Louisiana Purchase to the United States

Spanish concern – American Intruders

The final crisis for Spanish Texas came from problems associated with the retrocession of Louisiana to France (1800), its subsequent sale to the United States (1803), and the struggle for Mexican independence, begun in 1810 by Father Miguel Hidalgo y Costilla. Spain's response to the first two situations was a determination to defend its boundaries unimpaired, to keep out American intruders, and to colonize Texas with loyal Spanish subjects. Negotiations over the western boundary of Louisiana were complex and extended, but Spain and the United States, by the Neutral Ground Agreement of 1806, avoided a clash of arms when they temporarily accepted the Sabine River and Arroyo Hondo as borders of their respective lands. The boundary issue was not resolved until the Adam-Onís Treaty (1819–21).

Of particular concern to the Spanish was the fear of Anglo-American intruders as agents of political discord and illicit trade. One of the famous filibusters, Philip Nolan, was killed by Spanish forces in 1801 near the site of present Waco. His fate, however, did not frighten away subsequent adventurers, who responded to the lure of wealth and recognized Spain's tenuous hold on the vast, relatively unpopulated land that was Texas.

The most complete census data for Spanish Texas in the early nineteenth century are for 1804, the first year after the sale of Louisiana to the United States. It is quite possible that this systematic count resulted from the need to assess the strength and numbers of the Spanish and Hispanicized population in the face of aggressive Americans to the east.

The following population figures were compiled between January and December 1804: Nacogdoches, 789; Presidial Company of San Antonio de Béxar 413; Mission San Juan Capistrano, 74; Mission San Antonio de Valero, 121; Presidio (Settlement) of La Bahía, 399; Presidial Company of La Bahía, 301; Missions La Bahía, Rosario, and Refugio, 224; Mission San Francisco de la Espada, 107; Villa San Fernando de Béxar and Presidio (Settlement) of Béxar, 1,177. Total: 3,605.

Although the Spanish-speaking population included merchants and a few artisans such as tailors and blacksmiths, the vast majority of Texans were stock raisers and small farmers. The figures do not include unsettled Indians or black slaves; as Randolph B. Campbell has demonstrated, there were virtually no black bondsmen in Spanish Texas on the eve of Mexican independence.

The aggregate Hispanic population of Texas in 1810, when Father Hidalgo issued his Grito de Dolores, was probably fewer than 5,000. Spain's response to the first two situations was a determination to defend its boundaries unimpaired, to keep out American intruders, and to colonize Texas with loyal Spanish subjects. Negotiations over the western boundary of Louisiana were complex and extended, but Spain and the United States, by the Neutral Ground Agreement of 1806, avoided a clash of arms when they temporarily accepted the Sabine River and Arroyo Hondo as borders of their respective lands.

A more serious challenge to Spanish rule in Texas came from the activities of José Bernardo Gutiérrez de Lara and Augustus W. Magee. On the border of Texas and the United States, Gutiérrez and Magee organized a disparate band of filibusters who captured Nacogdoches, La Bahía, and San Antonio.

MEXICAN WAR OF INDEPENDENCE

After the fall of the capital, Gutiérrez issued a "Declaration of Independence of the State of Texas" and drafted a Centralist constitution. His Centralist stand and the execution of Spanish prisoners, including the governor of Texas, brought dissension between Gutiérrez and his Anglo-American followers and divided the motley collection of revolutionaries.

The insurgents suffered a disastrous defeat at the hands of Joaquín de Arredondo at the battle of Medina on August 18, 1813. Subsequently, Arredondo's bloody purges, which substantially depopulated the province, left an indelible mark on Texas and assured that it would remain a Spanish province for another seven years.

Those years, however, were frightfully destructive. Although Royalists successfully repelled all military invasions, the defenders of Texas had also become its predators. In the words of Antonio María Martínez, the last governor of Spanish Texas, the king's soldiers had "drained the resources of the country, and laid their hands on everything that could sustain human life."

In 1821, under the leadership of Agustín de Iturbide and Vicente Guerrero, a successful independence movement in Mexico brought Texas under that new nation. The Mexican War of Independence marked the close of an era in Texas history in which the Franciscan padres had founded and refounded missions at approximately forty different sites. Presidios, numbering ten, had extended from Central Texas eastward to the site of present Robeline, Louisiana, and southward to Chambers Country. Municipalities ranged from Laredo to San Antonio and Nacogdoches. Ranches and farms dotted the landscape. The majority of the population was probably matzo.

After Mexican independence, Hispanics were soon outstripped in numbers by Americans. Modern Texas, however, reflects its Spanish origins in many ways. Given the rich heritage and history of Spanish Texas, it seems preposterous that Jane Long , because she bore an Anglo child at Bolivar Point in December 1821, should continue to be nicknamed the "the Mother of Texas." That honor in large measure belongs to thousands of Indian and Spanish women, the earlier mothers of Texas.

Bibliography

Carlos E. Castañeda, *Our Catholic Heritage in Texas* (7 vols., Austin: Von Boeckmann-Jones, 1936–58; rpt., New York: Arno, 1976).

Donald E. Chipman, *Spanish Texas, 1519–1821* (Austin: University of Texas Press, 1992). Rupert N. Richardson, *Texas: The Lone Star State* (New York: Prentice-Hall, 1943; 4th ed., with Ernest Wallace and

Adrian N. Anderson, Englewood Cliffs, New Jersey: Prentice-Hall, 1981). Robert S. Weddle, *San Juan Bautista: Gateway to Spanish Texas* (Austin: University of Texas Press, 1968).

SOURCE; Donald E. Chipman, *Handbook of Texas Online*, s.v. : http://www.tshaonline.org/handbook/online/articles

ANGLO-AMERICAN COLONIZATION

Anglo-American colonization in Mexican Texas took place between 1821 and 1835. Spain had first opened Texas to Anglo-Americans in 1820, less than one year before Mexico achieved its independence. Its traditional policy forbade foreigners in its territory, but Spain was unable to persuade its own citizens to move to remote and sparsely populated Texas. There were only three settlements in the province of Texas in 1820: Nacogdoches, San Antonio de Béxar, and La Bahía del Espíritu Santo (later Goliad), small towns with outlying ranches.

The missions near the latter two, once expected to be nucleus communities, had been or were being secularized (i.e., transferred to diocesan from Franciscan administration), while those near Nacogdoches had been closed since the 1770s. Recruiting foreigners to develop the Spanish frontier was not new. As early as the 1790s, Spain invited Anglo-Americans to settle in Upper Louisiana (Missouri) for the same reason. The foreigners were to be Catholic, industrious, and willing to become Spanish citizens in return for generous land grants. Spain expected the new settlers to increase economic development and help deter the aggressive and mobile Plains Indians such as the Comanches and Kiowas. Mexico continued the Spanish colonization plan after its independence in 1821 by granting contracts to empresarios who would settle and supervise selected, qualified immigrants.

Anglo-Americans were attracted to Hispanic Texas because of inexpensive land. Undeveloped land in the United States land offices cost $1.25 an acre for a minimum of 80 acres ($100) payable in specie at the time of purchase. In Texas each head of a family, male or female, could claim a headright of 4,605 acres (one league-4,428 acres of grazing land and one labor-177 acres of irrigable farm land) at a cost about four cents an acre ($184) payable in six years, a sum later reduced by state authorities.

Beginning in 1824 when the Mexican Republic adopted its constitution, each immigrant took an oath of loyalty to the new nation and professed to be a Christian. Because the Catholic Church was the established religion, the oath implied that all would become Catholic, although the national and state colonization laws were silent on the matter. Religion was not a critical issue, however, because the church waited until 1831 to send a resident priest, Michael Muldoon, into the Anglo-Texan communities. This was inconvenient for those wishing to marry because there was no provision for civil ceremonies, and only priests had authority to perform nuptial rites. Anglo-Texans unwilling or unable to seek a priest in Catholic communities received permission from the authorities to sign a marriage bond, a practice common in the non-Anglican foothills of Virginia and the Carolinas before 1776, promising to formalize their union when a priest arrived.

Two other reasons brought Anglo-American settlers to Texas. Through the 1820s, most believed that the United States would buy eastern Texas from Mexico. Many thought that that portion of Texas had been part of the Louisiana Purchase and that the United States had "given" it away to Spain in exchange for Florida in the 1819 Adams-Onís Treaty, which established the Sabine River boundary.

The Texas pioneers expected annexation would stimulate immigration and provide buyers for their land. A second attraction was that Mexico and the United States had no reciprocal agreements enabling creditors to collect debts or to return fugitives. Therefore, Texas was a safe haven for the many Mississippi valley farmers who defaulted on their loans when agricultural prices declined at the end of the War of 1812 and bankers demanded immediate payment. Faced with seizure of their property and even debtors' prison in many states, men loaded their families and belongings into wagons and headed for the Sabine River, where creditors could not follow and there was opportunity to start over.

PROPOSED COLONIZATION OF TEXAS

Against this background, Moses Austin traveled from Missouri to Spanish San Antonio in 1820 to apply for an *empresario* grant to bring Anglo-American families to Texas. Like so many others, he had suffered losses in banking failures and hoped to restore his family's fortunes by charging small fees-12½ cents an acre-for land within his assigned area. He had become a Spanish citizen in 1798, when he moved from Virginia to the St. Louis area where he acquired an empresario grant to develop a lead mine and import workers.

The Spanish governor at San Antonio gave tentative approval to Austin's plan, subject to review by his superiors, to bring 300 moral, hardworking, Catholic families from the former Spanish territory of Louisiana. Although the authorities wanted him to settle close to San Antonio, Austin opted for a still-to-be-defined area along the lower Colorado River, where he hoped to establish a port. On his return to Missouri he *became ill and died at home in June 1821, leaving the plan with his eldest son, Stephen Fuller Austin.*

Austin's advertisements for colonists *coincided with Mexican independence* and the presumption that a republic would be organized. Stimulated by these events, some families began moving immediately to the Red River near future Texarkana and across the Sabine along the old Spanish road leading to Nacogdoches. There they remained as squatters, some with intentions of joining the Austin colony, but others engaged in trading with the Indians and Mexicans.

The younger Austin visited San Antonio in mid-1821 as his father's heir and received permission to explore the lower Colorado for a site for the colony. His roaming convinced him that the Brazos watershed should be added to his grant. Upon returning to Texas in early 1822, Austin discovered he must go to Mexico City to confirm the contract with the national government, even though his first settlers were on their way with only vague instructions about where to settle.

Soon after he reached the capital, a coup established an empire, and the resulting turmoil delayed Austin for a year. In April 1823 he finally received a contract under the Imperial Colonization Law, which had been passed in January. Because the empire collapsed in April and the republic was reestablished, Austin's empresario contract was the only one issued under this law. The reinstated republican Congress immediately approved the imperial contract, and *Austin rushed back to Texas to organize his colony.*

The Imperial Colonization Law specified that colonists must be Catholic, so Austin's first 300 families were affected. The 1824 National Colonization Law and the 1825 Coahuila and Texas State Colonization Law said only that foreigners must be Christian and abide by the laws of the nation, thereby implying they would be members of the established church. Protestant preachers occasionally visited Texas, but they seldom held public services. In 1834 the state decreed that no person should be molested for political or religious beliefs as long as he did not disturb public order. This was as close as Texans came to freedom of religion and speech before 1836.

Slavery was also an issue. Mexicans abhorred slavery as allowed in the United States, but pragmatic politicians shut their eyes to the system in their eagerness to have the Anglos produce cotton in Texas. National and state laws banned the *African* slave trade, but allowed Anglo-Americans to bring their family slaves with them to Texas and buy and sell them there until 1840. Grandchildren of those slaves would be freed gradually upon reaching certain ages.

The state inferred in 1827 that it might emancipate slaves earlier, and the immigrants took the precaution of signing indenture contracts with their illiterate servants binding them for ninety-nine years to work off their purchase price, upkeep, and transportation to Texas. Mexican officials recognized the subterfuge as debt peonage, and black slaves continued to arrive in Texas. The most serious threat to Anglo slaveholders occurred when President Vicente Ramón Guerrero emancipated all slaves on September 15, 1829, in commemoration of independence. Austin's Mexican friends quickly secured an exemption from the law for Texas.

Austin, the most successful Texas impresario, made four six-year contracts between 1823 and 1828 for a potential 1,200 families. They were to be settled between the watersheds of the Brazos and Colorado rivers and as far as the Lavaca River below the Old San Antonio Road, as well as eastward to the San Jacinto River (but not including Galveston Island) and a small area around the site of present-day Austin. A fifth contract issued in 1831 for 800 families to be settled along the Brazos above the old Spanish Road was challenged by Sterling Clack Robertson, who had an expired prior claim. The ensuing conflict made accurate tallies difficult.

Empresarios did not own the land within their grants, nor could they issue titles; the state appointed a land commissioner to give deeds only after 100 families had been settled. Surveyors laid off leagues and labores along the watercourses and roads, after which colonists could choose vacant tracts. The settlers paid fees to the state, the surveyor, the land commissioner, and the clerk, who wrote the deeds on stamped paper and recorded the payments.

Austin's plan to restore his family's well-being by selling land was denied because empresarios could not receive fees. The state gave them a bonus of 23,000 acres for each 100 families settled. By 1834, at the virtual end of the empresario system, Austin had settled about 966 families and earned 197,000 acres of bonus land that he could locate where he chose. He could sell the land, but only to those willing to live in Texas.

Austin, as the pioneer empresario in Texas, was burdened with more duties than later contractors. With no published compendium of the Mexican laws, administrative and judicial authority rested with Austin, and the result was a mix of Mexican decrees with pragmatic Anglo-American implementation. Local settlements within his colony elected alcaldes, similar to justices of the peace, and constables.

Austin sat as superior judge until 1828, when sufficient population permitted the installation of an ayuntamiento at San Felipe, the capital of the colony. This council, with elected representatives from the settlements, had authority over the entire Austin colony and acted like a county government. As population grew, other settlements within the colony qualified for ayuntamientos. These councils settled lawsuits, regulated the health and welfare of the residents by supervising doctors, lawyers, taverns, and ferries, surveyed roads, and sold town lots. Capital cases were referred to authorities in Monterrey and later Saltillo. The remoteness of the court disturbed Anglo-Texans, who wanted accessible courts.

Austin also commanded the local militia to defend the colony against Indians and to keep the peace. His contract area had only a few small Indian villages belonging to such sedentary groups as the Bidais and Coushattas, who wanted only to trade. Less friendly were the seasonally migrant remnant coastal tribes of Karankawas or the inland Tonkawas, who foraged for game and targeted the settlers' livestock. Pioneers along the Colorado River suffered most. Austin led several punitive expeditions between 1823 and 1826; he also negotiated moderately successful treaties with these declining tribes. North and west of the Austin colony Indians continued to resist the flow of immigrants well beyond the colonial period.

Other men besides Austin wanted empresario contracts in Texas, and a few were in Mexico City in 1822. Because of the changing political scene and the slow passage of the colonization laws, they had to wait until 1825, after the passage of national and state colonization laws passed in August 1824 and March 1825. The national law prohibited foreigners from settling within twenty-six miles of the Gulf of Mexico or within fifty-two miles of the Sabine River border without special executive permission. To encourage immigration, settlers were free from national taxes for four years. Land ownership was limited to eleven leagues. Owners had to be residents of Mexico. Preference was given to native Mexicans in the selection, and the national government could use any portion of land needed for the defence and security of the nation.

The state colonization law detailed how to apply for land, how much would be given to heads of families, including females or single persons, and the fees to be paid. The law granted freedom from tithes and the *alcabala*, an internal excise tax, for ten years. Within three weeks four contracts were signed: Haden Edwards[qv] could settle 800 families in the Nacogdoches area, Robert Leftwich 800 along the Brazos valley above the Old San Antonio Road, Green DeWitt 400 on the Guadalupe River, and Frost Thorn 400 north of Nacogdoches.

DeWitt, who developed the area around Gonzales, was the second most successful empresario in Texas. He settled 189 families before his contract expired in 1831. His colony suffered Indian attacks and also controversy with his neighbor, Mexican native empresario Martín De León.

De León, a native of a small town 100 miles southwest of the site of present Matamoros, moved his family north across the Nueces River and into the province of Texas in the early 1800s. He lived by catching mustangs and wild cattle and raising mules, then selling the animals in San Antonio or even trailing them to Louisiana.

In April 1824, before the passage of the national colonization law, he received permission from the provincial deputation in San Antonio to establish a town for forty-one families about twenty miles northeast of La Bahía on the banks of the Guadalupe River. No boundaries were mentioned. By October, De León and twelve families had arrived at a cypress grove, the site of the town of Guadalupe Victoria (now Victoria, Texas), named for the first president of Mexico, who took office that year.

Unaware of the colonization grant to De León in San Antonio, the state assigned the same area of the Guadalupe valley with specific boundaries to DeWitt in April 1825. When DeWitt's settlers arrived, trouble was inevitable. Because the colonization laws gave preference to native Mexicans, De León petitioned the state for redress, and the authorities told DeWitt in October 1825 to respect De León's prior claims but failed to establish boundaries.

The state named land commissioners for both De León's and DeWitt's colonies. The commissioners issued titles in 1831, the year DeWitt's six-year contract expired permanently.

The boundaries remained unresolved. Eventually sixteen non-Hispanic families, some of whom were Anglo-Americans with Irish roots, received headrights in De León's otherwise Hispanic community.

De León also quarreled with neighboring empresarios James Power and James Hewetson,[qqv] natives of Ireland and residents of the United States and Mexico. They received a state contract in June 1828 to settle 200 families-half Mexican and half Irish-in the twenty-six-mile coastal reserve between the mouth of the Guadalupe and the mouth of the Lavaca River, an area that received approval from the president. In 1829 their boundary was extended south to the Nueces River.

Two hundred titles were issued to Europeans, but because many were single men the colonial contract was left incomplete, since the law specified families. Nearby, two other Irish natives, residents of Matamoros, secured a contract in 1828 to bring 200 European families to the Nueces above the Power-Hewetson grant. John McMullen and James McGloin's colony was known as the Irish Colony; most of its residents clustered around San Patricio, where the land commissioner issued eighty-four titles.

Of more importance to the development of Anglo-Texan communities were the large grants made in 1825 to Edwards and Leftwich that were adjacent to the Austin colony on the east and north. Edwards's contract specified land for up to 800 Anglo-American families in a large area of East Texas from northwest of Nacogdoches, including the forks of the Trinity, west to the Navasota River, thence southeast along the Trinity River valley to upper Galveston Bay. The tract did not include Galveston Island or the twenty-six-mile-wide coastal reserve forbidden to foreigners.

The eastern boundary was the fifty-two-mile-wide border reserve along the Sabine River running north from the Gulf of Mexico to the thirty-second parallel. The state instructed Edwards to respect the property of long-time residents in the Nacogdoches area, some of whom had been there since the 1780s.

Edwards, insensitive to Hispanic culture, reached Nacogdoches in October 1825 and threatened to dispossess those who had no proof of ownership unless they paid him for the land. The Spanish process to acquire land on the remote frontier was lengthy and expensive, and almost nobody had deeds. But besides Hispanics, there were a number of Anglo-American hunters and traders who had moved there in the 1790s, as well as new squatters who had arrived since 1821.

Although local alcaldes were authorized for the scattered settlements, there was no ayuntamiento until 1828, so all official business had to be conducted in San Antonio. In general, Hispanics and old-time Anglos opposed Edwards and complained to the political chief in San Antonio; some newcomers supported the empresario, while others remained aloof. Tensions mounted during 1826, and in response to complaints against Edwards the state abrogated his contract in October and banished him from Texas. Anglo-Texans, even those not involved in the controversy, were disturbed that a contract, almost sacred in Anglo culture, could be canceled. Did that mean that the provisions of the colonization laws might change regarding slavery? Were other empresario contracts vulnerable?

At the time of the cancellation, Haden Edwards was recruiting colonists in Mississippi, but his brother, Benjamin W. Edwards, rallied supporters to protest the order. In November they seized the alcalde (an Anglo old-timer) and tried others at a drumhead court; after a few days all were released. The following month the insurgents declared independence from Mexico by decreeing the Republic of Fredonia. Benjamin Edwards forged an alliance with some disaffected Cherokee leaders who were unable to secure titles to their villages northwest of Nacogdoches, and Edwards promised to divide Texas between the red men and the whites.

The Indians reconsidered, however, and failed to support Edwards, who fled towards Louisiana in January 1827, when soldiers from San Antonio accompanied by Austin's militia approached Nacogdoches. Thus ended the quixotic Fredonia Rebellion, which aroused fears of widespread Anglo-Texan revolt among Mexican leaders. The state banished the ringleaders, promised to send a land commissioner to issue titles to families who had settled the area in good faith, and established a permanent garrison at Nacogdoches to guard against ruffians and filibusters from the United States.

Almost as troublesome was the Leftwich contract, which, unknown to the authorities, was intended as a profit-making undertaking for the benefit of stockholders. This empresario grant eventually was known as Robertson's colony. Leftwich, who was in Mexico City when Austin was there, represented seventy Tennessee investors called the Texas Association. The land assigned was in the upper Brazos valley just north of Austin's grant and touching that of Edwards on the east. When Leftwich returned to Nashville in 1825 with the empresario contract in his name instead of the company's, the investors had to buy out his interest. They sent several agents to Coahuila-Texas to get the contract in their name, but the suspicions raised by the Fredonia Rebellion deterred their efforts. With Austin's help, the Nashville Association (as it was now called) was finally recognized as successor to Leftwich in October 1827. But the stockholders failed to send settlers until October 1830, only six months before the end of their contract.

Shareholder Sterling C. Robertson and six others reached the garrison at Fort Tenoxtitlán, one of the new military posts established in 1830; Tenoxtitlán guarded the old Spanish road crossing on the Brazos River. Nine families trailed Robertson, and when they reached Nacogdoches the commandant detained them. The newly passed Law of April 6, 1830, prohibited the entrance of Anglo-Americans into Texas unless they had a passport to Austin's or DeWitt's colony. Austin had quickly secured an exemption from the restriction for his and DeWitt's colonies when he discovered an ambiguous phrase seeming to allow immigration to "established" colonies. This he interpreted as those with more than 100 families in residence. The authorities acquiesced. The Law of April 6 also forbade the immigration of slaves, but even the authorities admitted that this restriction was impossible to enforce.

Passage of the restrictive law was the result of heightened suspicions that the United States intended to seize Texas. Between 1825 and 1829, the United States minister to Mexico, Joel R. Poinsett, unsuccessfully pressured the new nation to sell eastern Texas, and his successor, Anthony Butler, had similar instructions. Mexican leaders feared a rebellion of Anglos and ***annexation*** to the United States, just as Spain had lost Baton Rouge and Mobile in the early 1800s. Thus the Fredonian Rebellion inspired an official inspection tour in 1828 from San Antonio to Nacogdoches which revealed that Anglo-Americans greatly outnumbered native Mexicans in Texas. Although there was no obvious subversive activity, the Anglos continued to speak only English and conducted legal matters primarily in Anglo tradition. The authorities concluded that Mexico might lose Texas if more Anglo-Americans were allowed to enter and that native Mexicans must be encouraged to settle in the frontier state to "Mexicanize" it.

The government dispatched troops to strategic entrances to Texas in late 1830 to enforce the law and also to aid the newly installed customs collectors in levying national import duties. The special exemption from the tariff for Texas pioneer settlers had expired. It was hoped that the new garrisons would produce native Mexican communities and that the tariff would pay for the troops needed to preserve Texas. However, this understandable response to save Mexican hegemony inflamed Anglo-Texans, who inherited their grandfathers' distaste for standing armies and troops billeted in residential communities, and who had come to believe that their exemption from tariffs was permanent.

Robertson's colonists surreptitiously passed Nacogdoches and reached the Brazos in November. Amid a flurry of letters to officials, Austin offered the unlucky settlers sanctuary, a step approved officially in September 1831. Meanwhile Robertson asked Austin, who was leaving to take his seat in the state legislature at Saltillo, to intercede for an extension of the six-year contract. Austin agreed, although he believed the case dead-the six years was up, the Law of April 6, 1830, prevented Anglo settlement, and a French immigration company had already applied for the old

Leftwich grant. In early February, however, Austin asked the state for a contract in his name and that of his associate, Samuel May Williams, to settle 800 European and Mexican families on the former Nashville grant, plus some additional land north and west. The application was approved on February 25, 1832.

This action caused trouble with Robertson, who felt betrayed. Austin defended his action by saying that Robertson's cause was hopeless. Austin wanted, he stated, a more dependable frontier neighbor than a French company on his northern frontier, which was still exposed to Indian raiders. Robertson appealed to the ayuntamiento in San Felipe for relief in November 1833 by offering testimony from various persons that the Nashville company had sent 100 families before the expiration of the contract in 1831. The council, influenced by anti-Austin sentiment, agreed that Robertson should be reinstated, and he left for Saltillo, where he presented his documents on April 2, 1834.

In May the governor returned the contract to Robertson despite arguments by Austin and Williams's attorney; Robertson's land commissioner began granting titles immediately. Sam Williams attended the 1835 session of the state legislature and won back the colony in May. Austin and Williams's new land commissioner began issuing deeds, while Robertson's commissioner continued to act until the end of the year. The resulting overlapping claims kept lawyers employed past mid-century.

The most controversial of all of the empresario grants, however, were those of David G. Burnet, Joseph Vehlein, and Lorenzo de Zavala, who sold their respective contracts to New York and Boston speculators. This action was contrary to the intent of the national and state colonization laws.

The sale was not only illegal, but also deleterious to the purchasers. As a result of it, small and large investors and innocent colonists lost their promised land and sometimes their lives. Burnet, a native of New Jersey and resident of Ohio, had traded with the Indians in Spanish Texas about 1818 but soon returned to Ohio. In December 1826 he received a contract to settle 300 families in the northwestern area of the former Edwards grant near Nacogdoches. At the same time, Vehlein, a Mexico City merchant, contracted through agents to settle 300 families in the southern portion of the canceled Edwards grant

Neither man visited the land and neither sent a single colonist. Burnet, a younger son of a prominent family, tried to recruit antislavery settlers in Ohio, but lacked the financial means to succeed. In 1829, with time running out, he unsuccessfully offered half of his future bonus land to wealthy and powerful individuals if they would send 300 families to his grant. Finally, in October 1830, he negotiated the sale of his grant to the Galveston Bay and Texas Land Company, as did Vehlein's agent. Joining in the sale was Lorenzo de Zavala, a native of Yucatán, a prominent politician and currently a political refugee in New York City. He had acquired the long fifty-two-mile-wide border reserve on the Sabine River as his empresario grant in 1828, when he agreed to settle 500 European or Mexican families. He, too, never saw his Texas land nor sent a colonist.

The Galveston Bay company issued stock entitling investors to scrip in denominations of leagues and labores. Some investors, such as James Prentiss, used their scrip to create separate ventures such as the Union Land Company and the Trinity Land Company. Those companies dispatched two schooners to Galveston Bay in January 1831 with a few European settlers recruited in New York City. Though the Law of April 6, 1830, prohibited sending Anglo-Americans, some American workmen accompanied the settlers. John Davis Bradburn, the commandant at Anahuac, allowed the unfortunate immigrants to land briefly. A few stayed and farmed, but most returned to the United States as best they could.

The entire undertaking was extortionate. Upon landing, each immigrant was to apply for a headright, of which he would receive 177 acres after three years of labor. The remaining 4,428 acres was signed over to the company. The company employed agents in Mexico to lobby for revocation of the restrictive law of 1830, a step that succeeded in 1834. Moreover, those squatters who had lived in East Texas since 1821 without titles discovered they were now colonists of the three empresarios. Although the state had approved naming a land commissioner in 1828, politics delayed implementation. Finally, in 1834, land commissioners arrived to give deeds to those long-time squatters and also the newcomers. One called himself the agent of the empresario and tried to collect fees until complaints to the state ended the illegal practice.

There were a number of lesser empresarios, including Benjamin R. Milam, Arthur G. Wavell, John Cameron, Stephen Julian Wilson, John Charles Beales, and Richard Exeter, as well as a few native Mexicans. All failed to complete their contracts within the allotted time. Most of their grants were located on the southwestern and northern perimeters of Texas.

Settlers continued to arrive in the established colonies and the area claimed by the Galveston Bay Company until the land offices were closed by the General Council on October 27, 1835, and confirmed by the Consultation the next month. By this time most residents, except in Robertson's colony on the northern frontier, no longer thought of themselves as colonists dependent on an empresario. De León was dead, DeWitt died in 1835, Austin was a sort of elder statesman, and few people in the Burnet, Vehlein, or Zavala grants were acquainted with the phantom empresarios.

By 1834 Texas was divided into three departments: Bexar, the Brazos, and Nacogdoches, each with its own political chief reporting directly to the governor and each having more than one town with an ayuntamiento. This gave the residents a feeling of self-government. Their main goal was achieving separate statehood from Coahuila. The old Austin colony continued to attract newcomers, including some of those unable to secure titles in other colonies. Its stability, its guaranteed deeds, and its easy accessibility by water from New Orleans through Galveston Bay and the Brazos river were drawing cards.

The Anglo-American settlers imported their culture to Texas *and resisted Mexicanizing* even after 1830, when purchase by the United States was increasingly unlikely. Rich or poor, Anglo immigrants were independent-minded, self-sufficient republicans suspicious of the traditional deferential society of Hispanic culture, even though Mexican reformers were struggling to build a republic.

The spirit of Jacksonian democracy pervaded even those not admiring President Andrew Jackson. Moreover, collecting national tariff duties in Texas in 1830 coincided with the growing anti-tariff movement in South Carolina, which resulted in the Nullification Crisis. Texans, like other agrarians, manufactured nothing and disliked import duties on necessities. It was not surprising, then, that American ship captains, supported by Anglo-Texan merchants, refused to pay the new duties and exchanged fire with the fort at the mouth of the Brazos River in December 1831.

At this same time, Colonel Bradburn, charged with enforcing Mexican laws regarding immigration and the tariff at Anahuac, arrested civilians and held some without bail for trial before the commandant general at Matamoros. Anglo-Texans believed Bradburn was acting arbitrarily. They did not understand that his actions were required under Mexican law, which lacked anything like a Bill of Rights. Angry men from the Brazos marched to confront Bradburn at Anahuac, while others loaded illegal cannons on a ship to join them.

At the mouth of the river, the Anglo-Texans forced the surrender of the fort. The first of the Anahuac Disturbances and the battle of Velasco took place in June 1832, just as the Federalist party's army defeated that of the conservative administration, thus ending a four-year-old civil war. The Texans claimed to be helping the Federalists, who were led by Antonio López de Santa Anna, and convinced authorities sent to investigate that they were not revolutionaries against Mexico. Their action resulted in the departure of all Centralist troops and customs collectors from Texas.

Wanting to capitalize on their support of Santa Anna, who was to be the new president, the Anglo-Texans drafted petitions for separate statehood, a better judicial system, and similar reform measures. In April 1833 Austin took the requests to Mexico City, where most were approved except for separate statehood. In a moment of despondency, Austin wrote an incriminating letter urging the leaders at San Antonio to act unilaterally on separation. For this act of sedition he was arrested in January 1834 and incarcerated in Mexico City until July 1835.

Texans remained relatively quiet during Austin's absence fearing for his life. After Santa Anna's installation as president in 1833, he left governing to his Federalist- reformer vice president. By 1835, however, Santa Anna reversed himself, became a Centralist dictator, and sent the army to punish his political enemies. Garrisons and customs collectors returned to Texas in January 1835 and the Anglo-Texan response was predictable. A second attack recaptured Anahuac in June, and when Austin reached Texas in September he surprised many by endorsing the resistance movement against the oppressive administration. Though at first supporting reform of the Mexican government, public opinion moved quickly to favoring a separate Texas nation.

There are no accurate figures detailing the number of Anglo-Americans who settled in Texas between 1821 and 1835. Although Mexican law required an annual census of all residents, the Anglo-Texans resisted such bureaucratic demands. Existing tallies reveal that in 1826 Austin had 1,800 people, including 443 slaves; DeWitt counted 159 whites and 29 slaves; and the lower Trinity River, then outside of any empresario grant, was populated by 407 settlers with 76 slaves

Subsequent extant records show Austin's colony with 2,201 people in 1828, 4,248 in 1830, and 5,565 in 1831, while DeWitt had only 82 persons, including 7 slaves, in 1828. In 1834 the English-speaking Col. Juan N. Almonte made an official inspection of Texas and estimated that 4,000 people lived in the Department of Bexar, 2,100 in the Department of the Brazos (Austin's colony), and 1,600 in the Department of Nacogdoches.

The 3,000 decrease in Austin's colony suggests that Almonte's figures were wrong. A few lists of residents in the Nacogdoches area in 1834–35 and scattered other records are extant, but such incomplete records only hint at the Texas population. Anglo-American colonization in Texas was obviously a success, although not what Mexican leaders envisioned. Neither side worked to understand, appreciate, nor resolve the cultural differences intrinsic to the union proposed by Mexico and accepted by Anglo-American immigrants.

Bibliography

Eugene C. Barker, *The Life of Stephen F. Austin* (Nashville: Cokesbury Press, 1925; rpt., Austin: Texas State Historical Association, 1949; New York: AMS Press, 1970).

A. B. J. Hammett, *The Empresario Don Martín de León* (Waco: Texian Press, 1973).

Mary Virginia Henderson, "Minor Empresario Contracts for the Colonization of Texas, 1825–1834," *Southwestern Historical Quarterly* 31, 32 (April, July 1928).

Margaret Swett Henson, *The Cartwrights of San Augustine* (Austin: Texas State Historical Association, 1993).

Margaret S. Henson, *Juan Davis Bradburn: A Reappraisal of the Mexican Commander of Anahuac* (College Station: Texas A&M University Press, 1982).

Margaret S. Henson, *Samuel May Williams: Early Texas Entrepreneur* (College Station: Texas A&M University Press, 1976).

Edward Albert Lukes, *De Witt Colony of Texas* (Austin: Jenkins, 1976).

Malcolm D. McLean, comp. and ed., *Papers Concerning Robertson's Colony in Texas* (19 vols., Arlington: University of Texas at Arlington Press, 1974–93).

Ernest Wallace and David M. Vigness, eds., *Documents of Texas History* (Austin: Steck, 1963). David J. Weber, *The Mexican Frontier, 1821–1846* (Albuquerque: University of New Mexico Press, 1982).

SOURCE; Margaret Swett Henson - *Handbook of Texas Online,* s.v.: http://www.tshaonline.org/handbook/online/articles

FORT ANAHUAC

It was the site of the first armed confrontation between Anglo-Texans and Mexican troops, on June 10–12, 1832. In November 1830 Col. Juan Davis Bradburn chose the site for the fort and its town on a bluff, called Perry's Point since 1816, overlooking the entrance to the Trinity River. The garrison was one of six new outposts located at strategic entrances to Texas and designed to enforce the Law of April 6, 1830. All the garrisons carried Mexican names. Anáhuac, the name of the ancient home of the Aztecs, was borrowed for the Chambers county fort. Bradburn brought plans and a cardboard fort with him.

The garrison lived temporarily in a fortified wooden barracks a half mile north of the bluff in the center of the site of modern Anahuac. The barracks was later used as the jail that held William B. Travis and others. Bricks for the walls and buildings of the permanent fort were made by convict soldiers on-site, beginning in March 1831. A Masonic and military ceremony marked completion of the foundation on May 14, 1831.

The exterior walls were 100 by seventy feet and enclosed two redoubts diagonally opposite on the southwest and northeast corners. Inside the perimeter was a reinforced-brick building about fifty by thirty-five feet. The southwest redoubt, overlooking Trinity Bay, was named Fort Davis (for Bradburn); it was manned by a maximum of fifty men and defended by a six-pound cannon, while its twin on the northeast guarded the land approach. The cavalry tethered its horses between the two redoubts.

An excavated passage connected the enclosure with the powder magazine on the east side, where two bulwarks named Hidalgo and Morelos (for martyrs of the Mexican independence movement) near the sites of the brick kilns, each with a sixteen-pound cannon, guarded the compound.

The garrison grew from forty men and four officers in November 1830 to a maximum of 285 men and ten officers in May 1832. After March 1832 about 100 of the men were stationed at Velasco, at the mouth of the Brazos, under Col. Domingo de Ugartechea. The troops were from the Eleventh and Twelfth battalions; the boatmen came from the battalion of

Pueblo Viejo de Tampico; La Bahía supplied twenty-five cavalrymen and one officer.

Texan insurgents under Col. Francis White Johnson attacked the fort on June 10–12, 1832, to rescue Travis. The troops dismantled the fort when they left in July 1832, and a fire in November gutted the wooden parts. The wooden calaboose was burned in December 1832, and practical residents removed bricks for fireplaces and foundations.

In January 1835 Capt. Antonio Tenorio arrived with about forty troops to reopen the fort, but it was in such disrepair that he asked his superiors for wood to make repairs. The wood arrived in May but was burned by irate Texans. Tenorio had no artillery when Travis and his volunteers attacked on June 29, so his troops fled into the woods. He capitulated the next day, and the small garrison sailed to Harrisburg and retreated to Bexar.

The fort was never used again; the land became private property. In 1938 the county surveyor made field notes of the existing foundations. Erosion caused by rechanneling the Trinity River sometime after the 1930s caused the remains of the southwestern redoubt to fall into the water. Chambers County acquired the site for a park in 1946, and officials ordered it cleared and the rubble buried for safety reasons and to prevent vandalism. An amateur excavation was made in 1968 before preservation laws went into effect, but no in-depth archeological study has been made of the site.

Bibliography

Margaret S. Henson, *Anahuac in 1832: The Cradle of the Texas Revolution* (Anahuac, Texas: Fort Anahuac Committee of the Chambers County Historical Commission, 1982).

Margaret S. Henson, *Juan Davis Bradburn: A Reappraisal of the Mexican Commander of Anahuac* (College Station: Texas A&M University Press, 1982).

James Wright Steely, comp., *A Catalog of Texas Properties in the National Register of Historic Places (Austin: Texas Historical Commission, 1984).*

Source; Margaret Swett Henson; *Handbook of Texas On Line*, s.v. http://www.tshaonline.org/handbook/online/articles

FORT LIPANTITLÁN

Fort Lipantitlán (meaning "Lipan land") was conceived about 1825 by José M. J. Carbajal. The site, now in northwestern Nueces County, was that of camping grounds of the Lipan Apache Indians on the west bank of the Nueces River about three miles upstream from the old town of San Patricio, which is on the east side of the river.

At the site a number of ancient trails beaten by game animals, Indians, and explorers crossed. An old presidio was also reportedly there as early as 1734 but had completely vanished. Mexican general Manuel de Mier y Terán, acting on the orders to restrict Anglo immigration into Texas, commissioned the fort and placed Capt. Enrique Villareal[qv] in command. He served until 1835, when he was relieved by Capt. Nicolás Rodríguez.

Author John Linn wrote, "The fort was a simple embankment of earth, lined within by fence-rails to hold the dirt in place, and would have answered tolerably well, perhaps, for a second-rate hog pen." After its construction it was garrisoned with from eighty to 120 men; however, many times the complement was much less.

Each of the four parapets was designed for one cannon, but it is doubtful if the fort was ever fully armed. Evidently several buildings and at least one barracks were built surrounding the embankments. James McGloin, in his account of the battle of Lipantitlán (November 4, 1835), makes reference to burning several houses, including a barracks.

The Mexican armies, coming north by land out of Mexico, headed to Goliad, Refugio, and East Texas during the Texas Revolution and crossed the Nueces at either the De Leon Crossing at the fort or Paso de Santa Margarita near San Patricio. During this period Mexican troops were in almost constant occupancy of the fort. Captain Rodríguez was in command of the Mexican forces when Capt. Ira J. Westover and a force of about seventy Texans defeated the Mexicans on November 4, 1835. Since the Texans did not occupy the fort after their victory, Mexican forces continued to use the old fort on occasion.

Fort Lipantitlán played an important role in the years immediately after the war, when Federalist forces under Gen. Antonio Canales[qv] sought refuge on the Nueces River to regroup and to seek assistance from Americans. Gen. James Davis[qv] repulsed Mexican troops under Canales at Lipantitlán on July 7, 1842. After the Mexican War put an end to Mexican armies in Texas, Lipantitlán was abandoned and grew up in brush.

In the mid-1980s digs on private land uncovered a number of artifacts in a rather large camp adjacent to the old fort where families and army women lived. A population of 300 or more in the camp is considered likely. The site of the old fort is a state park, but no traces of the earthen embankments remain.

The archeological digs confirmed Indian presence before the Spanish and Mexican eras, as well as occupancy by Texas forces in 1842. Aboriginal ceramics, Indian artifacts, personal items, all types of military buttons and ordnance (Spanish, Mexican, and Texan), money, and miscellaneous hardware were among the hundreds of artifacts recovered.

Bibliography

Eugene C. Barker, ed., *The Austin Papers* (3 vols., Washington: GPO, 1924–28).

William Campbell Binkley, ed., *Official Correspondence of the Texan Revolution, 1835–1836* (2 vols., New York: Appleton-Century, 1936).

Bethel Coopwood, "The Route of Cabeza de Vaca," *Quarterly of the Texas State Historical Association* 3 (October 1899, January, April 1900).

Charles Adams Gulick, Jr., Harriet Smither, et al., eds., *The Papers of Mirabeau Buonaparte Lamar* (6 vols., Austin: Texas State Library, 1920–27; rpt., Austin: Pemberton Press, 1968).

Hobart Huson, *Refugio: A Comprehensive History of Refugio County from Aboriginal Times to 1953* (2 vols., Woodsboro, Texas: Rooke Foundation, 1953, 1955).

Hobart Huson, *Captain Philip Dimmitt's Commandancy of Goliad, 1835–1836* (Austin: Von Boeckmann-Jones, 1974).

Hobart Huson, *Souvenir Program, Refugio County Centennial Celebration* (Refugio: Refugio *Timely Remarks*, 1936?).

John J. Linn, *Reminiscences of Fifty Years in Texas* (New York: Sadlier, 1883; 2d ed., Austin: Steck, 1935; rpt., Austin: State House, 1986).

Joseph Milton Nance, *Attack and Counterattack: The Texas-Mexican Frontier, 1842* (Austin: University of Texas Press, 1964).

Source; Keith Guthrie; *Handbook of Texas Online*, s.v. http://www.tshaonline.org/handbook/online/articles/FF/qbf32.html

FORT TENOXTITLÁN

Fort Tenoxtitlán, constructed in 1830 in what is now northeastern Burleson County, was part of a chain of military garrisons designed to Mexicanize Texas and stanch immigration from the United States pursuant to the Law of April 6, 1830. On June 25, 1830, Lt. Col. José Francisco Ruiz was dispatched from Bexar in command of 100 cavalrymen of the presidial company of Álamo de Parras with orders from Gen. Manuel de Mier y Terán to establish a fort at the strategic point halfway down the Old San Antonio Road, where the thoroughfare crossed the Brazos River en route to Nacogdoches.

Ruiz reached the Brazos on July 13 and established temporary headquarters on the east bank about a half mile below the Old San Antonio Road. On October 17, 1830, the garrison moved to a permanent site on a high bluff on the west bank of the Brazos twelve miles above the San Antonio crossing, opposite the spot where the present Brazos-Robertson county line strikes the river. The small spring-fed creek nearby was subsequently known as Dam Creek, probably because its water was diverted into the settlement. Although Mier y Terán, who envisioned Tenoxtitlán as a future capital of Texas, issued elaborate instructions from Matamoros for the design of the fort, most were eventually disregarded; the fortifications themselves were likely of conventional log construction.

One of the garrison's most important duties was to assist in the transportation of military funds from Bexar to Nacogdoches. Despite the ban on American settlement, the nearby farming community included an undetermined number of American immigrants; as early as July 1831, for example, Francis Smith operated a thriving general merchandise store at the fort, trading manufactured goods to the Indians for beaver pelts and buffalo robes. On December 31, 1830, the *ayuntamiento* of San Felipe de Austin, acknowledging the importance of the garrison, established a commission to construct a road from San Felipe to Tenoxtitlán.

In late October 1830 Maj. Sterling C. Robertson of the Texas (or Nashville) Association, appeared at Tenoxtitlán requesting permission to select a settlement site for fifty American families accompanying him, provided by the colonization contract that his group had made with the province of Coahuila and Texas. Three months later official announcement of the provincial government's invalidation of this contract reached the fort. However, Colonel Ruiz, Texas-born himself and sympathetic to the American settlers, evaded orders to apprehend the colonists and turn them over to the authorities in Nacogdoches, thus permitting them to scatter into various parts of Texas.

On July 13, 1832, despondent over the failure of his grand scheme to settle Mexicans in the Texas wilderness, Mier y Terán committed suicide. Thereupon the demoralized Colonel Ruiz decided to abandon Tenoxtitlán. He began evacuation of the garrison and entire Mexican settlement to Bexar on August 22, 1832. By December only a handful of Americans remained in occupation of the site. A trading post and settlement continued in the vicinity for many years but disappeared after 1860.

In 1936 a granite commemorative marker was erected by the Texas Centennial Commission near the site of the fort, fourteen miles northeast of Caldwell off Farm Road 1362. Another was erected in 1970 five miles east of Caldwell on State Highway 21. Tenoxtitlán, or "Prickly Pear Place" was the Aztec name for what became known as Mexico City.

Bibliography

Burleson County Historical Society, *Astride the Old San Antonio Road: A History of Burleson County, Texas* (Dallas: Taylor, 1980).

Malcolm D. McLean, "Tenoxtitlán, Dream Capitol of Texas," *Southwestern Historical Quarterly* 70 (April 1967)

Source; *Charles Christopher Jackson, Handbook of Texas Online*, s.v. http://www.tshaonline.org/handbook/online/articles/FF/qbf49.html (accessed May 31, 2010).

FORT TERAN

Fort Teran was a Mexican military encampment or station established in 1831 at a Neches River crossing that Spanish government representatives in Nacogdoches at the beginning of the nineteenth century had referred to as the "pass to the south." Three important trails crossed the Neches River at this point, underscoring the significance of this strategic site. The fort was named in honor of Gen. Manuel de Mier y Terán, commandant general of the eastern division of the Provincias Internas (which included Texas), and constructed at this location as part of a program to control the flow of smugglers and illegal immigrants into Texas.

Construction of a fort on the Neches River was a result of General Terán's inspection tour of East Texas in 1829. He observed that immigrants and smugglers were coming into Texas from Louisiana by using unguarded trails such as the Coushatta Trace, the Alabama Trace, and the Nacogdoches-Orcoquisac Road, all of which crossed the Neches River at the future site of Fort Teran. When Terán returned to Mexico, he helped to draft *Anastasio Bustamante*'s Law of April 6, 1830, forbidding American immigrants to settle in Mexican territory.

Responsibility for enforcing this law was assigned to a "director of colonization," and Terán was the first to hold this office. His program for closing Texas to immigrants from the United States included establishing garrisons on the Neches and several other rivers. He chose Peter Ellis Bean, a colonel in the Mexican army, to construct Fort Teran on the Neches. On September 25, 1831, Bean departed from Nacogdoches to establish the fort. Apparently construction proceeded very slowly, since the military commandant at Nacogdoches, José de las Piedras, on April 19, 1832, reported the need for additional carpenters and other craftsmen to assist in building the fort. When completed, this project consisted of approximately ten wooden cabins to provide housing for Colonel Bean and his small garrison. The Mexican government, however, found itself unable to support its Texas forts adequately, and later in 1832 transferred most of the troops.

After Fort Teran was abandoned by the remaining troops in 1834, the population in the immediate area was about a dozen persons. Samuel T. Belt opened a trading post at the fort site. A post office operated there in 1856–66, and this small community, sometimes called Fort Turan, continued as a trading and shipping point until the railroads came to Tyler and Angelina counties in the 1880s. Until 1878 steamboats continued to land near the fort, which was at the head of navigation on the Neches River. When Texas counties were organized after 1845, Fort Teran was used as a point of reference in describing the boundaries of Angelina and Jasper counties.

Bibliography

Juan N. Almonte, "Statistical Report of Texas," *Southwestern Historical Quarterly* 28 (January 1925).

Wallace W. Atwood, *The Physiographic Provinces of North America* (Boston: Ginn, 1940).

Eugene C. Barker, *Readings in Texas History* (Dallas: Southwest Press, 1929).

David J. Weber, *The Mexican Frontier, 1821–1846* (Albuquerque: University of New Mexico Press, 1982).

James E. Wheat, "The Story of Fort Teran on the Neches," *It's Dogwood Time in Tyler County*, March 1951.

James E. and Josiah Wheat, "Tyler County under Mexico," *It's Dogwood Time in Tyler County*, March 1966.

Amelia W. Williams and Eugene C. Barker, eds., *The Writings of Sam Houston, 1813–1863* (8 vols, Austin: University of Texas Press, 1938–43; rpt., Austin and New York: Pemberton Press, 1970).

Source; Howard N. Martin; *Handbook of Texas Online*, s.v. http://www.tshaonline.org/handbook/online/articles

OLD STONE FORT

Old Stone Fort is the modern name of a historic building, called La Casa Piedra by the Spanish, that was the first mercantile house and a frequent seat of civil government in early Nacogdoches. In 1779 Antonio Gil Ibarvo laid out the town near the intersection of the Old San Antonio Road and La Calle del Norte and built a stone house to use in the trading business. It was constructed of the native iron ore that occurs in the area abundantly.

The interior walls were made of ten-by-fourteen-inch sun-dried adobe blocks. Hand-hewn black walnut was used for sills and casements. The structure measured seventy feet along the Old San Antonio Road and twenty-three feet along Fredonia Street, as this lateral street was later called. With the second story the building was twenty feet tall; it remained the tallest structure in Nacogdoches for nearly a century. Originally each story had two main rooms, although subsequent owners rearranged the interior partitions and added a lean-to at the back.

The Stone House was not officially a government building despite its many public and civic uses in later decades. Ibarvo used it as a trading center in which he stored goods from the United States to trade to Indians for skins and hides. Since he was also the military and civil leader of the community, the Stone House assumed a quasi-public character that it never lost. Ibarvo faced accusations of smuggling in 1791 and was forced to leave Nacogdoches in 1796 but retained ownership of the Stone House until he sold his interest to José Luis de la Bega in 1805. Bega in turn sold it to William Barr within a year

During these years the government used the structure as if it were government property. In 1792 Juan Antonio Córtez used the Stone House as an office to issue land titles; in 1800 Ramón Músquiz used it as a military headquarters for his operations against the filibusters, and after Philip Nolan was killed, some of his men were jailed there briefly; Zebulon M. Pike's soldiers were billeted there on their return from Mexico; Governor Antonio Cordero y *Bustamante* resided in the building while Col. Simón de Herrera negotiated the Neutral Ground agreement with Gen. James Wilkinson; and Manuel Maria de Salcedo, governor of Texas in 1810, occupied the house when he visited East Texas.

The Gutiérrez-Magee expedition entered Texas in 1812, and their first stop was the Stone House in Nacogdoches. There they proclaimed Texas to be theirs. They remained in Nacogdoches for three months before moving on to La Bahía. While in the Stone House they prepared to publish the first Texas newspaper, the *Gaceta de Tejas*, which was subsequently printed in Natchitoches, Louisiana. Another newspaper, the Nacogdoches *Mexican Advocate*, was published in the Stone House during the time of the invasion of Texas and proclamation of independence by James Long[qv]. Though neither of these attempts to take Texas from Mexico succeeded, both were proclaimed at the Stone House. Barr's association with the Magee expedition cost him not only the ownership of the structure but his life as well. His business partner, John G. Davenport, succeeded to ownership.

The first *alcalde* of Nacogdoches, James Dill, established his office in the building and added to its uses as a civic center. *Empresario* Haden Edwards[qv] used the Stone House as the headquarters for his grant and later for his abortive Fredonian Republic, and the Mexican garrison stationed in Nacogdoches in the aftermath of the rebellion, commanded by Col. José de las Piedras, also headquartered there. The Stone House later became the primary objective of the Texans from the Ayish Bayou country in the battle of Nacogdoches (1832). Committees of Safety and Correspondence met there during the Texas Revolution, and the house served as a reception center for many Americans, including David Crockett.

John M. Durst purchased the Stone House from John Davenport, heir of Samuel Davenport, in 1829, and sold it to Juan Mora and Vicente Córdova, district judge and district attorney of Nacogdoches, in 1834. Thus the Stone House became a courthouse for the community, and in September 1837 the republic's first official court in East Texas met there with Judge Robert M. Williamson presiding. After Córdova's unsuccessful rebellion against the republic in 1838, the Stone House was purchased by John S. Roberts and his wife, Harriet. They used the building for commercial purposes until it was sold to William and Charles Perkins in 1901.

During much of that time Roberts operated a saloon in the Stone House, and it became an eyesore and a source of embarrassment to the town.

When the Perkins brothers dismantled the building in 1902 to build a modern commercial building, they donated the materials of the Stone House to the Cum Concilio Club. In 1907 this ladies' organization used the materials to construct a building on Nacogdoches public school property, and in 1936 the state of Texas moved the materials to the campus of Stephen F. Austin State Teachers College, where a replica of the original structure was erected for the Texas Centennial celebration. Because of its many uses by Spanish, Mexican, and American military forces, the reconstructed building became known as the Old Stone Fort.

Bibliography

Jordan Holt, The Edwards Empresarial Grant and the Fredonia Rebellion (M.A. thesis, Stephen F. Austin State University, 1977).

Archie P. McDonald, comp., Nacogdoches: Wilderness Outpost to Modern City, 1779–1979 (photocopy, Dolph Briscoe Center for American History, University of Texas at Austin).

Archie P. McDonald, *The Old Stone Fort* (Austin: Texas State Historical Association, 1981).

James G. Partin, A History of Nacogdoches and Nacogdoches County, Texas, to 1877 (M.A. thesis, University of Texas, 1967).

SOURCE; Archie P. McDonald - *Handbook of Texas Online*, s.v. http://www.tshaonline.org/handbook/online/articles/OO/cco3.html

The Alamo

San Antonio de Valero Mission (originally referred to as San Antonio de Padua) was authorized by the viceroy of Mexico in 1716. Fray Antonio de Olivares, who brought with him Indian converts and the records from San Francisco Solano Mission near San Juan Bautista on the Rio Grande, established the mission at the site of present San Antonio in 1718 and named it San Antonio de Valero in honor of Saint Anthony de Padua and the Duke of Valero, the Spanish viceroy.

The present site was selected in 1724; the cornerstone of the chapel was laid on May 8, 1744. Founded for the purpose of Christianizing and educating the Indians, the mission later became a fortress and was the scene of many conflicts prior to the siege of 1836. Its activity as a mission began to wane after 1765, and it was abandoned in 1793, the archives being removed to nearby San Fernando Church. In 1803 the Second Flying Company of San Carlos de Parras, a company of Spanish soldiers from Álamo de Parras, Coahuila, Mexico, occupied the abandoned mission, using its buildings as barracks for a number of years. From this association probably originated the name Alamo

According to some historians, the name was derived from a grove of cottonwood trees growing on the banks of the acequia, *álamo* being the Spanish word for "cottonwood." The Alamo was occupied by Mexican forces almost continuously from 1803 to December 1835, when the fortress under Gen. Martín Perfecto de Cos was surrendered to Texan forces.

On February 23, 1836, Mexican forces under the command of Gen. Antonio López de Santa Anna besieged Col. William B. Travis and his Texas garrison in the Alamo. The siege of the Alamo lasted thirteen days and was climaxed on March 6 with a complete loss of all the combatant Texans after the fall of the Alamo, the building was practically in ruins, but no attempt was made at that time to restore it.

The Republic of Texas, on January 13, 1841, passed an act returning the church of the Alamo to the Catholic Church. After Texas was annexed to the United States, it was declared that the Alamo was property of the United States, and in 1848 the United States government took over the building and grounds and until the Civil War used them for quartermaster purposes. For some time the Alamo was claimed by the city of San Antonio, the Catholic Church, and the United States government. The United States government finally leased the property from the Catholic Church and made some improvements. During the Civil War the Confederates used the building, but after the close of the war the United States government again took over and used it until 1876.

Under an act of April 23, 1883, Texas purchased from the church the Alamo property and placed the Alamo in the custody of the city of San Antonio on condition that the city should care for the building and pay a custodian for that purpose. This system continued until January 25, 1905, when the Texas legislature passed a resolution ordering the governor to purchase that part of the old Alamo fortress occupied by a business concern. It was further ordered that the governor should deliver the property thus acquired, with the property then owned by the state (the chapel of the Alamo), to the Daughters of the Republic of Texas.

A controversy over custody of the Alamo developed between the Daughters of the Republic of Texas and the De Zavala chapter of that organization at San Antonio, and for a time there was a controversy between the Daughters of the Republic of Texas and Governor O. B. Colquitt concerning restoration. Several appropriations for funds to improve the Alamo have been made, the largest being in connection with the celebration of the Texas Centennial. In the 1990s the Alamo was in custody of the Daughters of the Republic of Texas and remained the center of disputes over the custody, presentation, and boundaries of the site.

Bibliography

Frederick Charles Chabot, *The Alamo: Mission, Fortress and Shrine* (San Antonio: Lenke, 1935).

Anne A. Fox,. Anna B. Story, The Alamo from Its Founding to 1937 (M.A. thesis, University of Texas, 1938).

Henry Ryder Taylor, *History of the Alamo and of Local Franciscan Missions* (San Antonio: N. Tengg, 1908).

SOURCE; Amelia W. Williams - *Handbook of Texas Online*, s.v. http://www.tshaonline.org/handbook/online/articles/AA/uqa1.html

HENRY PERRY - FILIBUSTER

(17??–1817). Henry Perry was probably a native of Connecticut and possibly related to the noted military leader Daniel Perry. He joined the Gutiérrez-Magee expedition at Natchitoches in 1812, but it is unclear if he resigned from the United States Army with Magee or was an associate of Samuel Kemper. He was made a captain under Major Kemper. He was promoted to major at Bexar on June 16, 1813, succeeding Reuben Ross[qv] as commander of the Anglos. He led them against Ignacio Elizondo the next day and defeated the Spanish forces at Alazán Creek on June 20.

During the struggle for command of the expedition between Gutiérrez and José Álvarez de Toledo y Dubois, Perry supported the latter, who took command on August 4. After the defeat of the insurgent army at the battle of Medina on August 18, 1813, Perry fled to Natchitoches. He rejoined the United States Army in Louisiana and served at least from December 15, 1814, to June 15, 1815, probably in the quartermaster corps.

In July 1815 he joined a scheme under Juan Pablo Anaya in New Orleans to capture La Bahía and Bexar before joining fellow insurgents near Mexico City. The call for volunteers was so public that on September 1, 1815, President James Madison had to issue a proclamation against filibustering. Perry assembled 300 volunteers at Belle Isle, near the mouth of the Atchafalaya River, but remained without funds to sail to Copano Bay.

He secured a small vessel in November to ferry his troops to Galveston Bay, where they set up camp on what is now called Bolivar Peninsula. On a second voyage, the schooner foundered at the entrance to Galveston Bay and sixty men drowned. Perry ordered his men inland to a wooded eminence (known as Perry's Point through 1830) overlooking the mouth of the Trinity River below the site of modern Anahuac.

About March 1816 Perry abandoned the point and joined Luis Michel Aury[v] on Galveston Island. Perry commanded the Anglo-Americans until the arrival of Francisco Xavier Mina in November 1816; he and his men then served under Col. Guilford Dudley Young, a veteran of the War of

1812. Perry left with the Mina expedition for Soto la Marina, Tamaulipas, on April 7, 1817. After Mina captured Soto la Marina, Perry became convinced that Mina would ultimately fail and was disturbed that Texas had not been cleared of royalists.

Guided by Manuel Costillo of Camargo, he led forty-three men overland for Texas. On June 18, 1817, they reached La Bahía, where Perry demanded that Juan Ignacio Pérez surrender the garrison. When the royalists refused and prepared to attack, Perry and his men fled northeastward. On June 19 the royalist army surrounded them in a nearby wood, where most were killed or wounded. Perry was wounded but refused to surrender, preferring death by his own hand.

Bibliography

Henry Stuart Foote, *Texas and the Texans* (2 vols., Philadelphia: Cowperthwait, 1841; rpt., Austin: Steck, 1935).

Julia Kathryn Garrett, *Green Flag Over Texas: A Story of the Last Years of Spain in Texas* (Austin: Pemberton Press, 1939).

Margaret S. Henson, *Juan Davis Bradburn: A Reappraisal of the Mexican Commander of Anahuac* (College Station: Texas A&M University Press, 1982).

Harris Gaylord Warren, *The Sword Was Their Passport: A History of American Filibustering in the Mexican Revolution* (Baton Rouge: Louisiana State University Press, 1943).

SOURCE; Margaret Swett Henson - *Handbook of Texas Online*, s.v. http://www.tshaonline.org/handbook/online/articles/PP/fpe42.html

History of the State of Texas

The history of the state of Texas begins in 1519, when the first European explorers found the region was populated by various Native American tribes. During the period of 1519 to 1848, all or parts of Texas were claimed by six countries: Spain, France, Mexico, the Republic of Texas, the United States of America—as well as the Confederate States of America in 1861-65.

The first European base was established until 1682, when René-Robert Cavelier, Sieur de La Salle established a French colony, Fort Saint Louis, near Matagorda Bay. The colony was killed off after three years, but its presence motivated Spanish authorities to begin activity. Several missions were established in East Texas; they were abandoned in 1691. Twenty years later, concerned with the French presence in neighboring Louisiana, Spanish authorities again attempted to colonize Texas.

Over the next 110 years, Spain established numerous villages, *presidios*, and missions in the province. A small number of Spanish settlers arrived, in addition to missionaries and soldiers. Spain signed agreements with colonizers from the United States. When Mexico won its independence from Spain in 1821, Mexican Texas was part of the new nation. To encourage settlement, Mexican authorities allowed organized immigration from the United States, and by 1834, over 30,000 Anglos lived in Texas, compared to only 7,800 Mexicans

Angry at the government in Mexico City, Texans fought and won the Texas Revolution in 1835-36. Texas now became an independent nation, the Republic of Texas. Attracted by the rich cotton lands and ranch lands, tens of thousands of immigrants arrived from the U.S. (bringing slaves) and from Germany as well. In 1845, Texas joined the United States, becoming the 28th state. Determined to protect slavery, Texas declared its secession from the United States in 1861 to join the Confederate States of America.

Only a few battles of the American Civil War were fought in Texas; most Texas regiments served in the east. When the war ended the slaves were freed and Texas was subject to Reconstruction, a process that left a residue of bitterness among whites and a second- class status for blacks in a Jim Crow system of segregation.

Cotton and ranching dominated the economy, with railroad construction after 1870 a major factor in the formation of new cities. Toward the end of the 19th century timber became an important industry in Texas as well. In 1901 a petroleum discovery at Spindletop Hill, near Beaumont, created the most productive oil well the world had ever seen. The wave of oil speculation and discovery that followed came to be known as the "Oil Boom", permanently transforming and enriching the economy of Texas

EARLY EUROPEAN EXPLORATION

The first European to see Texas was Alonso Álvarez de Pineda, who led an expedition on behalf of the governor of Jamaica, Francisco de Garay, in 1519. While searching for a passage between the Gulf of Mexico and Asia, Álvarez de Pineda created the first map of the northern Gulf Coast. This map is the earliest recorded document of Texas history.

Between 1528 and 1535, four survivors of the Narváez expedition, including Álvar Núñez Cabeza de Vaca and Estevanico, spent six and a half years in Texas as slaves and traders among various native groups. Cabeza de Vaca was the first European explorer to explore the interior of Texas.

(Source; Wikipedia)

French colonization of Texas: 1684-1689

From 1685 until 1689, a French colony, Fort Saint Louis, existed near what is now Inez, Texas. Explorer Robert Cavelier de La Salle intended to found the colony at the mouth of the Mississippi River, but inaccurate maps and navigational errors caused his ships to instead anchor 400 miles west, off the coast of Texas near Matagorda Bay.

The colony faced numerous difficulties during its brief existence, including hostile Native Americans, epidemics, and harsh conditions. Mindful of his original mission, La Salle led several expeditions to find the Mississippi River. These did not succeed; instead, La Salle explored much of the Rio Grande and parts of East Texas. During one of his absences in 1686 the colony's last ship was wrecked, leaving the colonists unable to obtain supplies from the French colonies in the Caribbean Sea.

As conditions deteriorated, La Salle realized the colony could only survive with help from French settlements in Illinois Country. His last expedition ended along the Brazos River in early 1687 when La Salle and five of his men were murdered by rivals in the group. Although a handful of men reached Illinois, help never arrived. The remaining members of the colony were killed or captured during a Karankawa raid in late 1688. Although the colony lasted only three years, its existence established France's claim to possession of the region that is now Texas, and later supported the claim by the United States to the region as part of the Louisiana Purchase.

Spain learned of La Salle's mission in 1686. Concerned that the French colony could threaten Spain's control over New Spain and the southern portion of North America, Spanish authorities funded multiple expeditions to find and eliminate the settlement. The unsuccessful expeditions helped Spain to better understand the geography of the Gulf Coast region. When the Spanish finally discovered the remains of the French settlement in 1689, they buried the cannons and burned the buildings. Years later, Spanish authorities built a *presidio* at the same location. When the Spanish *presidio* ceased to exist, the site of the French settlement was forgotten. It was rediscovered in 1996, and the area is now an archaeological site.

FRENCH COLONIZATION OF TEXAS, 1684-1689

Although Álvarez de Pineda had claimed the area that is now Texas for Spain, the area was essentially ignored for over 160 years. Its initial settlement by Europeans occurred by accident. In April 1682, French nobleman René-Robert Cavelier, Sieur de La Salle had claimed the entire Mississippi River Valley for France. The following year, he convinced King Louis XIV to establish a colony near the Mississippi, essentially splitting Spanish Florida from New Spain.

La Salle's colonization expedition left France on July 24, 1684 and soon lost one of its supply ships to Spanish privateers. A combination of inaccurate maps, La Salle's previous miscalculation of the latitude of the mouth of the Mississippi River, and overcorrecting for the Gulf currents led the ships to be unable to find the Mississippi. Instead, they landed at Matagorda Bay in early 1685, 400 miles (644 km) west of the Mississippi. In February, the colonists constructed Fort Saint Louis.

René-Robert Cavelier, Sieur de La Salle founded the French colony in Texas.

After the fort was constructed, one of the ships returned to France, and the other two were soon destroyed in storms, stranding the settlers. La Salle and his men searched overland for the Mississippi River, traveling as far west as the Rio Grande and as far east as the Trinity River. Disease and hardship laid waste to the colony, and by early January 1687, fewer than 45 people remained. That month, a third expedition launched a final attempt to find the Mississippi. The expedition experienced much infighting, and La Salle was ambushed and killed somewhere in East Texas.

The Spanish learned of the French colony in late 1685. Feeling that the French colony was a threat to Spanish mines and shipping routes, King Carlos II's Council of war recommended the removal of "this thorn which has been thrust into the heart of America. The greater the delay the greater the difficulty of attainment." Having no idea where to find La Salle, the Spanish launched ten expeditions—both land and sea—over the next three years.

The last expedition discovered a French deserter living in Southern Texas with the Coahuiltecans. The Frenchman guided the Spanish to the French fort in late April 1689. The fort and the five crude houses surrounding it were in ruins. Several months before, the Karankawa had become angry that the French had taken their canoes without payment and had attacked the settlement sparing only four children.

SPANISH TEXAS, 1690-1821

Establishment of Spanish colony

News of the destruction of the French fort "created instant optimism and quickened religious fervor" in Mexico City. Spain had learned a great deal about the geography of Texas during the many expeditions in search of Fort Saint Louis. In March 1690, Alonso De León led an expedition to establish a mission in East Texas. Mission San Francisco de los Tejas was completed near the Hasinai village of Nabedaches in late May, and its first mass was conducted on June 1. On January 23, 1691, Spain appointed the first governor of Texas, General Domingo Terán de los Ríos.

On his visit to Mission San Francisco in August, he discovered that the priests had established a second mission nearby, but were having little luck converting the natives to Christianity. The Indians regularly stole the mission cattle and horses and showed little respect to the priests. When Terán left Texas later that year, most of the missionaries chose to return with him, leaving only 3 religious people and 9 soldiers at the missions. The group also left behind a smallpox epidemic. The angry Caddo threatened the remaining Spaniards, who soon abandoned the fledgling missions and returned to Coahuila. For the next 20 years, Spain again ignored Texas.

After a failed attempt to convince Spanish authorities to reestablish missions in Texas, in 1711 Franciscan missionary Francisco Hidalgo approached the French governor of Louisiana for help. The French governor sent representatives to meet with Hidalgo. This concerned Spanish authorities, who ordered the reoccupation of Texas as a buffer between New Spain and French settlements in Louisiana. In 1716, four missions and a presidio were established in East Texas. Accompanying the soldiers were the first recorded female settlers in Spanish Texas.

ESTABLISHMENT OF THE MISSIONS, 1718-1719

The new missions were over 400 miles (644 km) from the nearest Spanish settlement, San Juan Bautista. Martín de Alarcón, who had been appointed governor of Texas in late 1716, wished to establish a way station between the settlements along the Rio Grande and the new missions in East Texas. Alarcón led a group of 72 people, including 10 families, into Texas in April 1718, where they settled along the San Antonio River. Within the next week, the settlers built mission San Antonio de Valero and a presidio, and chartered the municipality of San Antonio de Béxar, now San Antonio, Texas.

The following year, the War of the Quadruple Alliance pitted Spain against France, which immediately moved to take over Spanish interests in North America. In June 1719, 7 Frenchmen from Natchitoches took control of the mission San Miguel de los Adaes from its sole defender, who did not know that the countries were at war. The French soldiers explained that 100 additional soldiers were coming, and the Spanish colonists, missionaries, and remaining soldiers fled to San Antonio.

The new governor of Coahuila and Texas, the Marquis de San Miguel de Aguayo, drove the French from Los Adaes without firing a shot. He then ordered the building of a new Spanish fort *Nuestra Señora del Pilar de Los Adaes*, located near present-day Robeline, Louisiana, only 12 miles from Natchitoches. The new fort became the *first capital of Texas*, and was guarded by 6 cannon and 100 soldiers. The six East Texas missions were reopened, and an additional mission and presidio were established at Matagorda Bay on the former site of Fort Saint Louis.

Temporary Peace Negotiated with the Indians

In the late 1720s, the viceroy of New Spain closed the presidio in East Texas and reduced the size of the garrisons at the remaining presidios, leaving only 144 soldiers in the entire province. With no soldiers to protect them, the East Texas missions relocated to San Antonio.Spanish missions within the boundaries of what is now the state of Texas.

Although the missionaries had been unable to convert the Hasinai tribe of East Texas, they did become friendly with the natives. The Hasinai were bitter enemies of the Lipan Apache, who transferred their enmity to Spain and began raiding San Antonio and other Spanish areas. A temporary peace was finally negotiated with the Apache in 1749, and at the request of the Indians a mission was established along the San Saba River northwest of San Antonio. The Apaches shunned the mission, but the fact that Spaniards now appeared to be friends of the Apache angered the Apache enemies, primarily the Comanche, Tonkawa, and Hasinai tribes, who promptly destroyed the mission.

In 1762, France finally relinquished their claim to Texas by ceding all of Louisiana west of the Mississippi River to Spain as part of the treaty to end the Seven Years War. Spain saw no need to continue to maintain settlements near French outposts and ordered the closure of Los Adaes, making San Antonio the new provincial capital. The residents of Los Adaes were relocated in 1773. After several attempts to settle in other parts of the province, the residents returned to East Texas without authorization and founded Nacogdoches.

Mexican Control of Texas, 1821-1836

Spanish control of Texas was followed by Mexican control of Texas, and it can be difficult to separate the Spanish and Mexican influences on the future state. The most obvious legacy is that of the language; every major river in modern Texas, except the Red River, has a Spanish or Anglicized name, as do 42 of the state's 254 counties. Numerous towns also bear Spanish names. An additional obvious legacy is that of Roman Catholicism.

At the end of Spain's reign over Texas, virtually all inhabitants practiced the Catholic religion, and it is still practiced in Texas by a large number of people. The Spanish missions built in San Antonio to convert Indians to Catholicism have been restored and are a National Historic Landmark.

The Spanish introduced European livestock, including cattle, horses, and mules, to Texas as early as the 1690s. These herds grazed heavily on the native grasses, allowing mesquite, which was native to the lower Texas coast, to spread inland. Spanish farmers also introduced tilling and irrigation to the land, further changing the landscape.
Texas eventually adopted much of the Anglo-American legal system, but some Spanish legal practices were retained, including homestead exemption, community property, and adoption.

Texas was governed by Mexico from 1821 until 1836. This period began with the victory of Mexico over Spain during its war of independence in 1821. For the first several years of its existence, Mexican Texas operated very similarly to Spanish Texas. The 1824 Constitution of Mexico joined Texas with Coahuila to form the state of Coahuila y Tejas. The same year, Mexico enacted the General Colonization Law, which enabled all heads of household, regardless of race or immigrant status, to claim land in Mexico.

The first empresarial grant had been made under Spanish control to Stephen F. Austin, whose settlers, known as the Old Three Hundred, settled along the Brazos River in 1822. The grant was later ratified by the Mexican government. Twenty-three other empresarios brought settlers to the state, the majority from the United States of America.

Many of the Anglo-American settlers owned slaves. Texas was granted a one-year exemption from Mexico's 1829 edict outlawing slavery but Mexican president Anastasio Bustamante ordered that all slaves be freed in 1830. To circumvent the law, many Anglo colonists converted their slaves into indentured servants for life. By 1836 there were 5,000 slaves in Texas.

Also in 1830, Bustamante outlawed the immigration of United States citizens to Texas. Several new presidios were established in the region to monitor immigration and customs practices. Angry colonists held a convention in 1832 to demand that U.S. citizens be allowed to immigrate. The following year, their Convention of 1833 proposed that Texas become a separate Mexican state. Although Mexico implemented several measures to appease the colonists, Antonio Lopez de Santa Anna's measures to transform Mexico from a federalist to a centralist state provided an excuse for the Texan colonists to revolt.

The first violent incident occurred on June 26, 1832 at the battle of Velasco. On March 2, 1836, Texans signed the Texas Declaration of Independence. The Texas Revolution ended on April 21, 1836 when Santa Anna was taken prisoner following the battle of San Jacinto. Although Texas then governed itself as the Republic of Texas, Mexico refused to recognize its independence.

In 1825, Mexican authorities became concerned with the actions of empresario Haden Edwards in Nacogdoches. Edwards had threatened to confiscate the land of any Mexican already living in the area in which he planned to bring settlers unless the Mexicans could present written deeds to the property. Mexican authorities promptly told him that he did not have the authority to confiscate land and he should honor the claims of the previous settlers.

After multiple confrontations, on December 16, 1826, Edwards, his brothers, and 30 settlers issued a declaration of independence and called themselves the Republic of Fredonia. Other empresarios disassociated themselves from Edwards, and Austin sent 250 militiamen to Nacogdoches to help the Mexican forces quell the revolt. Edwards was finally forced to flee Mexican territory.

After hearing reports of other racial issues, the Mexican government asked General Manuel Mier y Teran to investigate the outcome of the 1825 colonization law in Texas. In 1829, Mier y Teran issued his report, which concluded that most Anglo Americans refused to be naturalized and tried to isolate themselves from Mexicans. He also noted that slave reforms passed by the state were being ignored.

Although many Mexicans wanted to abolish slavery, fears of an economic crisis if all of the slaves were simultaneously freed led to a gradual emancipation policy. In 1823, Mexico forbade the sale or purchase of slaves and required that the children of slaves be freed when they reached fourteen. Any slave introduced into Mexico by purchase or trade would also be freed. By 1825, however, a census of Austin's Colony showed 1,347 Anglo-Americans and 443 people of African descent, including a very small number of free Negroes. Two years later the legislature of Coahuila y Tejas outlawed the introduction of additional slaves into the state and granted freedom at birth to all children born to a slave. The new laws also stated that any slave brought into Texas should be freed within six months.

In 1829, slavery was officially outlawed in Mexico. Austin feared that the edict would cause widespread discontent and tried to suppress publication of it. Rumors of the new law quickly spread throughout the area and the colonists seemed on the brink of revolt. The governor of Coahuila y Tejas, Jose Maria Viesca, wrote to the president to explain the importance of slavery to the Texas economy, and the importance of the Texas economy to the development of the state. Texas was temporarily exempted from the rule. On April 6, 1830, Mexican president Anastasio Bustamante ordered Texas to comply with the emancipation proclamation or face military intervention.

To circumvent the law, many Anglo colonists converted their slaves into indentured servants for life. Others simply called their slaves indentured servants without legally changing their status. Slaveholders wishing to enter Mexico would force their slaves to sign contracts claiming that the slaves owed money and would work to pay the debt. The low wages the slave would receive made repayment impossible, and the debt would be inherited, even though no slave would receive wages until age eighteen. This tactic was outlawed by an 1832 state law which prohibited worker contracts from lasting more than ten years. A small number of slaves were imported illegally from the West Indies or Africa. The British consul estimated that in the 1830s approximately 500 slaves had been illegally imported into Texas. By 1836, there were approximately 5,000 slaves in Texas.

Exportation in the slave-owning areas of the state surpassed that of the non-slave-owning areas. A survey of Texas in 1834 found that the department of Bexar, which was comprised mostly of Tejanos, had exported no goods. The Brazos department, including Austin's colonies and those of Green DeWitt, had exported 600,000 pesos worth of goods, including 5,000 bales of cotton. The department of Texas, which included the eastern settlements, expected to export 2,000 bales of cotton and 5,000 head of cattle.

ANGLO-AMERICAN IMMIGRATION – RESTRICTIONS

David G. Burnet's empresarial contract was cancelled when he could not bring enough settlers. Burnet later became the interim president of the Republic of Texas.Bustamante implemented other measures to make immigration less desirable for Anglo-Americans. He rescinded the property tax law, which had exempted immigrants from paying taxes for ten years. He further increased tariffs on goods entering Mexico from the United States, causing their prices to rise.

The 1830s laws also brought settlement contracts under federal rather than state control. Colonies that did not have at least 150 inhabitants would be canceled. Among the affected colonies were the Nashville Company run by Sterling C. Robertson and the Bay and Texas Land Campy, run by David G. Burnet, Lorenzo de Zavala and Joseph Vehlein. Finally, *he prohibited further immigration to Texas from the United States*, although Anglos would still be welcome in other parts of Mexico. The ban and other measures did not stop U.S. citizens from migrating to Texas by the thousands. By 1834, it was estimated that over 30,000 Anglos lived in Texas, compared to only 7,800 Mexicans.

Anglos often viewed the Mexicans as foreigners and intruders. The feeling was often returned; Rafael Antonio Manchola, son-in-law of empresario Martin de Leon, served as the commander of the presidio at La Bahia from 1828 to 1830 and then as the alcade of Goliad. He warned the military commander for Texas that

"'No faith can be placed in the Anglo-American colonists because they are continually demonstrating that they absolutely refuse to be subordinate, unless they find it convenient to what they want anyway, all of which I believe will be very detrimental to us for them to be our neighbors if we do not in time, clip the wings of their audacity by stationing a strong detachment in each new settlement which will enforce the laws and jurisdiction of a Mexican magistrate which should be placed in each of them, since under their own colonists as judges, they do nothing more than practice their own laws which they have practiced since they were born, forgetting the ones they have sworn to obey, these being the laws of our Supreme Government.'"

Many Americans thought the United States had been cheated out of Texas. American land speculators believed they could make fortunes in the vast region of Texas, and American politicians believed Texas could help maintain a balance of power between free and slave states. In 1827, American president John Quincy Adams offered US$1 million for Texas. Mexican president Guadalupe Victoria refused. Two years later, Andrew Jackson increased the United States' offer to $5 million; President Vicente Guerrero again declined to sell.

In July 1829, Mexican authorities had other concerns, as General Isidro Barradas landed 2,700 Spanish troops to the eastern coast of Mexico, near Tampico in an attempt to reclaim the country for Spain. At the request of the government, Austin mustered a local militia to help defend Texas if the invasion were to reach the northern regions of the country. Yucatan governor Antonio López de Santa Anna led a force of Mexican troops to halt the invasion. Barradas surrendered as his troops suffered greatly from tropical diseases, and Santa Anna was hailed as a hero. During the invasion, the Mexican Congress had granted war powers to President Guerrero, making him essentially a dictator. This alarmed the Anglo colonists in Texas, who were accustomed to a separation of powers.

The Precursor to Revolt

Mier y Teran's 1829 report had recommended new garrisons in Texas which could oversee the Anglo colonists and encourage Mexicans to resettle in the area. The new garrisons were to be partly staffed by convicts. The first was established along Galveston Bay in 1831 at the site of present-day Anahuac. It became the first port in Texas to collect customs. A second custom port, Velasco, was established at the mouth of the Brazos River, while a third garrison established Fort Teran on the Neches River below Nacogdoches to combat smuggling and illegal immigration.

Mier y Teran further ordered the garrison at Bexar to abandon their fort and create a new presidio. Tenoxtitlan was established in 1830 on the west bank of the Brazos River, 100 miles above San Felipe. Shortly after the fort was completed, 50 immigrants from Tennessee arrived in the area under empresario Sterling C. Robertson. The settlers had arrived illegally, as Robertson's contract had been invalidated by Guerroro's 1830 laws. The garrison commander chose not to expel them, instead sending to Mexico for advice. Three months later he received instructions to expel the settlers immediately. He chose not to do so, allowing the Robertson colony to be saved.

The fort closed in 1832. After having received no replacements or supplies, the commander finally ordered all of the soldiers to return to San Antonio.Anahuac was placed under the control of Colonel Juan Davis Bradburn. Bradburn enforced the 1830 laws strictly, angering many colonists. He forbade the state commissioner from granting property titles to squatters and insisted on enforcing the law freeing any slave who set foot in Mexican territory.

In 1832, local men organized a militia, supposedly to protect the settlement from Indian attacks, although all Indians in the area were peaceful. Mexican law forbade residents from creating militias, so Bradburn arrested the ringleader, Patrick C. Jack. Citizens were very angry. In Brazoria, residents held a town meeting to decide what to do. William H. Wharton complained that there was little support within

Austin's colony to oppose Bradburn with military force; he and other advocates of armed conflict felt that their opposition from other settlers was as deep as that of the Mexican soldiers in the area. Bradburn eventually agreed to release Jack, and tensions cooled for a brief period.

In May 1832, Bradburn received a letter, ostensibly from a friend, warning that 100 armed men were stationed 40 miles away, intent on reclaiming the slaves. When Bradburn realized that the letter was a hoax, he arrested Travis for questioning. He intended to send Travis to Matamoros for a military trial on charges of attempted insurrection, with the goal being separation from Mexico. Conviction on this charge would certainly lead to Travis's execution. The settlers were outraged that the arrest did not require a warrant, a statement of charges, or trial by jury. Most were unfamiliar with Mexican law and assumed that the United States Bill of Rights still applied to them. Settlers attacked the Anahuac garrison to free Travis in an event that became known as the Anahuac Disturbances.

Additional settlers had gathered in Brazoria to transport several cannon to aid the group in Anahuac. Colonel Domingo de Ugartechea, who led the garrison at Velasco, at the mouth of the Brazos River, refused to allow the ship carrying the cannon to pass. On June 26, settlers initiated the Battle of Velasco; Ugartachea surrendered the following day.Several days later, Colonel Jose de las Piedras arrived from Nacogdoches to assist Bradburn. He removed Bradburn from his command, and the settlers dispersed.

INSURRECTION OF 1832

Santa Anna against Mexican president Bustamante

In 1832, Antonio Lopez de Santa Anna led an insurrection against Mexican president Bustamante. Although most of the Mexican Army supported the Bustamante administration, this led to a small civil war. Many of the Anglo settlers sided with Santa Anna and followed General José Antonio Mexía, who led soldiers in Texas against Bustamante. Mexia removed the commander at Matamoros from his post. In October, 55 delegates from Texas communities attended the Convention of 1832 in San Felipe. The delegates drafted three petitions to the Congress of Mexico. They wished for an annulment of Article 11 of the colonization law of 1830, which prohibited foreign settlement as well as customs reform, recognition of squatters as valid immigrants, and a separate state for Texas.

A second convention was held the following year. This one, attended by recent arrivals such as Sam Houston, appointed a commission to draft a constitution for a new state of Texas and selected delegates to represent Texas before the federal government. Austin was chosen to deliver the proposed constitution to Santa Anna's government in Mexico City. Although Austin pointed out that Texas had been given permission to form a separate state and had now grown to 46,500 inhabitants, the political chief of Bexar warned the government that the Anglos might be proposing separate statehood as part of a plan to join with the United States. Austin was arrested on November 21, 1833 on suspicion of treason.

The Mexican government attempted to address some of the Texans' concerns. Article 11 was repealed on November 21, 1833, allowing American immigrants to again flow into Texas. Five months later, Coahuila y Tejas separated Texas into three departments, San Antonio-Bexar, Brazos, and Nacogdoches, with political chiefs for each department and more representation in the state legislature. Furthermore, trial by jury was introduced, and English was authorized as a second language.

An Anglo American, Jefferson Chambers, was appointed superior circuit judge of Texas in 1835 and extensions were granted for settlement contracts that had not met their conditions for the number of settlers. Six English-speaking Texan communities were elevated to municipalities.

In March 1833, the capital of the state was transferred from Saltillo to Monclova. The following year, conservatives began urging Santa Anna to overturn the federal system and introduce centralism. Some legislators believed that centralism would be the only way to retain Texas, as newspapers in the United States continued to make statements about the forthcoming annexation of Texas.

When the national congress attempted to centralize the nation, a civil war ensued. As fighting erupted, Saltillo declared that Monclova had been illegally made the state capitol and selected its own governor. Texans in Saltillo recommended establishing a provisional government in Bexar during the unrest to strengthen the autonomy of Texas. Juan Seguin, political chief of Bexar, called for a town meeting to create a government but was forced to postpone it when Mexican troops advanced in the direction of Texas.

TEXAS WAR OF INDEPENDENCE

By the end of the year, however, Santa Anna began to exhibit centralist tendencies, and in 1835 *he revoked the Constitution of 1824* and began consolidating his power. In various parts of the country federalists revolted, and in May 1835 Santa Anna brutally crushed a revolt in Zacatecas; over 2000 noncombatants were killed.

The federalists, including Agustín Viesca, the governor of Coahuila y Tejas, were afraid that Santa Anna would march against Coahuila after subduing the rebels in Zacatecas, so they disbanded the state legislature on May 21, 1835 and authorized the governor to set up an office in a different part of the state. Viezca was arrested as he traveled to San Antonio. Under the pretext of being angry over Viezca's imprisonment the people of Anahuac organized a resistance under Travis. In actuality, they were angry that the two-year grace period on tariffs had ended and the Anahuac customs office had reopened. When Viezca escaped and reached Texas, no one recognized him as governor.

As protests spread across Texas, Mexican officials increasingly blamed the settlers from the United States for the discontent. As historian Alwyn Barr noted, many of the new settlers had "lived entirely within growing Anglo colonies ... and had made few adjustments to the Spanish traditions of Mexico.

THE TEXAS REVOLUTION

The *Texas Revolution* or *Texas War of Independence* was a military conflict between Mexico and settlers in the Texas portion of the Mexican state Coahuila y Tejas. The war lasted from October 2, 1835 to April 21, 1836. However, a war at sea between Mexico and Texas would continue into the 1840s. Animosity between the Mexican government and the American settlers in Texas (who were called Texians), as well as many Texas residents of Mexican ancestry, began with the Siete Leyes of 1835, when Mexican President and General Antonio López de Santa Anna abolished the Constitution of 1824 and proclaimed a new anti-federalist constitution in its place.The new laws were unpopular throughout Mexico, leading to violence in several states. War began in Texas on October 2, 1835, with the Battle of Gonzales. Early Texian successes at La Bahia and San Antonio were soon met with crushing defeat at the same locations a few months later.

The war ended at the Battle of San Jacinto where General Sam Houston led the Texian Army to victory over a portion of the Mexican Army under Santa Anna, who was captured shortly after the battle. The conclusion of the war resulted in the creation of the Republic of Texas in 1836.

The New Country of Mexico (Formed Oct.2.1824)

The Mexican War for Independence severed control that Spain had exercised on its North American territories, and *the new country of Mexico was formed from much of the individual territory that had comprised New Spain.* On October 4, 1824, Mexico adopted a new constitution which defined the country as a federal republic with nineteen states and four territories. The former province of Spanish Texas became part of a newly created state, Coahuila y Tejas, whose capital was at Saltillo, hundreds of miles from the former Texas capital, San Antonio de Bexar (now San Antonio, Texas).

The new country emerged from the war essentially bankrupt. With little money for the military, Mexico encouraged settlers to create their own militias for protection against hostile Indian tribes. Tejas was very sparsely populated and in the hope that an influx of settlers could control the Indian raids, the government liberalized immigration policies for the region. The first group of colonists, known as the Old Three Hundred, arrived in 1822 to settle an empresarial grant that had been given to Stephen F. Austin. Of the 24 empresarios, only one settled citizens from within the Mexican interior; most of the remaining settlers came from the United States.

The Mexican-born settlers in Tejas were soon vastly outnumbered by people born in the United States. To address this situation, President Anastasio Bustamante implemented several measures on April 6, 1830. Chief among these was a prohibition against further immigration to Tejas from the United States, although American citizens would be allowed to settle in other parts of Mexico.

Furthermore, the property tax law, intended to exempt immigrants from paying taxes for ten years, was rescinded, and tariffs were increased on goods shipped from the United States. Bustamante also ordered all Tejas settlers to comply with the federal prohibition against slavery or face military intervention.

These measures did not have the intended effect. Settlers simply circumvented or ignored the laws. By 1834, it was estimated that over 30,000 Anglos (short for anglophones, people whose first language is English) lived in Coahuila y Tejas, compared to only 7,800 Mexican-born citizens. By 1836, there were approximately 5,000 slaves in Tejas.

Texans were becoming increasingly disillusioned with the Mexican government. Many of the Mexican soldiers garrisoned in Tejas were convicted criminals who were given the choice of prison or serving in the army in Tejas. Many Texians were also unhappy with the location of their state capital, which moved periodically between Saltillo and Monclova, both of which were in southern Coahuila, some 500 miles (800 km) away; they wanted Tejas to be a separate state from Coahuila (but not independent from Mexico) and to have its own capital. Some American immigrants and Mexican citizens were accustomed to the rights they had in the U.S. that they did not have in Mexico.

For example, Mexico did not protect Freedom of Religion, instead requiring colonists to pledge their acceptance of Roman Catholicism; Mexican Law required a tithe paid to the Catholic Church In the Mexican interior, violence began to erupt between those who supported federalism and those who wanted a centralized government.

Origin of conflict (economic reasons)

Cotton was in high demand throughout Europe and most farmers in United States focused on cotton. This was the focus of the Texians too, who wanted to raise cotton for big profits. But Mexico's government demanded that the Texians produce corn, grain and beef, dictating which crops each Texian settler would plant and harvest. The other interest of the Texians was representation of their interests in the government.

They believed a closer location for the capital would help to stem corruption and facilitate other matters of government. Texians continued to lobby to overturn the laws of 1830. In April 1833, settlers called a convention to discuss proposed changes in immigration, judicial, and other political policies. The delegates also advocated separate statehood for Tejas and elected Austin to carry a proposed state constitution to Mexico City. The new Mexican President, Antonio Lopez de Santa Anna, approved many of the proposals, but refused to agree to separate statehood; Austin was jailed when he wrote a letter advocating that Texians act unilaterally on statehood.

The number of American immigrants entering Tejas quickly escalated. Santa Anna believed that the influx of American immigrants to Texas was part of a plot by the U.S. to take over the region. In 1834, because of perceived troubles within the Mexican government, Santa Anna went through a process of dissolving state legislatures, disarming state militias, and abolishing the Constitution of 1824. He also imprisoned some cotton plantation owners who refused to raise their assigned crops, which were intended to be redistributed within Mexico instead of being exported. These actions triggered outrage throughout Mexico

Mexican preparation

In early 1835, as the Mexican government transitioned from a federalist model to centralism, wary colonists in Texas began forming Committees of Correspondence and Safety. A central committee in San Felipe de Austin coordinated their activities The Texians staged a minor revolt against customs duties in June; these Anahuac Disturbances prompted Mexican President Antonio López de Santa Anna to send additional troops to Texas. In July, Colonel Nicolas Condelle led 200 men to reinforce Presidio La Bahía.

The following month, a contingent of soldiers arrived in Béxar with Colonel Domingo de Ugartechea. Fearing that stronger measures were needed to quell the unrest, Santa Anna ordered his brother-in-law, General Martín Perfecto de Cos to "repress with strong arm all those who, forgetting their duties to the nation which has adopted them as her children, are pushing forward with a desire to live at their own option without subjection to the laws". Cos landed at Copano Bay on September 20 with approximately 500 soldiers. Austin was released in July, having never been formally charged with sedition, and was in Texas by August. Austin saw little choice but revolution. A consultation was scheduled for October to discuss possible formal plans to revolt, and Austin sanctioned it

THE TEXAS OFFENSIVE

Before the consultation could happen, however, in accordance with Santa Anna's nationwide call to disarm state militias, Colonel Domingo Ugartechea, who was stationed in San Antonio, ordered the Texians to return a cannon given to them by Mexico that was stationed in Gonzales. The Texians refused. Ugartechea sent Lieutenant Francisco Castañeda and 100 dragoons to retrieve it. When he arrived at the rain-swollen banks of the Guadalupe River near Gonzales, there were just eighteen Texians to oppose him. Unable to cross, Castañeda established a camp, and the Texians buried the cannon and called for volunteers. The Texians stalled for several days until reinforcements arrived. Texians attacked early on October 2, 1835. The Battle of Gonzales ended with a Mexican withdrawal. Only one Texian was injured when he fell off his horse during the skirmish.

After learning of the Texian victory, Cos made haste for Béxar. He left with the bulk of his soldiers on October 5, but because he was unable to find adequate transportation, most of his supplies remained at La Bahía. Unaware of Cos's departure, on October 6 Texians in Matagorda decided to march on the Mexican garrison at Presidio La Bahía in Goliad. They intended to kidnap Cos and, if possible, steal the estimated $50,000 that was rumored to accompany him. On October 10, the Texians stormed the presidio, and the Mexican garrison surrendered after a 30-minute battle One Texian was wounded, and estimates of Mexican casualties range from one to three soldiers killed and from three to seven wounded Approximately 20 soldiers escaped. They warned the garrisons at Copano and Refugio of the advancing Texians; those garrisons abandoned their posts and joined the soldiers at Fort Lipantitlán, near San Patricio.

The Come and take it flag, used during the revolution

"Come and take it" was a slogan used in the Texas Revolution in 1835. In March 1831, Juan Gomez, a Lieutenant in the Mexican Army, worked alongside Tadeo Ortiz, a consul at Bordeaux, France, and granted a small cannon to the colony of San Antonio. The small bronze cannon was received by the colony and signed for by Randy Tumlinson. It was then transported to Gonzales, Texas and later was the object of Texas pride.

At the minor skirmish known as the Battle of Gonzales, a small group of Texans successfully resisted the Mexican forces who had orders to seize their cannon. As a symbol of defiance, the Texans had fashioned a flag containing the phrase along with a black star and an image of the cannon which they had received six years earlier from Mexican officials.The Texians confiscated over $10,000 in food, blankets, clothing, and other provisions. For the next three months, the provisions were parceled out among companies in the Texian Army.

Over the next several days, Texians continued to gather at La Bahia. Austin ordered that 100 men remain at Goliad, under the command of Captain Philip Dimmitt, while the rest should join the Texian Army in marching on Cos's troops in Béxar. Within days of his appointment, Dimmitt began advocating for an attack on Fort Lipantitlán.

THE BRIEF BATTLE OF LIPANTITLÁN

Dimmitt believed that Texian control of Fort Lipantitlán would "secure the frontier, provide a vital station for defense, create instability among the centralists, and encourage Mexican federalists". The Mexican soldiers at Fort Lipantitlán intimidated the settlers in San Patricio, leaving them afraid to openly support the federalists who defied Mexican president Antonio López de Santa Anna.

On October 31, Dimmitt sent a group of men under Adjutant Ira Westover to take the fort. They arrived at Fort Lipantitlán late on November 3 and took the undermanned fort without firing a shot. The next day, the Texians dismantled the fort. As they prepared to return to Goliad, the remainder of the Mexican garrison, who had been out on patrol, approached.

The Battle of Lipantitlán lasted only 30 minutes, and resulted in the retreat of the Mexican soldiers. Their departure left only one remaining group of Mexican soldiers in Texas, those under Cos at Béxar. The Texians controlled the Gulf Coast, so all communication with the Mexican interior would now be transferred overland. The long journey left Cos unable to quickly request or receive reinforcements or supplies.

SIEGE OF BEXAR

While Dimmitt supervised the Texian forces along the Gulf Coast, Stephen F. Austin worked to organize the men gathered in Gonzales into a cohesive army. On October 13, Austin led the newly formed Texian Army toward Bexar to engage Cos and his troops. One week later, the men reached Salado Creek and initiated a siege of Bexar. The Texians gradually moved their camp nearer Bexar, and on October 27 had made camp at Mission San Francisco de la Espada.

That afternoon Austin sent James Bowie and James Fannin with a contingent of men to find a closer campsite. The men realized that Mission Concepción was a good defensive spot. Rather than return immediately to Austin, as their orders specified, Bowie and Fannin instead sent a courier to bring Austin directions to Concepción. The next day, an angry Austin issued a statement threatening officers who chose not to follow orders with court-martial.

Cos had learned that the Texian army was temporarily divided and sent Ugartechea and troops to engage Bowie and Fannin's men. The ensuing Battle of Concepción, which historian J.R. Edmondson describes as "the first major engagement of the Texas Revolution",was the last offensive against the Texians that Cos would order. Although historian Alwyn Barr believed that the battle "should have taught ... lessons on Mexican courage and the value of a good defensive position",historian Stephen Hardin believes that "the relative ease of the victory at Concepción instilled in the Texians a reliance on their long rifles and a contempt for their enemies".

The Texian volunteers had little or no experience as professional soldiers, and by early November many had begun to miss their homes. As the weather turned colder and rations grew smaller, many soldiers became sick, and groups of men began to leave, most without permission. On November 18, however, a group of volunteers from the United States, known as the New Orleans Greys, joined the Texian Army.

Unlike the majority of the Texian volunteers, the Greys looked like soldiers, with uniforms, well-maintained rifles, adequate ammunition, and some semblance of discipline. The Greys, as well several companies of Texians who had arrived recently, were eager to face the Mexican Army directly. The Texian volunteers, however, were becoming discouraged with the siege. Within days Austin resigned his command to become a commissioner to the United States; Texians elected Edward Burleson as their new commander.

On November 26, Burleson received word that a Mexican pack train of mules and horses, accompanied by 50–100 Mexican soldiers, was within 5 miles of Bexar. After a near mutiny, Burleson sent Bowie and William H. Jack with cavalry and infantry to intercept the supplies In the subsequent skirmish, the Mexican forces were forced to retreat to San Antonio, leaving their cargo behind. To the disappointment of the Texians, the saddlebags contained only fodder for the horses; for this reason the battle was later known as the Grass Fight.

Although the victory briefly uplifted the Texian troops, morale continued to fall as the weather turned colder and the men grew bored. Burleson proposed that the army lift the siege and retreat to Goliad until spring. His war council was ambivalent until Colonel Ben Milam stood up and yelled "Who will go with old Ben Milam into San Antonio?"

Several hundred soldiers, including the New Orleans Greys, agreed to participate in the attack, which commenced on December 5. Milam and Colonel Frank W. Johnson led two columns of men into the city, and for the next few days they fought their way from house to house towards the fortified plazas where the Mexican soldiers waited. Milam was killed by a sharpshooter on December 7.

On December 9, Cos and *the bulk of his men withdrew into the Alamo Mission* on the outskirts of Bexar. Cos presented a plan for a counterattack; cavalry officers believed that they would be surrounded by Texians and refused their orders. Possibly 175–soldiers from four of the cavalry companies left the mission and rode south. Sanchez Navarro said the troops were not deserting but misunderstood their orders and were withdrawing all the way to the Rio Grande. The following morning, Cos called Sanchez Navarro to the Alamo and gave him orders to "go save those brave men. ... Approach the enemy and obtain the best terms possible" On December 11, the Texians officially accepted Cos's surrender.

Under the terms of the surrender, Cos and his men would leave Texas and no longer fight against the Constitution of 1824. With his departure, there was no longer an organized garrison of Mexican troops in Texas, and many of the Texians believed that the war was over. Johnson described the battle as "the period put to our present war". Burleson resigned his leadership of the army on December 15 and returned to his home. Many of the men did likewise, and Johnson assumed command of the 400 soldiers who remained. Soon after, a new contingent of Texians and volunteers from the United States arrived with more heavy artillery.

According to Barr, the large number of American volunteers "contributed to the Mexican view that Texian opposition stemmed from outside influences". Within several weeks of the Mexican surrender, Johnson and Dr. James Grant enticed 300 of the Texians to join them in preparing to invade Mexico, leaving Colonel James C. Neill to oversee the remaining 100 soldiers garrisoned at the Alamo. Although the Matamoros Expedition, as it came to be known, was but one of many schemes to bring the war to Mexico, nothing came of it. On November 6, 1835, the Tampico Expedition under José Antonio Mexía left New Orleans, intending to capture the town from the Centralists. The expedition failed. These independent missions drained the Texian movement of supplies and men

Provisional government

In Gonzales, the consultation scheduled for the month before finally got underway after enough delegates from the colonies arrived to signify a quorum. After bitter debate, they finally created a provisional government that was not to be separate from Mexico but only to oppose the Centralists. They elected Henry Smith as governor and Sam Houston was appointed commander-in-chief of the regular Army of Texas. There was no regular army yet; Austin's army was all volunteers, so Houston would have to build one. They had more land than money so land was chosen as an incentive to join the army; extra land would be given to those who enlisted as regulars and not as volunteers.

The provisional government commissioned privateers and established a postal system. A merchant was sent to the U.S. to borrow $100,000. They ordered hundreds of copies of various military textbooks. They gave Austin the option to step down as commander of the army in Béxar and go to the U.S. as a commissioner. On November 24, 1835, Austin stepped down as general. Elections were held, and Colonel Edward Burleson became Austin's successor.

Santa Anna's Offensive

Army of Operations

As early as October 27, Santa Anna had been making plans to quell the unrest in Texas. He stepped down from his duties as president to lead what he dubbed the Army of Operations in Texas, which would relieve Cos and put an end to the Texian revolt. Santa Anna and his soldiers believed that the Texians would be quickly cowed. The Mexican Secretary of War, José María Tornel, wrote: "The superiority of the Mexican soldier over the mountaineers of Kentucky and the hunters of Missouri is well known. Veterans seasoned by 20 years of wars can't be intimidated by the presence of an army ignorant of the art of war, incapable of discipline, and renowned for insubordination."

The units comprising the Army of Operations were generally operating at under full strength, and many of the men were raw recruits. A majority of the troops had been conscripted or were convicts who agreed to serve in the military instead of jail.The Mexican officers knew that the Brown Bess muskets they carried lacked the range of the Texian weapons, but Santa Anna was convinced that his superior planning would nonetheless result in an easy victory.As part of his preparations, Santa Anna orchestrated a warning to the American citizens who were flocking to Texas. At his behest, the Mexican Congress passed a resolution stating: Foreigners landing on the coast of the Republic or invading its territory by land, armed, and with the intent of attacking our country, will be deemed pirates and dealt with as such, being citizens of no nation presently at war with the Republic and fighting under no recognized flag.
All foreigners who shall import, by either sea or land, in the places occupied by the rebels, either arms or ammunition or any kind for their use, will be deemed pirates and punished as such.

In this time period, captured pirates were executed immediately. The resolution thus gave the Mexican Army *permission to take no prisoners in the war against the Texians*. Santa Anna also sent a strongly worded letter to Andrew Jackson, the United States president, warning that any

Americans found fighting the Mexican government *would be treated as pirates.*

The letter was not widely distributed, and it is unlikely that most of the American recruits serving in the Texian Army were aware of that there would be no prisoners of war.

By December 1835 6,019 soldiers had gathered at San Luis Potosi to march into Texas. Several of Santa Anna's officers argued that the Army of Operations should advance along the coast, so that they would be able to receive additional supplies via sea. Instead, Santa Anna ordered the army inland to Bexar, the political center of Texas and the site of Cos's defeat; Santa Anna wanted to restore the reputation of his family after his brother-in-law's embarrassing surrender. The long march would also provide an opportunity to train the new recruits. In late December, the army began the march north.

Progress was slow. There were not enough mules to transport all of the supplies, and many of the teamsters, all civilians, quit when their pay was delayed. The large number of *soldaderas*–women and children who followed the army–reduced the already scarce supplies. The soldiers were soon reduced to partial rations. After reaching Saltillo, the army halted for two weeks so that Santa Anna could recover from an illness. Officers took advantage of the break to train the men.

Many of the new recruits did not know how to use the sights of their guns, and many refused to fire from the shoulder because of the large recoil. The march into Texas resumed on January 26, and the army crossed the Rio Grande on February 12.Temperatures in Texas reached record lows, and by February 13 an estimated 15–16 inches of snow had fallen. A large number of the new recruits were from the tropical climate of the Yuca tán, and some of them died of hypothermia. Others contracted dysentery. Soldiers who fell behind were sometimes killed by Comanche raiding parties. Nevertheless, the army continued to march toward Bexar. As they progressed, settlers in their path in South Texas evacuated northward. The Mexican army ransacked and occasionally burned the vacant homes.

Goliad Campaign

The Goliad Campaign refers to a series of battles which occurred in 1836 as part of the Texas Revolution, which ultimately led to the *Goliad massacre*. Troops from the army of Mexico defeated Texian forces in several clashes, and eventually *massacred many of their prisoners of war, spreading outrage and resentment* among the population of the fledgling Republic of Texas.

General José Urrea marched into Texas from Matamoros, making his way north following the coast of Texas, thus preventing any foreign aid by sea and opening up an opportunity for the Mexican Navy to land much needed provisions. After surprising Colonel Frank Johnson and his troops at the Battle of San Patricio, Urrea's forces defeated a small Texian force at the Battle of Agua Dulce on March 2, 1836. Urrea then led his troops toward Goliad, where Colonel James Fannin commanded the only Texian Army troops outside *the Alamo*.

Fannin delayed his retreat, and his force of about 300 men was caught on the open prairie at a slight depression near Coleto Creek and made three charges at a heavy cost in Mexican casualties. Overnight, Urrea's forces surrounded the Texians, brought up cannon and reinforcements, and induced Fannin's surrender under terms the next day, March 20. About 342 of the Texian troops captured during the Goliad Campaign were executed a week later on Palm Sunday, March 27, 1836, under Santa Anna's direct orders, widely known as the Goliad Massacre.

According to Harbert Davenport, "The impact of the Goliad Massacre was crucial. Until this episode Santa Anna's reputation had been that of a cunning and crafty man, rather than a cruel one...together with the fall of the Alamo, branded *both Santa Anna and the Mexican people with a reputation for cruelty and aroused the fury of the people of Texas, the United States, and even Great Britain and France*, thus considerably promoting the success of the Texas Revolution."

Background

By 1835, about 30,000 Americans, mostly from the southern states, had emigrated to the Mexican province of Texas. The province had been opened to Catholic American immigrants (Mexico, like Spain, saw religion as part and parcel of citizenship) when Mexico became independent from Spain in 1821. Mexico became a Federal republic, like the United States, in 1824, and the province of Texas was joined with the state of Coahuila in 1827.

Many in Texas were working to get the Mexican Federal government to recognize Texas as a separate state within Mexico. They wanted their own legislature and local government. Some of these settlers had brought slaves with them to work on the ranches and expansive cotton fields (about 5000 slaves by 1835). When Mexico abolished slavery, a Tejano delegate to the legislature in Saltillo got a bill passed allowing Texas slave holders to avoid the law. As a result, a growing cotton economy flourished in Texas.

As Mexico sank into dictatorship under General Santa Anna, and the Federal Republic and constitution of 1824 were abolished, many Tejanos and Anglo Texians joined in a rebellion against the centralization of government power under Santa Anna. Some wanted a return to Federalism and the restoration of the 1824 constitution. Others wanted independence for Texas. Still others wanted Texas to be joined to the United States. When the conflict began in 1835, settlers proceeded to seize key towns and forts, including the Alamo in San Antonio and the Presidio La Bahia in the town of Goliad.

Texan Force and the Battle of Refugio

In Goliad, Colonel James Fannin commanded the Texan force of nearly 500 trained soldiers and militia. Fannin had chosen to keep his troops at Goliad mainly because it had a fort, from which he believed it would be easier to fight than out in the open. Fannin also believed that by occupying Goliad he could prevent Mexican commander Antonio López de Santa Anna from drawing supplies from the Gulf of Mexico. However, Fannin was called to assist Colonel William Travis at the Alamo.

On February 26, 1836, he attempted to march to San Antonio but turned back at the San Antonio River because of the inability to travel with the artillery and arms. Meanwhile, Mexican forces under General José de Urrea were quickly reaching Goliad, and they defeated three Texan forces at the Battle of San Patricio on February 27, the Battle of Agua Dulce on March 2, and the Battle of Refugio on March 12.

Battle of Refugio

The *Battle of Refugio* was fought from March 12–March 15, 1836, near Refugio, Texas. Mexican General José Urrea and *1,500 Centralista soldiers fought against Amon Butler King and his 28 American volunteers* and Lieutenant Colonel William Ward and his approximately 120 Americans. The battle, a part of the Goliad Campaign of the Texas Revolution, resulted in a Mexican victory and splintered Texan resistance.

Fannin sent Captain Amon Butler King on a mission to Refugio on March 11, to remove several noncombatant families out of the path of Urrea's army. On March 13, King was surrounded by elements of the Mexican army and sent out a plea for help to Fannin, who sent Lieutenant-Colonel William Ward and the Georgia Battalion to reinforce him. Urrea, meanwhile, heard of their presence and marched a flying column of 300 Mexican troops to Refugio, hoping to overtake the Texians.

On March 14, the two sides clashed and fought until dark, when Captain King led his men in an independent escape attempt. They were overtaken shortly *and surrendered* for lack of munitions. *Captain King and all but one man were executed in short order.*

The Georgia Battalion attempted to escape to Victoria, where they expected to link up with the balance of Fannin's command. After wandering lost on the coastal prairie for several days, the Georgia Battalion reached Victoria, only to find it in the possession of the Mexican army. Short on munitions and supplies, with no hope of rescue, the majority of Ward's men voted to surrender under good terms. Only then were they made aware that Colonel Fannin and his men had already surrendered following the Battle of Coleto. They were marched back to Goliad to face the same fate as the rest of Fannin's command.

The Battle of Coleto (Fannin's Retreat)

Fannin was ordered by General Sam Houston on March 11, 1836, to abandon Goliad and retreat to the Guadalupe River near Victoria. However, he had sent most of his carts and horses with Ward to Refugio and had no cavalry. He sent couriers to Ward, but all of them were intercepted by Urrea's cavalry. Fannin hoped to retreat to Victoria, but he hesitated for several days. Urrea, meanwhile, sent cavalry to surround and isolate Goliad. He linked up with several more units of Mexican infantry, bringing the total number of Mexican troops in the area to 1,500.

Battle of Coleto

The *Battle of Coleto*, also known as the Battle of Coleto Creek, the Battle of the Prairie, and the Batalla del encinal del Perdido, was fought on March 19 and 20, 1836, during the Goliad campaign of the Texas Revolution. In February, General José de Urrea led a branch of the Mexican army up the Gulf Coast of Mexican Texas toward Goliad, where a large contingent of soldiers from the Texian Army were garrisoned under Colonel James W. Fannin. Simultaneously, Mexican president Antonio Lopez de Santa Anna led a larger force into the Texan interior, where on March 6 his troops won the Battle of the Alamo. After learning of the Alamo's defeat, Texian general Sam Houston order Fannin to retreat from Goliad and join the rest of the army in Victoria.

On March 19, Fannin led his men on a leisurely retreat from Goliad. Mexican troops surrounded the Texians later in the day, before Fannin could reach the shelter of a grove of timber at Coleto Creek, some 400 yards (370 m) away. Texians formed a square in the middle of the prairie and attempted to defend their position. Although Mexican troops launched three separate attacks against the square, they could not penetrate the Texian position. As night fell, Mexican sharpshooters were able to wound and kill more Texians. With little water to give to the wounded or to cool their artillery, the Texians felt they were unable to withstand further fighting. On the morning of March 20, the Texians surrendered.

Urrea attempted to secure honorable terms for his Texian prisoners. However, Santa Anna had received authorization from the Mexican Congress to treat all captured Texian troops as pirates rather than prisoners-of-war. Against Urrea's pleadings, all of the Texians were sentenced to death. A few Texians escaped the Goliad Massacre on March 27, but most were killed.

Background

Colonel James Fannin was the commander of the Texan troops at Fort Defiance in late 1835 and early 1836. During the siege of the Alamo in February 1836 he attempted a march of 100 miles to relieve the Texan forces at the Alamo but due to poor preparation for the journey and word that general Urrea's Mexican forces were approaching Goliad, he turned back. After the Alamo fell to Santa Anna's forces the Texians received orders from General Sam Houston to fall back to Victoria. Fannin therefore abandoned the fort but proceeded without adequate supplies and without haste on his retreat.

Prelude

By 09:00 on March 19 they began their retreat from Goliad, during a period of heavy fog. The Texian force included the San Antonio Greys, the Red Rovers, the Mustangs commanded by Burr H. Duval, a militia from Refugio commanded by a man named Frazer, Texian regular soldiers commanded by Ira Westover, and the Mobile Greys. Nine heavy artillery pieces with different calibers were ordered by Fannin to be taken by the Texians, along with 1000 muskets, but he neglected to ensure that a good amount of food and water was transported. Carts loaded with heavy equipment were being pulled by hungry and tired oxen. Urrea did not realize the Texians had left until 11:00.

The two hour lead was removed, when a Texian cart crossing the San Antonio River broke, a cannon had to be brought out of the river, and Fannin ordered that the oxen be allowed to graze for a period of time after the Texians had proceeded about a mile past Manahuilla Creek, resulting in the retreat being stopped. John Shackelford, Burr H. Duval, and Ira Westover opposed Fannin's decision to allow the oxen to graze, arguing that they should continue their retreat until they reached the protection of the Coleto Creek timber. Shackelford would state that Fannin argued that the Mexican army against them was poor, and that Urrea would not follow them.

In an effort to catch Fannin's troops Urrea left his artillery, and some of his men, in Goliad. He began his pursuit with, according to Mexican sources, 80 cavalrymen and 360 infantrymen. Mexican mounted scouts determined the location of the Texians, and reported the size of the force, which Urrea concluded was smaller than he originally thought. As a result, he ordered 100 of his soldiers to go back to Goliad to help secure Presidio La Bahía. He also ordered the artillery he left in Goliad to be brought to him, and that the artilley would be escorted by some of the soldiers he was sending back.

Meanwhile, Horton's 30 cavalrymen were serving as advance guards, and were positioned to cover all sides of the Texian force. The rear guard was not alert, and did not detect the Mexican cavalry that was approaching the Texians. Shortly after they resumed their march another Texian cart broke down, and its cargo had to be transferred to another one, delaying the retreat again. Shortly after Fannin had sent Horton to scout the Coleto Creek timber that was in sight, the Mexican cavalry overtook the Texians. As the Texians tried to get to high ground 400 to 500 yards away from the position they were in when the cavalry overtook them, the ammunition cart broke.

The Texian soldiers formed a square against the Mexicans. The high grass of the prairie meant the Texian view of the Mexicans was impaired. The Texians had little water. Each Texian soldier received three to four muskets. The square was three ranks deep. The front line contained the San Antonio Greys and Red Rovers, whilst Duval's Mustangs and Frazer's Refugio militia formed part of the rear line. The left flank was covered by Westover's regulars, whilst the right was protected by the Mobile Greys. In the corners of the square, the artillery had been positioned. Fannin stood in the rear of the right flank. In addition, a number of sharpshooters were deployed around Abel Morgan's hospital wagon, which could no longer be moved after the ox that was moving it was killed by Mexican fire.

The Mexican soldiers then attacked the square. The left of the Texian square was confronted by the rifle companies under Morales, and the right was assaulted by the grenadiers and part of the San Luis Battalion. The Mexican formations involved in this attack on the right of the square was under the personal supervision of Urrea. The Jiménez Battalion under Col. Mariano Salas fought the front, and Col. Gabriel Núñez's cavalry was ordered against the rear of the square. By sunset, when Urrea ordered the Mexicans to cease any more major attacks against the square due to a lack of Mexican ammunition, the majority of the action of 19 March was over.

The Mexicans had assaulted the square three times. Making effective use of their bayonets, multiple muskets, and nine cannons, the Texians had prevented the Mexicans each time from breaking the square. Urrea said that he was impressed with the fact that the Texians had managed to maintain the square against the three charges, and he was also impressed with the Texian weapon fire. Dr. Joseph H. Barnard, a Texian, recorded that by sunset seven Texians had been killed. He also recorded that sixty

Texians, *including Fannin, had been wounded.* Forty of the sixty had been wounded several times.

After sunset, Urrea ordered Mexican sharpshooters to be positioned in the tall grass around the square, and that they fire at the Texians. Before Texian sharpshooters were able to remove the threat posed by the Mexican sharpshooters, by firing at the flash caused by the Mexican guns, the Mexican sharpshooters were able to inflict more Texian casualties. As a result of all the fighting that occurred on 19 March, the Texians had suffered at least ten dead and sixty wounded, whilst the Mexicans suffered an unspecified high amount of casualties.

The fighting of 19 March had not demoralised the Texian soldiers. They were encouraged by the thought that Horton would succeed in getting Texian reinforcements from Guadalupe Victoria to Fannin. However, Horton had not been able to break through the Mexican defences.

During the day's fighting the Texian soldiers that were retreating to Guadalupe Victoria after the earlier battle of Refugio were close enough to Fannin to hear gunfire. However, they were exhausted and hungry, and did not move to the square. Urrea stationed three detachments of Mexican troops around the square, to prevent the Texians in the square from escaping, and during the night Mexican false bugle calls were sounded to keep the Texians alert.

The Texians' lack of water, and the inability to light fires in the square, meant the wounded Texians could not be treated. The pain being experienced by the wounded resulted in the general decrease in morale amongst the Texian soldiers during the night. The poor weather during the night further lessened the morale of the soldiers.

The lack of water also meant that the artillery could not be used effectively the next day, because water was needed to cool and clean the cannons. The fighting of 19 March had also left many Texian artillerists casualties, and ammunition for the cannons was low. All these factors contributed to the conclusion by Fannin and other officers during the night that they could not sustain another day of fighting.

An idea for the Texians to escape to a more defendable position under cover of darkness, before Urrea received reinforcements, was rejected because it was decided that those who were too injured to escape, which included friends and relatives of unwounded Texians, should not be left behind. It was therefore decided that the Texians should attempt to make another stand from their current position the next day.

As a result, during the night, the Texians dug trenches and erected barricades of carts and dead animals. Urrea, meanwhile, had been reinforced with munitions, fresh troops, and two or three artillery pieces from Goliad. He positioned the Mexican artillery on the slopes overlooking the Texian square.

At 06:15 on March 20, the Mexicans were grouped for battle. After one or two rounds were fired by Mexican artillery Fannin and his officers re-iterated their conclusion that the Texians could not take another day's fighting, and decided to seek honourable terms for surrender. They drafted terms of surrender, which included statements that the Texian wounded would be treated, that they would be gain all the protection expected as prisoners of war, and that they would be paroled to the United States of America.

However, Santa Anna had stated earlier that any Texian can only be allowed to surrender unconditionally. As a result, Urrea could not guarantee that all the terms would be followed by Santa Anna. He stated that he would talk to Santa Anna on behalf of the terms of surrender presented by the Texians. The document of surrender was signed by Benjamin C. Wallace, Joseph M. Chadwick, and Fannin. As a result of the signing, the battle of Coleto ended.

THE GOLIAD MASSACRE

Those Texians that could walk were sent to Goliad, under Mexican escort. It would take until about March 23 until those Texians that could not walk were transported to Goliad. During that time, Texian physicians were told that wounded Mexicans were a priority to treat, as opposed to the wounded Texians. Fannin arrived in Goliad on March 22. Urrea, meanwhile, had moved onto Guadalupe Victoria, from where he wrote to Santa Anna a letter recommending that the Texian prisoners should be treated with clemency. However, Santa Anna did not follow Urrea's recommendation, instead ordering the Mexican commander Jose Nicolas de la Portilla at Goliad to execute the Texian prisoners.

Coleto Creek

On March 18, Urrea's advance scouts were viewing Goliad. Not until the morning of March 19 did Fannin finally begin his retreat from Goliad. Weighted down with cannon and 500 extra guns, Fannin burned his extra supplies in an attempt to lighten the load. The Texians had traveled only six miles (10 km) from their fort when, on March 19, the Mexican army engaged the Texians on an open prairie. The Texians were less than one mile (1.6 km) from the safety of the tree line of Coleto Creek.

That afternoon, Urrea's cavalry encircled the Texians. After several hours of fighting, the Mexicans had suffered an estimated 200 casualties and the Texians nine killed and about sixty wounded. Fighting was halted that day at dark. Fannin could have probably have cut his way to safety, but he refused to abandon the wounded. The next morning, seeing Urrea receive one hundred more men and three more artillery pieces, Fannin agreed to surrender. Fannin met to discuss terms with Colonel Juan José Holzinger (often misspelt as "Holsinger"). Fannin was unaware General Santa Anna had decreed execution for all rebels, but General Urrea negotiated surrender "at the disposal of the Supreme Mexican Government", falsely stating that no prisoner taken on those terms had lost his life.

The Texans were marched back to Goliad and held as prisoners, each believing that they were going to be set free in a matter of weeks.

Knowing the prisoners' probable fate, General Urrea departed Goliad, leaving command to Colonel Jose Nicolas de la Portilla, and later writing to Santa Anna to ask for clemency for the Texians. Urrea wrote in his diary that he "...wished to elude these orders as far as possible without compromising my personal responsibility." On March 26, 1836, 19:00, Portilla received orders from Santa Anna in triplicate to execute the prisoners. At around 8 a.m. on Palm Sunday, March 27, 1836, Colonel Portilla had the 342 Texians marched out of Fort Defiance into three columns on the Bexar Road, San Patricio Road, and the Victoria Road.

Once the columns reached their selected location, the Mexican *soldados* formed into two ranks on one side of the captives. The Texians were then fired on at point-blank range only a few hundred yards from the fort. The wounded and dying were then clubbed and stabbed. Those who survived the initial volley were run down by the Mexican cavalry. Fannin's men wounded in the Battle of Coleto were shot or bayoneted where they lay. *Colonel Fannin was the last to be executed, after seeing his men butchered.*

Their bodies were stacked into piles and burned. There were twenty-eight Texians who did manage to escape by feigning death and other means. Three known survivors escaped to Houston's army and participated in the Battle of San Jacinto. In some accounts of the Goliad Massacre, a Mexican woman, Francisca (Francita, Panchita or Pancheta) Alavez, sometimes referred to by other names (Alvarez or Alavesco), rescued about 20 Texian soldiers and became known as "The Angel of Goliad."

Other people known to rescue some prisoners were: Juan Holzinger (saved two German Texians captured among Capt. Amon B. King's men and at Victoria he saved twenty-six of Lt. Col. William Ward's troops by claiming to need them to transport cannons across the San Antonio River), Colonel Garay, Father Maloney (also referred as Molloy), Urrea's wife and an unnamed girl.

Written account of the massacre

One survivor of the massacre, a young German named H. Von Ehrenberg, wrote an account of the murders on December 3, 1853. He said the Texan prisoners and American volunteers numbered about 400, while the Mexican captors totaled 700, in addition to cavalry and smaller groups of Mexican soldiers he saw gathered on the prairie. *He described the slaughter:*

"Kneel down!" Now burst in harsh accents from the lips of the Mexican commander. No one stirred. Few of us understood the order, and those who did would not obey. The Mexican soldiers who stood at about three paces from us, leveled their muskets at our breasts. Even then we could hardly believe that they meant to shoot us, for if we had we should assuredly have rushed forward in our desperation, and weaponless though we were, some of our murderers would have met their death at our hands.

The sound of a second volley, from a different direction than the first just then reached our ears, and was followed by a confused cry, as if those at whom it had been aimed, had not all immediately been killed. A thick cloud of smoke was wreathing toward the San Antonio River. The blood of my lieutenant was on my clothes, and around me lay my friends convulsed in their last agony. I saw nothing more. Unhurt myself, I sprang up and, concealed by the thick smoke, fled along the hedge in the direction of the river, the noise of the water for my guide.

On I went, the river rolled at my feet, the shouting and yelling behind. "*Texas forever*!" And without a moment's hesitation, I plunged into the water. The bullets whistled round me as I swam slowly and wearily to the other side, but none wounded me. Whilst these horrible scenes were occurring on the prairies, Col. Fannin and his wounded companions were shot and bayoneted at Goliad, only Dr. Shackleford and a few hospital aids having their lives spared, in order that they might attend the wounded Mexicans.

By kind permission of: Presidio La Bahia, Goliad, Texas.

SURVIVORS ACCOUNTS OF GOLIAD[1]

Joseph Barnard
Andrew Boyle
Samuel Brown
Dillard Cooper
John C. Duval
Nicolas Eastland
Herman Ehrenberg
Joseph Fields
Isaac D. Hamilton
Benjamin Holland
William L. Hunter
"Kentuck"
Mrs. George W. Cash
Capt. Miller's Secretary
Abel Morgan
Jack Shackelford
Charles B. Shain
Joseph Spohn
Lewis M.H. Washington

Source; by kind permission of Presidio La Bahia, Goliad, Texas.

---

[1] A spurious account in the late 19th, century allegedly by a Lt. Harrison, has been postulated to be a fictional account and thus has not been included within these pages.

THE ANGEL OF GOLIAD – FRANCITA ALAVEZ

Perhaps the most compassionate figure of the Texas Revolution, Senora Alazez. Historical accounts refer to the "Angel of Goliad" as a lady of Mexican birth named Francita, Francisca, Panchita, or Pancheta/Panchita with surname Alavez, Alvarez, or Alevesco

Multiple survivor accounts of the Massacre at Goliad mention with honor and reverence and credit their survival to a Mexican lady immortalized with the term "The Angel of Goliad" is credited with *directly sparing the lives of a number of men at Goliad.* Francita Alavez was not the wife, but the traveling companion of Col. Telesforo Alavez, an officer in General Urrea's army. At Goliad, she plead and saved the lives of a number of men including fifteen year old Benjamin Hughes. Later at Victoria, she assisted in the escape of Issac Hamilton. He had escaped the Goliad Massacre, but was captured again at Victoria.

Her selfless and sincere concern for the enemy Texians displayed deep and humane feeling beyond common bonds and was recalled later by some of those whom she comforted during their imprisonment. Numerous accounts, including that in the current Handbook of Texas, relate that Señora Alavéz returned to Mexico City with the Captain and was there abandoned by him upon which she returned to Matamoros penniless where she was aided by Texians who knew of her humanitarian efforts.

History tells no finer story than that of the "Angel of Goliad", the Mexican lady whose merciful heart, unyielding courage, and almost unbelievable exertions induced Urrea's officers to evade, and partially disobey, Santa Ana's orders to shoot all prisoners, and to mitigate the rigors of the prisoners' lot. In the words of Dr.Joseph Barnardd, one of the beneficiaries of her mercies: [Linn: Reminiscences of Fifty Years in Texas, 171-175].

"I must not here omit the mention of Senora Alvarez, whose name ought to be perpetuated to the latest times for her virtues, and whose action contrasted so strangely with that of her countrymen, and deserves to be recorded in the annals of this country and treasured in the heart of every Texan.

Various accounts place Francita Alavéz with the movements of Captain Alavéz in carrying out his assignments at Copano Bay, Goliad, Victoria and Matamoros where she aided Texian prisoners at all locations. Numerous accounts, including that in the current Handbook of Texas, relate that Señora Alavéz returned to Mexico City with the Captain and was there abandoned by him upon which she returned to Matamoros penniless where she was aided by Texians who knew of her humanitarian efforts. From there she is said to have disappeared from the history books.

Though four generations of Texans have delighted to praise her, they have been singularly incurious as to her name, personality, and subsequent fate. She came into Texas with Urrea's army and was swept out again with the Mexican retreat from San Jacinto. From March through May of the year of Texan independence her virtues shone resplendently against the grim cruelty of Santa Anna; and then, insofar as Texas and Texans were concerned, she stepped gently out of their hearts and lives. Not even the beneficiaries of her mercies took the trouble to learn, or, at least they failed to record, her name.

Source; by kind permission of Presidio La Bahia, Goliad, Texas.

REMINISCENCES OF ANDREW BOYLE

It was in the year 1870 that Andrew A. Boyle (previously known as Andrew Michael O'Boyle), a son of San Patricio and a Texian soldier when he was eighteen, lay in his bed three months before his death in Los Angeles, then a small town in the Southern California desert. He had summoned his daughter to fetch her pen and paper, for he had a story that he had carried in his memory but had neglected to put on paper during his action-filled life. He realized that now was his last chance to make his contribution as an eyewitness to the Goliad Massacre on Palm Sunday, March 27, 1836, from which he had been spared by a twist of fate-more accurately, a kind Providence. *From The Forgotten Colony: San Patricio de Hibernia by Rachel Bluntzer Hébert*

Reminiscences of the Texas Revolution **by Andrew A. Boyle**

These recollections of the Texas revolution were dictated by Andrew A. Boyle in 1870, just before his death, to his daughter, Mrs. W. B. Workman. For the manuscript The Quarterly is indebted to his granddaughter, Miss Gertrude Dardin Workman. Although Mr. Boyle's memory of details was inaccurate, the paper is important in two particulars: (1) it adds another witness to the list of Texan participants who have unanimously testified that Fannin did not surrender at discretion, as General Urrea claimed, and (2) it gives *a first hand account of the execution of the wounded prisoners at Goliad.*

On the seventh day of January, 1836, at San Patricio de las Nueces, I enlisted in Captain Westover's battery (the first company of regular artillery in the Texas army). Our command was soon after ordered to Goliad, where it was incorporated with the forces commanded by Colonel Fannin. Colonels Bowie and Crockett, then in command of the Alamo, sent a courier to Colonel Fannin in the latter part of February, asking him for reinforcements

A hundred men[1] were at once detailed, and had crossed the San Antonio river on their way to the assistance of the doomed garrison, when they were recalled on account of a report brought in by a scout named "Comanche," of the advance of the Mexican army under General Urrea, toward San Patricio.

The main body of the enemy, under Santa Anna, had marched directly from Laredo upon San Antonio. Our commander, by the advice of "Comanche," determined to march to San Patricio, leaving one company in garrison at Goliad. The character of the scout was notoriously bad, and Colonel Fannin was informed of the fact, but gave no heed to the warning, although two of us volunteered to go to San Patricio and ascertain the truth of the report.

Three days' rations were distributed, and everything was in readiness to commence the march the next morning, when an American named Ayres arrived from the Old Mission, some fifteen miles distant in the direction of San Patricio, and brought reliable news of the arrival of the Mexicans at that place, and of their maltreating all Americans there, bearing themselves with special insolence toward the women.

Source ; by kind permission of Presidio La Bahia, Goliad, Texas.

Colonel Fannin immediately dispatched Captain King with a party of twenty men to remove all American women and children to our fort. Captain King was surrounded by a superior force of the enemy, but cut his way through them and retreated to the Old Mission Church, from which point he sent a messenger to Colonel Fannin, stating his position and asking for reinforcements.

Fannin sent Colonel Ward, with his Georgia Battalion, to King's assistance. On Ward's arrival at the Mission Church, a difference of opinion arose as to who should command the whole force. Not being able to come to any agreement, they separated; Ward retreating in a southeasterly direction for some distance, and then striking for the San Antonio river with the intention of joining us. King got out of the church, and after a skirmish with the Mexicans, retreated on the direct road to Goliad.

---

[1] Contemporary statements place this number between three and four hundred men.

He and his men were taken prisoners, tied together with rawhide, and shot immediately. We heard of the surrender and killing of King and his men, and the retreat of Colonel Ward in the direction above mentioned, and were in daily expectation of Ward's command arriving at Goliad. About the 8th or 9th of March we heard of the fall of the Alamo and the killing of Colonels Bowie and Crockett and all their men.[2] Colonels Bowie and Crockett having refused all propositions for surrender or capitulazition, the garrison held out until reduced to seven men, who asked for quarter and were refused.On the 17th of March the enemy appeared on the opposite side of the river. We sent over a skirmishing party of one company (under Captain Shackelford's command, I think), who had an engagement with the enemy, we watching from the ramparts with the most intense anxiety.

They were recalled by Colonel Fannin, after the enemy's retreat to the Old Mission Church. On the following day the enemy appeared in force at the same place, and orders were given by Colonel Fannin to bake bread sufficient for several days, and carry dried beef sufficient for the same length of time. The guns were taken down from the bastions, and orders were also given to be ready to march before daylight in the morning.

From cause unknown to me, we did not evacuate the fort until between 8 and 9 o'clock next morning. We marched down the river and crossed at a ford below, which was effected without difficulty. Our object in crossing at the lower instead of the upper ford in front of the Old Mission, was to avoid, if possible, an action with the enemy (he outnumbering us at least six to one), and to get into the interior of Texas and join Houston's army. We continued our march until we crossed a creek called Manawee *(Manahuilla)*, distant from the crossing about three miles. We traveled slowly, our cannon and baggage wagons being drawn by oxen. A halt was called and we ate some breakfast.

After breakfast, the march was continued; nothing new transpiring until about half past twelve o'clock; the Mexican army was then descried on our left and rear; their cavalry approached us rapidly, seemingly with the intention of cutting us off from the timber of the Coleto creek; they fired a few shots at us, when Colonel Fannin exclaimed (I was standing close by him at the time), "That's the signal for battle; I won't retreat another foot."

---

[2] Ramon Martinez Caro, Santa Anna's secretary, says that after the Alamo was taken five men who had hidden themselves during the action surrendered to General Castrillon, but were shot by Santa Anna's order.

We then unlimbered our pieces, being six in number, formed ourselves into a hollow square, placing the baggage wagons, hospital wagon and magazine in the centre; we remained in this position five or ten minutes, when Colonel Fannin, seeing clearly the main object of the enemy was to cut us off from the timber, ordered us to limber up again and continue the march. We left the road, marching in an oblique direction to the left toward the nearest timber; when within as well as I can recollect, three quarters or one mile of the timber, the enemy's infantry overtook us and we were obliged to halt.

We formed as previously, our little force then not numbering more than 311 men, maintained an action from half past one o'clock, P. M., and fought until near dark, when the enemy retreated, leaving twenty-five of us killed and wounded. I had been shot in the right leg at about half past three in the afternoon. Our real trouble commenced after the retreat of the enemy, and arose principally from the want of water, from which the wounded especially suffered severely.

A few of our men dug for water while the rest were throwing up intrenchments, as we expected to renew the battle on the following day. In the fight just finished we had killed our oxen and used the carcasses for breastworks. I lay that night near Colonel Fannin, who had been slightly wounded in the thigh. I remember his good-naturedly offering me his "good leg for a pillow."

In the morning the Mexicans again advanced, largely reinforced from General Santa Anna's division, and well supplied with artillery. After firing a few round shot, all of which passed over our heads, they hoisted a white flag, which we answered. A consultation of officers was held, at which it was concluded to capitulate, as preferable to attempting to prolong a hopeless struggle. Our wounded men were on our hands, and suffering; we had no means of caring for them, and Colonel Fannin strongly expressed his determination not to abandon them.

Two officers from each army then met in parley and agreed upon articles of capitulation, guaranteeing our lives and personal property. We agreed to give up all government property in our possession, and to remain prisoners of war until honorably exchanged or sent to the United States, upon parole never to return to Texas. These articles were signed by both parties, and the surrender was completed. Those of our command able to march were at once taken to Goliad, the wounded waiting two or three days for Mexican carts.

Our sufferings were intense, on account of the heat of the sun, thirst, and want of medical attention. Upon our arrival at Goliad we--the wounded---were placed in the hospital; the rest of the command was guarded in the yard of the fort. Just one week after the surrender, all the wounded men were marched out of the fort in separate divisions and shot.

Soon after, a Mexican officer came into the hospital, and ordered me to tell all those able to walk to go outside. I interpreted for him, and the men commenced gathering up their blankets. In the meantime, four Mexican soldiers came in and began to carry out those who were too severely wounded to walk. I was assisted by two comrades who were but slightly wounded. As we passed the door, an officer told me we were all to be shot. This I told the men.

The wounded were placed in the corner of the yard upon which the church door fronts. A company of soldiers formed in front of us and loaded their pieces with ball cartridge. Then a file of men under a corporal took two of our number, marched them out toward the company, and after bandaging their eyes, made them lie with their faces to the ground, after which, placing the muzzles close to their heads, they shot them as they lay.

At this time an officer, apparently of distinction, came into the yard and asked in a loud voice, in English, whether any one named Boyle was there or not. I was near him as he entered, and answered at once. He then ordered an officer to take me to the officers' hospital and have my wound attended to, saying that he would call upon me there.

Source; by kind permission of Presidio La Bahia, Goliad, Texas.

When I arrived at the hospital the Mexican officers seemed kindly disposed to me, and gave me a pair of "armas de pelo" to lie on. Mr. Brooks, Aid to Colonel Fannin, was there at the time, with his thigh badly shattered near the hip. I found him entirely ignorant of what had been going on. Upon being informed he said, "I suppose it will be our turn next."

In less than five minutes four Mexicans carried him out, cot and all, placed him in the street not fifteen feet from the door, in a position in which I could not avoid seeing him, and there shot him. His body was instantly rifled of his gold watch, stripped, and thrown into a pit at the side of the street. Colonel Ward and his command, who had been captured between the Lavaca and Navidad rivers a few days after our surrender, were also shot. The whole number of men thus barbarously executed was, according to Mexican report, four hundred and seventeen.[3]

A few hours after the murder of Mr. Brooks, the officer who had asked me in the yard came into the hospital. Addressing me in English he said: "Make your mind easy, Sir; your life is spared." I asked if I might inquire the name of the person to whom I was indebted for my life. "Certainly," said he, "my name is General Francisco Garay, second in command of Urrea's division." He had taken my name and description from my sister, Mary, at whose house he had been quartered while his division occupied San Patricio, and by whom and my brother Roderick he had been kindly treated.

She and my brother had refused all remuneration from him, only asking that if I should ever fall into his hands I should be kindly treated. The General informed me that he himself was on the eve of departure to join General Urrea, but that he had given orders to General Portillo, commandant of the garrison, to furnish me a passport whenever I should call for it. With this he took his leave. The passport was obtained without difficulty in pursuance of the order given by General Garay, and I secured passage in an ox-cart to the Mission and thence to San Patricio, where I remained.

We knew nothing of the battle of San Jacinto until about the 28th of April, although we had noticed Mexican troops traveling towards the Rio Grande. A dragoon rode up one day and asked me to sell him two bits' worth of dried meat; I offered to give him all the meat he might want if he would answer a few questions.

He consented, and I learned for the first time that a battle had been fought on the 21st near San Jacinto Creek, and that the result had been disastrous to the invading army; that General Santa Anna had been taken prisoner, and that the Americans had seemed inclined to give no quarter, charging with the cry of "Alamo and Fannin." The remains of the Mexican forces

[3] The real number seems to have been between three hundred and twenty and three hundred and thirty.

engaged, as well as General Urrea's division, which had been stationed at Brazoria, were in full retreat. The effect of such glad news upon my feelings may be imagined.

General Garay arrived a few days afterwards, and called to see us as he passed hastily through town. At his request, I accompanied him to Matamoras. Upon arriving there, he explained that stringent orders had been given to the effect that no American, who had been at any time a prisoner in Mexican hands, should be suffered to remain in Texas. He also informed me that all prisoners were to be closely confined, but that he would allow me the freedom of the city, upon my giving my parole not to attempt to escape.

About three weeks afterward, the General invited me to accompany him to the City of Mexico, stating that I should no longer be considered as a prisoner, in case I accepted his offer, which was accompanied with the most profuse offers of friendship and assistance. Notwithstanding the gratitude which I naturally felt toward the preserver of my own life, I was compelled to decline, on account of my anxiety to see my relatives in the United States.

At my urgent solicitation, General Garay then released me from my parole, and left me free to control my own movements. I concluded to start on foot for Brazos Santiago, but experienced great difficulty in procuring from the alcalde the necessary permit to leave the city. I was afraid to apply to him directly, and all the American and Irish residents strongly remonstrated against my doing so.

I finally succeeded in passing myself as the son of an old Irishman who had obtained a passport for New Orleans, and had myself included in it; afterwards I had a separate document made out for myself. The next day I took passage on a brig at Brazos Santiago, and six days after, landed at New Orleans. I at once visited the Texas Consul in that city, Mr. Bryan,[4] but found that he could do for me nothing more than to furnish a free pass to Texas Being out of money and in rags, I was compelled to seek employment. I engaged with a painter for two dollars and a half a day, and went to work painting St. Mary's Market, though I had never painted except in water colors. I worked eleven days, at the expiration of which

[4] Bryan was not officially a consul, because at that time the United States had not recognized the independence of Texas, but he discharged many of the duties of a consul.

time I drew my money, purchased some clothes, and accepted Mr. Bryan's offer. He procured me passage an a schooner for the mouth of the Brazos river, where I landed in a few days. General Burnet, the first President of the Republic of Texas, then living at Velasco, gave me a letter to General Rusk, at that time commanding the army on the Guadalupe River

I walked to General Rusk's camp, a distance of a hundred and fifty miles, in five days. He was in daily expectation of the advance of the Mexicans, but excused me, on account of my impaired health, from further service in the army. I had a severe attack of fever and ague in Victoria, where General Rusk's headquarters were. As soon as I recovered, I went to Columbia, then the seat of government, and obtained a passport for New Orleans.

Los Angeles, December 15th, 1870.

From The Forgotten Colony: San Patricio de Hibernia by Rachel Bluntzer Hébert

Source; by kind permission of Presidio La Bahia, Goliad, Texas.
Reprinted from Boyle, Andrew A. - The Retreat from Goliad. (Pamphlet from Reminiscences of the Texas Revolution in the Texas Historical Association), The Union National Bank, Houston, TX, 1933.

Here ends the reminiscences of Andrew O'Boyle. He had left Texas not to return; it was the end of an era in his life. He had dropped the "Michael" and substituted the initial "A" and he had dropped the "O" from O'Boyle. Henceforth he was to be known as Andrew A. Boyle, a prosperous merchant and trader in New Orleans; so much so, that he was able to open stores elsewhere. While he was opening one in Red River in upper Louisiana, he found his errant father, brought him to New Orleans and got Anne and Mary to keep house for him.

In 1846 he married Maria Christie of New Orleans. Two children were born to their marriage, a daughter Maria and a son, John, in 1847 and 1849 respectively. Two months after the birth of John, Boyle was returning from Mexico with a box of silver coins, and while transferring the box from a small boat to a streamer, the small boat capsized, and the box sank. Boyle remained at the port for several days while divers tried unsuccessfully to recover the money. Meanwhile, the rumor that Boyle had lost his life reached his wife.

The young Mrs. Boyle died of "grief" before he could return to her. Baby John died soon afterwards. The grief-stricken businessman left New Orleans in 1851, then journeyed to Boston where he purchased a stock of shoes to take to San Francisco to open a store—Hobart and Boyle. Seven years later he moved to Los Angeles, then a small town. From Thomas Rubio he bought a tract of land east of the city. This land under a bluff and containing a vineyard later became known as Boyle Heights.

Active in the early life and progress of Los Angeles, Boyle was elected to the Los Angeles City Council. His insistence in limiting the number of years the city's franchise could be held by a private company eventually led to the municipal ownership of the city's water system. Water is essential to the growth of any city and so it was to Los Angeles. Having seen his daughter Maria happily married in 1867, Andrew Boyle expired February 9, 1871, at his home with his daughter Maria present, then Mrs. W.H. Workman of Los Angeles. It is the destiny of some sojourners on this planet earth to have a life so filled with adventure, escapes, hardships, and grief intermingled with joy, that it would fill the life span of many older than 52-year old Andrew Boyle. As his life attests, the Irish who came to Texas under the auspices of McMullen and McGloin scattered but left their mark on distant places.

From The Forgotten Colony: San Patricio de Hibernia by Rachel Bluntzer Hébert. Source; by kind permission of; Presidio La Bahia, Goliad, Texas

From *Reminiscences of John J. Linn: Fifty Years in Texas*

A. M. O'Boyle, a member of Captain Westover's company, was saved by Colonel Gary, of the Mexican Army. When the men under General Urrea crossed the river at San Patricio, Colonel Gary stopped at the first house, which was owned by Mr. Edward O'Boyle. Mrs. O'Boyle, being at home, was asked by Colonel Gary to be permitted to stay over-night, as it was late in the day, to which Mrs. O'Boyle consented. Next morning, when the colonel was about leaving, he thanked her for her kind hospitality and said that any service he could return he would do so cheerfully. She remarked that he was quite welcome, and knew of nothing, only that she had a brother in Fannin's command at Goliad; that if, in the fortunes of war, he could do him a favor was all she asked, to which he agreed and asked for the name, which he wrote down in his memoranda.

Colonel Gary was a Greek by birth, whom an adventurous nature had led from the classic East to seek his fortune on the shores of the distant occident. On the morning that the prisoners were led forth to execution he rode up to the line and asked if there was a man present named Michael O'Boyle. O'Boyle replied in the affirmative, when Colonel Gary instructed him to follow him, which he did as well as his wounded ankle would permit. Colonel Gary placed him with Miller's men, and saved his life, thus keeping the promise made to the sister of O'Boyle.

DILLARD COOPER'S REMEMBRANCES OF THE FANNIN MASSACRE

From Rangers and Pioneers of Texas by A.J. Sowell 1884, as reprinted from the American Sketch Book 1881.

According to James T. DeShields in Tall Men With Long Rifles, Cooper died in extreme poverty in the 1890's in Llano, TX stating "during his latter years the pitiful pension of $150.00 a year, provided by the great and opulent state of Texas, barely sufficed to buy food and medicines for the aged hero and his faithful wife. Napoleon was not far wrong when he said 'Republics are ungrateful.'" ... several ... cried out for mercy. I remember one, a young man, who had been noted for his piety, but who had afterwards become somewhat demoralized by bad company, falling on his knees, crying aloud to God for mercy, and forgiveness. Others, attempted to plead with their inhuman captors, but their pleadings were in vain ... On my right hand, stood Wilson Simpson, and on my left, Robert Fenner ... while some of them were rending the air with their cries of agonized despair, Fenner called out to them, saying: "Don't take on so, boys; if we have to die, let's die like brave men ... At that moment, I glanced over my shoulder and saw the flash of a musket ...

On the morning of the 27th of March, 1836, about daylight, we were awakened by the guards, and marched out in front of the fort, where we were counted and divided into three different detachments, We had been given to understand that we were to be marched to Capono, and from there shipped to New Orleans. The impression, however, had in some way been circulated among us, that we were to be sent out that morning to hunt cattle; though I thought at the time that it could not be so, as it was but a poor way, to hunt cattle on foot.

Our detachment was marched out in double file, each prisoner being guarded by two soldiers, until within about half a mile southwest of the fort, we arrived at a brush fence, built by the Mexicans. We were then placed in single file, and were half way between the guard and the fence, eight feet each way. We were then halted, when the commanding officer came up to the head of the line, and asked if there were any of us who understood Spanish. By this time, there began to dawn upon the minds of us, the truth, that we were to be butchered, and that, I suppose, was the reason that none answered. He then ordered us to turn our backs to the guards. When the order was given not one moved, and then the officer, stepping up to the man at, the head of the column, took him by the shoulders and turned him around.

By this time, despair had seized upon our poor boys, and several of them cried out for mercy. I remember one, a young man, who had been noted for his piety, but who had afterwards become somewhat demoralized by bad company, falling on his knees, crying aloud to God for mercy, and forgiveness. Others, attempted to plead with their inhuman captors, but their pleadings were in vain, for on their faces no gleam of piety was seen for the defenseless men who stood before them. On my right hand, stood Wilson Simpson, and on my left, Robert Fenner. In the midst of the panic of terror which seized our men, and while some of them were rending the air with their cries of agonized despair, Fenner called out to them, saying: "Don't take on so, boys; if we have to die, let's die like brave men."

At that moment, I glanced over my shoulder and saw the flash of a musket; I instantly threw myself forward on the ground, resting on my hands. Robert Fenner must have been instantly killed, for he fell with such force upon me as almost to throw me over as I attempted to rise, which detained me a few moments in my flight, so that Simpson, my companion on the right, got the start of me. As we ran towards an opening in the brush fence, which was almost in front of us, Simpson got through first, and I was immediately after him. I wore, at that time, a small, round cloak, which was fastened with a clasp at the throat.

As I ran through the opening, an officer charged upon me, and ran his sword through my cloak, which would have held me, but I caught the clasp with both hands, and tore it apart, and the cloak fell from me. There was an open prairie, about two miles wide, through which I would have to run before I could reach the nearest timber, which was a little southwest of the place from where we started.

I gained on my pursuers, but saw, between me and the timber, three others, who were after Simpson. As I neared the timber, I commenced walking, in order to recover my strength, before I came near them. When he first started, we were all near together, but as Simpson took a direct course across the prairie, I, in order to avoid his pursuers, took a circuitous course. There were two points of timber projecting into the prairie, one of which was nearer to me than the other. I was making for the furthest point, but as Simpson entered the timber, his pursuers halted, and then ran across and cut me off,

I then started for the point into which Simpson had entered, but they turned and cut me off. I then stopped running and commenced walking slowly between them and the other point. They, no doubt, thinking I was about to surrender myself, stopped, and I continued to walk within about sixty yards of them, when I suddenly wheeled and ran into the point for which I had first started. They did not attempt to follow me, but just as I was about to enter the timber, they fired, the bullets whistling over my head caused me to draw my head down as I ran.

As soon as I entered the timber, I saw Simpson waiting and beckoning to me. I went towards him, and we ran together for about two miles, when we reached the river.We then stopped and consulted as to the best way of concealing ourselves. I proposed climbing a tree, but he objected, saying that should the Mexicans discover us, we would have no way of making our escape. Before we arrived at any conclusion, we heard some one coming, which frightened us so, that I jumped into the river, while Simpson ran a short distance up it, but seeing me, he also jumped in.

The noise proceeded from the bank immediately above the spot where Simpson was, and I could see the place very plainly, and soon discovered that two of our companions had made their escape to this place. They were Zachariah Brooks, and Isaac Hamilton. In the fleshy part of both Hamilton's thighs were wounds, one made by a gun-shot and another by a bayonet.

We all swam the river, and traveling up it a short distance, arrived at a bluff bank, near which was a thick screen of bushes, where we concealed ourselves. The place was about five miles above the fort. We did not dare proceed further that day, as the Mexicans were still searching for us, and Hamilton's wounds had become so painful as to prevent his walking, which obliged us to carry him. We remained there until about 10 o'clock that night, when we started forth, Simpson and myself carrying Hamilton, Brooks, though severely wounded, was yet able to travel. We had to proceed very cautiously and rather slowly.

Fort La Bahia being southeast of us, and the point we were making for, was about where Goliad now stands. We proceeded, in a circuitous route in a northeasterly direction.

We approached within a short distance of the fort, and could not at first account for the numerous fires we saw blazing. We were not long in doubt, for the sickening smell that was borne towards us by the south wind, informed us too well that they were burning the bodies of our companions. And, here, I will state what Mrs. Cash, who was kept a prisoner, stated afterwards; that some of our men were thrown into the flames and burned alive. We passed the fort safely, and reached a spring, where we rested from our journey and from whence we proceeded on our travels.

But the night was foggy, and becoming bewildered, it was not long before we found ourselves at the spring from which we started. We again started out, and again found ourselves at the same place; but we had too much at stake to sink into despondency.

So once more took our wounded companion, thinking we could not miss the right direction this time; but, at last when day began to break, to our great consternation, we found we had been traveling around the same spot, and were for the third time back at the identical spring from which we had at first set forth. It was now impossible to proceed further that day, as we dared not travel during the day, knowing we should be discovered by the Mexicans. We therefore concealed ourselves by the side of a slight elevation, amidst a thick undergrowth of bushes.

By this time, we began to grow very hungry, and I remembered an elm bush that grew at the entrance of the timber where we were concealed, which formed an excellent commissary for us, and from the branches of which we partook, until nearly every limb was entirely stripped. About 9 o'clock that morning, we heard the heavy tramp of the Mexican army on the march; and they not long after that passed within a stone's throw of our place of concealment. It seems indeed, that we were guided by an overruling providence in not being able to proceed further that night, for as we were not expecting the Mexican army so soon, we would probably have been overtaken and discovered by them, perhaps in some prairie, where we could not have escaped.

We remained in our hiding place the rest of the day, and resumed our journey after dark, still carrying our wounded companion. Whenever the enemy passed us, we had to conceal ourselves; and we laid several days in ponds of mud and water, with nothing but our heads exposed to view. When in the vicinity of Lavacca, we again got ahead of the Mexicans; and, after traveling all night, we discovered, very early in the morning of the ninth day, a house within a few hundred yards of the river. We approached it, and found the inhabitants had fled.

When we entered the house, we discovered a quantity of corn, some chickens, and a good many eggs lying about in different places. Our stomachs were weak and revolted at the idea of eating them raw, so we looked about for some means of striking a fire, first searching for a rock, but failing to find one, we took an old chisel and ground it on a grindstone for about two hours, but could never succeed in getting the sparks to catch. We then concluded to return and try the eggs raw.

We had taken one, and Simpson was putting on his shoes, which he had taken off to rest his feet, which were raw and bleeding, and had just got one on when he remarked: "Boys, we would be in a tight place if the Mexicans were to come upon us now." So saying, he walked to the window, when to his horror, there was the whole Mexican army not more than a mile and a half off, and fifteen or twenty horsemen coming at full speed within a hundred yards of us.

We took up our wounded man and ran to the timber, which was not far off, Simpson leaving his shoe behind him. We got into the timber and concealed ourselves between the logs of two trees, the tops of which having fallen together, and being very thickly covered with leaves and moss, formed an almost impenetrable screen above and around us

We had scarcely hidden ourselves from view, when the Mexicans came swarming around us, shouting and hallooing through the woods, but did not find us. We heard them from time to time, all throughout the day and next night. The next morning, just before day, the noise of the Mexicans ceased, and we concluded they had left. Simpson then asked me to go with him to get his shoe, as it would be difficult for him to travel without it, and I consented to do so. We went out to the edge of the timber and stopped some time to take observations before proceeding further. Seeing nothing of the Mexicans, we proceeded to the house, found the shoe, and possessing ourselves of a couple of ears of corn, and a bottle of water, we returned to our companions.

We had no doubt that the Mexicans had gone, so we sat down and drank the water and ate an ear of corn, when Brooks asked Simpson to go with him to the house, saying he would get a chicken, and we could eat it raw. They started, and had hardly got to the edge of the timber when I heard the sound of horses feet, and directly afterwards the Mexicans were to be seen in every direction.

I was sure they had captured Simpson and Brooks. Soon I heard something in the brush near us, but did not know whether it was the boys or Mexicans, but it turned out to be the boys, who crept undercover, and, in a few minutes, four Mexicans came riding by, passing within a few feet of where we were lying, with our faces to the ground.

After going into the woods a short distance they turned and passed out again, but it was not long after when six of them came riding quite close, three on each side of us, and leaning down and peering into our hiding place. It seemed to me they could have heard us, for my own heart seemed to raise me almost from the ground by its throbbings.

I felt more frightened than I ever had been before; for at the time of the massacre, every thing had come on me so suddenly that my nerves had no time to become unstrung as they now were. The Mexicans passed and repassed us, through the day, so we dared not move from our hiding place. A guard was placed around us the following night, the main body having, no doubt, gone on, and left a detachment to search for us. I think they must have had some idea of our being some of Fannin's men, or they would scarcely have gone to that trouble.

About 10 o'clock that night we held a consultation, and I told my companions it would not do to remain there any longer, as the Mexicans were aware of our place of concealment, and would surely discover us the next day. We all decided then to leave, and they requested me to lead the way out. I told them we would have to crawl through the timber and a short piece of prairie, until we crossed the road near which the Mexicans were posted; that they must be careful to remove every leaf and stick in the path, and to hold their feet up, only crawling on their hands and knees, as the least noise would betray us to the enemy.

I was somewhat acquainted with the locality; for we were now not far from Texana, and I had some times hunted along these woods. Thus I led the way. Hamilton's wounds were so painful that we could move only slowly, and we must have been two hours crawling about 200 yards. When we at length passed the timber and reached the road, I stopped to make a careful survey of the situation.

I could see the Mexicans placed along the road, about a hundred yards on each side of us. The moon was shining, but had sunk towards the west, which threw the shadow of a point of timber across the road, and concealed us from view. It would have been hard to discover us from the color of our clothes, as the earthy element with which they were mixed had entirely hidden the original fabric.

We continued to crawl, until we reached a sufficient distance not to be discovered, when we rose up and walked. Although Hamilton had, with a great deal of pain, managed to crawl, yet it was impossible for him to walk, and his wounds had by this time become so much irritated and inflamed that he could scarcely bear to be carried. We traveled that night only a short distance, and hid ourselves in a thicket near a pond of water. Brooks had been trying to persuade me to leave Hamilton; but, although our progress was impeded by having to carry him, I could not entertain the idea for a moment.

I indignantly refused, but still he would seize every opportunity to urge it upon me. He said it would be impossible for us to escape, burdened as we were with Hamilton. I could only acknowledge the truth of this, for it was a desperate case with us. The foe was around us in every direction. Brooks, finding that I was not to be persuaded, then attempted to influence Simpson.

On the tenth day out, they took the bottle and went to the pond near by, for water. As they were returning, (I suppose Brooks did not know he was so near the place they left us), both Hamilton and myself heard Brooks urging Simpson to leave him. He told him if we remained with Hamilton, we would certainly lose our lives; but there was some slight chance of escaping, if we left him, and that Hamilton's wounds had become so much worse that he was bound to die, unless he could have rest; and, as we were doing him no good, and ourselves a great deal of injury by carrying him, it was, our duty to leave him.

Now Brooks had never carried him a step; Simpson and myself having done that; yet Brooks was the first who had ever proposed leaving him; and, although there was a great deal of truth in what he was saying, yet I felt quite angry with him, as I heard him trying to persuade Simpson. Hamilton did not say a word to them when they came in, but sat with his face buried in his hands a long time.

At length, he looked up, and said: "Boys, Brooks has told you the truth; I can not travel any further, and if you stay with me, all will be killed. Go and leave me, boys; if I have rest I may recover, and if I ever should get off safe, you shall hear from me again." He spoke so reasonably, and we were so thoroughly convinced of the truth of what he said, after a brief consultation, we decided to depart without him. Hamilton had known Brooks in Alabama; he called him to him, and gave him a gold watch and $40 in gold, telling him to give it to his mother.

We then bade Hamilton farewell, all of us shedding tears as we parted, but when we turned to go, my resolution failed me, and I could not find it in my heart to leave him. I said: "Boys, don't let us leave him." But Simpson and Brooks said that we could do neither him nor ourselves any good by remaining, and that they were determined to go. I told them I would remain with him, and do the best I could for him.

So they started off without me; but Hamilton insisted so much that I should leave him, that I again bade him farewell, and followed and soon overtook the others. The reason that we started off in the day, was that it was raining quite hard, and we thought there would not be much danger in traveling, but we had not gone more than half way through the next prairie. when the weather cleared up, and we saw the whole Mexican army encamped at Texana, about two miles off; but they did not discover us, and we succeeded in reaching the timber on the Navidad.

In the evening we walked out to a slight eminence which overlooked the prairie, to reconnoiter. While gazing across the prairie, we could see three men on horseback, but so indistinct were they, that we could not at first tell whether they were Americans or Mexicans. As they approached, we hid in the undergrowth; and as they passed, we saw that they were Mexican couriers returning to the command.

At eight we again started forth, and coming out on the prairie, we discovered a road, which we concluded had been made by the refugees in their retreat from the enemy. During all this time we had nothing to eat but leaves and herbs, and the two ears of corn that we got at the house on Lavacca river. On the twelfth day, we reached the Colorado, at Mercer's crossing. As we were very tired, we sat down on the bank to rest a little, before attempting to swim over.

While sitting there, a dog on the opposite side of the river began to bark. When we heard that well-known sound, our very souls thrilled with joy, and that was the first time since the awful day of the massacre that a smile had ever illuminated our faces. We looked at each other, and then burst into a great big laugh. We were all good swimmers, but I some times took the cramp while swimming, so we concluded to cross on a log. We procured a dead mulberry pole, and hanging on to it, one at each end, and one in the middle, we crossed over to the land of freedom, and a land where we found plenty to eat. After recruiting a little, we procured horses, with the intention of joining Houston's army; but before we reached there, San Jacinto had been fought and won.

It was more than a year before I ever heard any thing of Hamilton. He remained in the same place where we left him nine days, some times lying in the pond of water, which assuaged the pain of his wounds. At the end of that time he was so much improved that he essayed to walk to Texana, and succeeded in doing so. He said the best eating he ever had in his life, was when he first entered Texana, and ate the meat from the rawhides the Mexicans had left

The next morning he took a skiff, and made his way down to Dimmitt's landing. He had scarcely reached there when he was taken prisoner by a Mexican soldier. Not long after, other soldiers came in, and tying Hamilton on a mule, started for camp. He suffered so much from his wounds that he fainted several times, on the way. Whenever this occurred, they would untie him, lay him on the ground, and throw water into his face until he revived, when they would again mount him on the mule and proceed on their way. Hamilton remained in their hands for sometime and gradually grew well of his wounds.

There was a Mexican who waited on him, who seemed much attached to him, and Hamilton was led to place much confidence in him. One morning, this Mexican told him that if he wanted to live another day, he must make his escape that night, as he had learned that he and two other prisoners were to be shot before morning. Hamilton then arranged a plan for the escape of himself and two of his companions, which was a success, after many trials and tribulations.

Source by kind permission of Presidio La Bahia, Goliad, Texas.

NICHOLAS W. EASTLAND AND CAPT. HORTON'S TROOP

From *Recollections of Early Texas: Memoirs of John Holland Jenkins*

It seems but justice to insert a few words by way of tribute to judge N. W. Eastland, now a resident of Bastrop County, who has given and suffered much for Texas. He is now eighty-two years old, having served under Andrew Jackson in the Seminole War. He came to Texas in 1833, and has been a devoted and active participant in all of her struggles and triumphs since then. A gentleman of fine military education, of varied experience, and rare conversational powers, he is even now, despite his extremely advanced age, agreeable and entertaining, giving in detail and from personal experience, many important incidents connected with our past history.

Two son's and a brother went out from the old man's house to battle for Texas, and were killed. His oldest son, Robert M. Eastland, a man of sterling qualities, was killed in the Dawson Massacre, but made a brave and heroic struggle for his life. Nat [Nathaniel W.] Faison, who is still living, saw him fall. He says Robert Eastland was first struck by a grapeshot, which broke his leg just above the ankle. In this condition he leaned against a mesquite tree for support and fought on, loading and shooting several times, until the Mexicans crowded on him and killed him.

By the way, judge Eastland had quite an interesting experience about the time of Fannin's massacre in 1836, which is worthy of note. He joined Fannin's body of men near Goliad, just as the officers were about to hold a council as to what measures of action to adopt. Having been an old friend and roommate of Fannin at West Point, he was invited to be present at this council of officers, by Fannin, whose tent served as council chamber.

The discussion was hot and earnest. At last a vote was taken as to whether they should take immediate action-that is go on that night, evacuate Goliad, and attack the Mexican force of cavalrymen, then about six miles off, or wait until daylight for further action. The latter motion prevailed, although a number of the officers, together with Eastland, were in favor of the former plan. Alas! If they had pursued that plan, Fannin's unfortunate band might have been spared their terrible fate, for the Mexican forces would have been surprised; besides, during the night they received heavy reinforcements under General Urrea, thus increasing the fearful odds that awaited the brave band of Texans, whose names will be cherished among the many martyrs to Texas' liberty.

They waited for daylight, and after evacuating Goliad started east to attack the army. Judge Eastland belonged to the advanced guard, a company of cavalrymen under Albert C. Horton, whose name is not unknown to Texas history, and whose conduct some have criticized, though unjustly, for he acted as he was compelled by circumstances. "Truly," a small incident brings wonderful results," and Eastland's experience, then, is a case in point. His horse having foundered, his company left him, and he thought he would have to remain with the main army. An old soldier seeing his trouble, however, informed him that nothing was better medicine for a foundered horse than running, and advised him to try it, saying he might thereby not only cure the animal but also overtake his company. So putting out at full speed he did accomplish both results, and was once more at his proper post in Horton's company.

As they left Goliad, they came first to the fresh trail showing the march of a larger reinforcement of Mexicans, and very soon could see Mexican spies going to and fro. Everything indicated not only the close proximity of the enemy, but eager interest and action. The company of scouts halted at a fire they found in their march, when suddenly a loud and terrible roaring was heard. Some said it was the noise of an approaching storm from the north, but it proved to be the tread of the Mexican army bearing upon Fannin's doomed band.

This was indeed a critical situation. Horton's small company of men found themselves separated from the main army of Texans by a force of three or four thousand Mexicans. They stood helpless, but unhurt, until two or three charges had been made. They could see some sign of the brave fight made by Fannin's men, as ever and anon from the din and smoke of battle a Mexican horse would come charging back without a rider, but they knew the terrible odds of the struggle, and felt that the fate of their fellow soldiers was sealed.

" ... heard our companions shrieking in the most agonizing tones "O God! O God! Spare us" ... It was then we knew what was to be our fate ... Major Wallace ... sprung, and struck the soldier on his right a severe blow with his fist: they being at open files, the soldier at the other file attempted to shoot him; but being too close, was unable. The soldier then turned his gun, and struck the writer a severe blow upon the left hand. I then seized hold of the gun and wrenched it from his hand ... the central one raised his gun to fire—I still ran towards him in a serpentine manner in order to prevent his taking aim—I suddenly stopped—dropped my piece, fired, and shot the soldier through the head and he fell instantly dead ... immediately ran and leaped into the river."

On or about the 12th of March, orders were issued by Gen. Houston, to destroy the town and fort of La Bahia, and the forces to fall back to Gonzales, to unite with him and concentrate all the Texian forces. Prior to the receipt of these orders, Captain King's company was ordered to the Mission, (a distance of about twenty-five miles) in order to relieve some families who were in danger of failing into the enemy's hands.

Upon their arrival at the Mission they were met and attacked by a large body of the enemy, and after a gallant and well-sustained fight, retreated in an orderly and judicious manner to the church, where they sustained themselves against a very superior number of Mexicans and Indians with but small loss to themselves, but a severe loss to the enemy, until relieved by the Georgia battalion under Col. Ward, who had been sent to their aid. The separation of our forces caused us to delay our retreat.

An express was sent to Col. Ward, at the Mission, to fall back and join the forces at Goliad with all possible despatch, or should he be cut off by the enemy, (of whose advance from San Patricio we had intelligence) to make good his retreat through the Gaudaloupe bottoms, and join the main army at Victoria.

On the 16th our scouts brought intelligence that a body of the enemy, 1500 strong, were on the San Antonio road. Many of the cannon having been dismounted preparatory to a retreat, we immediately remounted them, as we anticipated an attack that night. About 12 o'clock the picket guard gave the alarm, and retreated into the fort; it however proved to be only the enemy's spies reconnoitring. On the 17th the enemy forded the San Antonio river and showed themselves at the old Mission, a distance from our fort of four miles. This day we destroyed the whole town of La Bahia by fire, battering down all ruined walls, so as to secure us a full sweep of the enemy, should they attack the fort.

March 18—The enemy still hovering round the old Mission, a council of war was called, when it was decided, that inasmuch as our ammunition was not sufficient to sustain a siege, and as our provisions were short, and as we were well aware of the overwhelming force of the enemy, it was advisable to fall back to Victoria. This night we made every preparation for an early retreat in the morning and by daylight every one was in marching order: before day a scouting party was despatched to ascertain the position of the enemy, who returned shortly after daylight and reported the lower road being clear. Col. Horton was then ordered by Col. Fannin to post all, advance, rear, right and left guard.

March 19th, at about half-past six, A. M., took the line of retreat towards the lower ford; and about 9 A. M., got our baggage and cannon across. We had nine pieces of brass artillery, consisting of one six inch howitzer, three short sixes, two long and two short fours, with several small pieces for throwing musket balls. We then commenced our advance towards Victoria. We had advanced several miles without receiving any intelligence of the enemy by our videttes, and at about 10 A. M. halted to graze our cattle, and take some refreshment on the outskirts of some timber we had just passed.

We tarried about three fourths of an hour, when we again took up the line of march. We had advanced about four miles into the prairie, when we had intelligence of the enemy's approach. Col. Horton's cavalry, who were ordered in the rear, had neglected to remain in that position and, in consequence, the enemy had advanced within the distance of from one to two miles, here they were discovered by the infantry in the rear and almost simultaneously they were descried upon both flanks, evidently with the design of surrounding us.

The enemy had now formed a semicircle on our right and left, and as we had no means of moving our artillery but by exhausted and worn-out men, were fast surrounding us. Captains Hurst and Holland of the artillery were ordered to the rear, to keep up a retreating fire, under cover of which the army advanced about a mile and a half in the face of the enemy. *sdct*

It now became necessary for us to take a position, as we were entirely surrounded, our cavalry cut off from us and escaped, leaving us now two hundred and fifty effective men, consisting of the following companies:

New Orleans Greys-Captain Pellis.
Red Rovers-Captain Jack Shackleford, from Alabama.
Mustangs-Captain Duval.
Mobile Greys-Captain McManaman.
Regulars--Captain Westover.
1st Company Artillery-Captain Hurst.
2d do. do. -Captain Holland.
3d do. do. -Capt. Schrusnecki, a Polish engineer

We were about three hundred yards to the left of the road, in a valley with an elevation towards the road, of about six feet in the whole distance; we were unfortunately obliged to take that very disadvantageous position, in consequence of our having pursued our advance so far in order to gain the woods. We drew our wagons into a cluster, formed ourselves into an oblong circle around them, and posted our artillery in positions to defend it: the circle was about 49 feet of shortest central diameter, and about 60 feet of longest diameter.

It was now 1 o'clock, P. M., at which time we were attacked on all sides by the enemy, with a brisk fire of musketry: we were ordered not to fire, until the word of command was given, in order to draw the enemy within rifle-shot. We reserved our fire for about ten minutes, and several were wounded in our ranks previous to our firing.

At the request of the officers, the artillery was permitted to open fire. The wind was blowing slightly from the N. E., and the smoke of our cannon covered the enemy, under which they made a desperate charge, but were repulsed with a very severe loss. Our cannon was loaded with cannisters of musket balls, and the howitzer with grist. In this manner, the action was kept up with great fury by the enemy; charge after charge being made by the cavalry and infantry, and always repelled by a heavy loss on their part. Our men behaved nobly; and, although surrounded by overwhelming numbers, not a change of countenance could be seen.

Thus was the battle kept up and upon the repulse of each charge, column upon column of the enemy were seen to fall, like bees before smoke. Here would be seen horses flying in every direction without riders, and there dismounted cavalry making their escape on foot, while the field was literally covered with dead bodies. It was a sorry sight to see our small circle: it had become muddy with blood; Col. Fannin had been so badly wounded at the first or second fire as to disable him, the wounded shrieking for water which we had not to give them. The fight continued until dusk, when the enemy retreated, leaving us masters of the field, with ten men killed and wounded, while the enemy lay around heap upon heap. We possessed a great advantage over the Mexicans, they having no artillery, and we having nine brass pieces, with which we kept up an incessant fire of musket balls.

It now became prudent to take measures as to our next procedure: accordingly, the officers were all summoned to Col. Fannin, where he lay wounded, and the question was, whether we should maintain our present position, or retreat. It was carried that we should sustain ourselves as long as possible; consequently, we commenced heaving up a redoubt, some three feet above the mean level of the prairie, exclusive of the dyke.

The night was now very dark and cloudy, drizzling with rain and misty fog: the enemy encompassed us, and kept up a continual sound to charge, so that we appeared to be surrounded with bugles. We had with us 1000 spare muskets, which we loaded, and each man took an equal share, our cannon ammunition being nearly exhausted. Daylight broke upon us in this situation, and some of our men went out about a hundred yards, and brought into camp two Mexican prisoners, both badly wounded.

From them we ascertained that the number opposed to us was 1900 men, and that a reinforcement of two brigades of artillery would be there that morning, if they had not already arrived. We had no sooner received this intelligence, than this very artillery opened their cannon upon us: they had placed themselves behind a small hillock, and were entirely under cover. We could neither touch them with our cannon, nor charge , as they had so placed their cavalry, that the moment we should quit our artillery, they would cut us to pieces.

We accordingly met in council, to devise means and measures: it was accordingly decided that we should send a flag of truce to the enemy, and if possible, obtain a treaty, if upon fair and honorable terms. Accordingly, Capt. F. J. Desauque, the bearer of the express from General Houston, Capt. B. H. Holland, of the artillery, and an ensign, were despatched with a flag of truce: the flags met midway between the two armies, and it was decided that the two commanders should meet to decide the matter; in pursuance of which Col. Fannin was conveyed out, and met Gen. Urea, Governor of Durango, commander of the Mexican forces; and the following treaty was concluded upon, and solemnly ratified: a copy of it in Spanish was retained by General Urea, and one in English by Colonel Fannin:

Seeing the Texian army entirely overpowered by a far superior force, and to avoid the effusion of blood, we surrender ourselves prisoners of war, under the following terms:Art. 1st, That we should be received and treated as prisoners of war, according to the usages of civilized nations.

Art. 2d. That the officers should be paroled immediately upon their arrival at La Bahia, and the other prisoners should be sent to Copano, within eight days, there to await shipping to convey them to the United States, so soon as it was practicable to procure it: no more to take up arms against Mexico, until exchanged.

Art. 3d. That all private property should be respected, and officers' swords should be returned on parole or release.
Art. 4th. That our men should receive every comfort, and be fed as well as their own men.

Signed, Gen. UREA, Col MORATEAS, Col. HOBZINGER, on the part of the enemy; and on our part, signed by Col. FANNIN,and Maj. WALLACE.

The officers were then called upon to deliver up their side arms, which were boxed up, with their names placed by a ticket upon each, arid a label upon the box stating they should soon have the honor of returning them; and it was their principle to meet us now as friends, not as enemies. Col. Fannin and the men were that afternoon marched back to La Bahia; the wounded, together with, the captain of each company, and our surgeons were left on the field to dress the wounded, which was completed on the 21st, when we were all conveyed back to the fort, where we found the men in a most miserable state.

They were brutally treated: they were allowed but very little water to drink, in consequence of it having to be brought from the river; and but a small piece of meat, without salt, bread or vegetables. On the 23d, Major

Miller arid ninety men were brought into the fort prisoners: they had just landed at Copano, from the United States.

On the 25th, the Georgia battalion was also brought in: it had been surprised and captured between Victoria and Demill's point, and marched back, and confined with us. Here we were now nearly 500 strong, guarded by 1000 Mexicans, without being allowed the slightest liberty in any respect. The Mexicans had always said that Santa Anna would be at La Bahia on the 27th, to release us.

Accordingly, on that day, we were ordered to form all the prisoners: we were told that we were going to bring wood and water, and that Santa Anna would be there that day. We were ordered to march all the officers at the head of the file, except Col. Fannin, who lay wounded in the Hospital. As we marched out of the sally port, we saw hollow squares formed ready to receive us: we were ordered to file left, and marched into a hollow square of double filed cavalry, on foot, armed with carbines, commonly called scopets, and broadswords.

This square was filled and closed, and the head of the remaining files wheeled off into another square, and so on, until all were, strongly guarded in squares. The company, of which the writer of this was one, was ordered to forward, and no more was seen of our unfortunate comrades: we marched out on the Bexar road, near the burying-ground; and as we were ordered to halt, we heard our companions shrieking in the most agonizing tones *"O God! O God! Spare us"* and nearly simultaneously, a report of musketry. It was then we knew what was to be our fate.

The writer of this then observed to Major Wallace, who was his file leader that it would be best to make a desperate rush. He said, "No---we were too strongly guarded." He then appealed to several others, but none would follow. He then sprung, and struck the soldier on his right a severe blow with his fist: they being at open files, the soldier at the other file attempted to shoot him; but being too close, was unable.

The soldier then turned his gun, and struck the writer a severe blow upon the left hand. I then seized hold of the gun and wrenched it from his hand, and instantly started and ran towards the river. A platoon of men (I have been since informed, by two others who made their escape by falling when fired upon among the dead bodies of their comrades) wheeled and fired upon me, but all missed.

I then had a chain of sentinels to pass at about 300 yards' distance; they were about thirty yards apart, three of them closed to intercept my retreat, the central one raised his gun to fire—I still ran towards him in a serpentine manner in order to prevent his taking aim—I suddenly stopped—dropped my piece, fired, and shot the soldier through the head and he fell instantly dead.

I ran over his dead body, the other two firing at me but missing, and immediately ran and leaped into the river, and while swimming across was shot at by three horsemen, but reached the opposite banks in safety; and after wandering six days without food in the wilderness, succeeded on the tenth of April in joining General Houston's army, after having been retaken by the enemy once, but succeeded in making my escape in company with a wounded man who had got off from La Bahia, by falling among the dead as before stated, I am happy to state, that six more succeeded in saving their lives and regaining their liberty by the same stratagem. The number of the enemy according to their own account, killed at the battle of Cotelo, varies from, nine to eleven hundred.

Source ; by kind permission of Presidio La Bahia, Goliad, Texas.

WILLIAM LOCKHART HUNTER

[William Lockhart Hunter was born in 1809 in Virginia. He came to Texas with the New Orleans Grays under Robert Morris and participated in the Siege and Battle of Bexar. After surviving the Goliad Massacre and the Texas War of Independence, he returned to the Goliad region and practiced law. He was a justice in both Goliad and Refugio counties and served both in the Texas legislature. He overlapped in the legislature with survivor Dr. Barnard and Hunter, Barnard and John C. Duval are said to have met and shared experiences about the massacre until their deaths. Hunter married Eunice Fedelia Cook of IL and died in 1886.
No. 1650 Historical Notes. Lamar Papers

[Houston? Texas, 183-?] FROM WM L HUNTER. Born Virginia- came to Texas 1835 with the New Orleans Grays the first that came in the Revolution—Hunter was at Bexar at the capture of Coss—afterwards went with Grant's command (300 men) to Goliad- Genl. Houston he addressed the troops there. This was about the middle of January-1836 Houston's object in addressing the troops was to dissuade them from marching westward- His speech created dissentions among the soldiers- Grant marched to Refughio with the exceptions of Houston's Regulars, who remained at Goliad & Houston also- Houston remained two or three days and left for the East- It was during this visit that he made arrangement to have himself returned to the Convention as a member.

Soon after the departure of Houston, Fannin came with his Battalion to Copano & thence to Refughio soon to Goliad Previous to Fannin's arrival at Refughio, Grant had set out for Sanpatricio, with a diminished force; 75 men 3 pieces cannon; at Sanpatricio he left Johnson with a smal company whilst he went west for the purpose of procuring horses- he had got some horses & was on his return, when he surprised by Urea, west Nueces. *[Note in document: Hunter was mistaken in the name; Urea had not yet taken command; the person Hunter refers to Vital Fernandez, Commanding Genl Matamoras. It was with him that Grant corresponded]*

Urea had deceived Grant. Texas at this time was contending only for the constitution of 1824- Urea professed to be attached to that party, and made

Grant believe that he only wanted a good chance to avow himself on that side- He and Grant. had had some correspondence upon the matter-so that Grant, expected when he met Urea to have been received as a friend rather than trated as an enemy. Urea however, Mexican like, intercepted Grant & attempted to capture him; Grant would not be taken; he himself was killed in the fight, and nearly all of his men, with the exception of a few-

This was in the month of February, whilst Fannin was at Goliad, he recd. a letter from Houston, then at Gonzales, to aid the families in moving; burn the town; blow up the fort, and to retreat- Fannin immediately sent Capt. King with a Company to Refughio, to take off some families there who had applied to him for protection Capt King had a skermish there, & sent to Col. Fannin for a reenforcemt- Fannin Col. Ward with upwards 100 men to Kings rescue-

Two days after Ward left Goliad, Fannin recd. a second order from Houston apprising him of the fall of the Alamo; & ordering to retreat immediately- Fannin on receiving this order, dispatched a runner after Ward & King to return & retreat with him. The runner was killed by the Mexicans; Fannin sent another runner (Edwd Perry who was also captured & was taken before Urea, who examined his papers, and then gave him leave to go on & deliver Fannins despatch to Ward, on the Condition that Perry would return to Urea again as soon as he should deliver the papers- Urea told Perry to say to Ward that he had better leave the Church & retreat, or he would batter it down- Perry was taken in the night Urea was then encompassing the church. Ward on receiving the news from Perry, retreated unperceived by the Mexicans- Fanning in the mean while, hearing nothing from Ward & King, Commenced his retreat on the 19th March, about 9 O'Clock in the morning The two orders which Fannin recd from General Houston were seen and heard by Hunter, from whom this information is now derived The letters were read by Fannin to Capt. Pettus of N. O. Grays, when Houston *[Hunter?]* was present. [Endorsed:] Information Recd. from Hunter respecting Fannin Houston & Grant & Ward.

Escape of Judge W. L. Hunter:
From *Reminiscences of John J. Linn: Fifty Years in Texas*

Hunter made an extraordinary escape from the "Fannin massacre." He was a member of the "New Orleans Grays"; he was shot down at the first fire, and remained for a considerable time unconscious. Upon reviving he could not move his body, as a dead comrade had fallen upon him. Being very weak from the loss of blood, he extricated himself with difficulty, and discovered that he had been stripped of his clothing, retaining only undershirt and drawers. He summoned all his strength for one supreme effort to reach the river, and nearly failed in doing so. He submerged himelf in the water, and remained in that position all day. At night he crossed the river and struck out in an easterly direction.

He came to a small stream the next morning, upon the banks of wbich he remained nearly all day, suffering excruciating pains from his wounds, and being rendered weak from the loss of blood and hunger. He finally made anotlier start, and soon came to another stream, and in following the course of this he came across his own tracks where he had crossed before. He then took down the creek, and came to a house, near the Coletto, where he found some Mexicans who could speak a few words of Englisli, and received of them some clothing and food. These people treated him with the utmost kindness and did all that they could to alleviate his pains. The owner of the jacal, Juan Reynea, had previously lived at the Goliad crossing, but had removed to avoid the unwelcome visits of the soldiers, who were continually passing between Goliad and Victoria. *sdct*

With the aid of these Good Samaritans Hunter speedily recovered sufficient strength to resume his journey, when Señor Reynea himself accompanied him to the house of Mrs. Margaret Wright, wife of David R. Wright, five miles above Victoria, on the Guadalupe River. This good old mother in Israel died very recently, in the city of Victoria, at the advanced age of eighty-seven years. She nursed Hunter with a mother's care, and sheltered him from the Mexicans until after the battle of San Jacinto. This statement I have from the lips of Judge Hunter himself, who now resides in Goliad, near the spot or that most terrible episode in his life.

Source ; by kind permission of Presidio La Bahia, Goliad, Texas.

THE ESCAPE OF KENTUCK

Told by DeWitt Colonist Creed Taylor *to James T. DeShields and related in Tall Men With Long Rifles.*

I personally knew most of the men who escaped at Goliad, and have often listened to their recitals of that bloody affair. One of the most interesting being the escape of a young man, called "Kentuck." This fine young fellow was one of the volunteers from Kentucky; but of his past I knew nothing, nor did he impart his real name, being known to the boys in the camp only by the nickname of "Kentuck." He was a man of small stature, rather handsome of features, and probably twenty-two or twenty-three years old. He was a jolly good fellow and fairly well educated. Many old Texans will remember this unique and somewhat noted character of the early days of Texas. I can vouch for the truth of his story.

To enable the reader to appreciate the terrible ordeal to which "Kentuck" was subjected on that occasion, it is important to describe his wardrobe. All except his underwear was thick and heavy, consisting of jeans, pants and vest, a blanket overcoat and rawhide boots. From his guards he learned the Mexican game of monte, in which he soon excelled his teachers. He won about 300 silver dollars, the total weight of which was about eighteen pounds; and for the lack of a bag in which to carry this money, he had distributed it in the several pockets of his garments. On the night of the twenty-six of March, the prisoners were told that on the next day they were to be shipped from Copano Bay to New Orleans; and on the following morning, Palm Sunday, they were marched out single file, between two files of guards, and in separate divisions; two squads to the riverbank, and one a few hundred yards east-ostensibly for marching to the coast.

"Kentuck" was in one of the divisions marched to the riverbank. They were made to form single file a few feet from the river's brink, and the Mexican file next to the river passed through them to the other side. Now the guards, outnumbering them two to one, were all on one side, and the prisoners naturally turned and faced them. The enemy's purpose could no longer be concealed. Instantly the prisoners became aware that they were to be shot, and poor, emaciated, half-starved fellows, they were powerless to resist; else they might have rushed upon the file of soldiers, disarmed them and escaped, as did the Mier men at Hacienda Salado in 1842.

At that moment they heard the firing upon the other squads and the terrible screams of the wounded and dying men. "Kentuck" and his comrades were ordered to 'face about'-turn their backs to their guards-but many of them, including "Kentuck," scorned the order, and continued to look straight into the eyes of their murderers.

When the order was given to "Fire," many of the firing squad appeared to be more nervous than their unfortunate victims. Some of the Mexicans had not the courage to shoot brave men who could look into their eyes in the moment of death; and the man who faced "Kentuck" did not fire, though nearly all others were killed.

As his slain comrades fell, "Kentuck" also fell back into the river. He tried to swim across the stream, but encumbered by his heavy garments and his pockets loaded with silver, he made poor headway. The current bore him downstream and held him long enough to saturate his garments and fill his boots with water, and be came near drowning. Finally be reached the opposite bank, climbed up, and ran for his life. His heavy garments weighed him down, the heavy silver in his pockets beat upon him, and the sloshing water in his boots caused his feet to slide within them. Thus impeded and exited he thought he was scarcely moving. Looking back over his shoulder he saw some of the Mexicans were pursuing him.

They could not bring their muskets to bear on him as they swam across the river, but he could hear their swords clanking. Gradually they gained upon him, but when almost near enough to strike him, as luck would have it, in twisting about trying to elude them, some money jostled out of his pocket, and his pursuers halted and scuffled for the coins.

This enabled "Kentuck" to gain considerable distance and it gave him an idea. As his pursuers again drew near, he threw down a handful of money, for which they again halted and scuffled. Thus handful after handful of his dollars went until all of his pockets were empty. He threw off his overcoat, which greatly lightened his burden; and then his vest, his pursuers stopping and scuffling for each of these. His hard running in the wet rawhide boots had stretched them, so that be managed to kick them off, and thereby caused his pursuers another delay and scuffle.

"Kentuck" said that he did not mind the loss of his money and his garments so much, but be soon realized the importance of the loss of his boots. The prairie grass had recently been burned and the sharp rubble penetrated his socks and pierced his feet, causing great pain; but he was running for his life and did not relax his speed.

His pants were burdensome and he dropped them; but, being wet, they did not slip easy from his extremities, causing him to tangle up and fall; but he did not halt, rolling over and over until be freed himself, then bounding on again with renewed effort. He had no more "graft" with which to bribe his pursuers, and as they gained on him, "Kentuck" said be thought that at last he was a gone "coonskin"; but just then be reached a dense chaparral thicket into which he plunged and hid from his pursuers who beat around for a while and then gave up the pursuit.

After a short breathing spell, "Kentuck" pushed on, making his way in great pain through the thorny chaparral and prickly pears. His under garments were reduced to shreds and his flesh fearfully torn. In this plight be dragged himself on, and emerged into the open again, fell in with a comrade who had also feigned death, and escaped from the slaughter.

These comrades hurried on, avoiding roads as far as possible for fear of being overtaken by some party of the enemy. All families on their route had fled eastward, but they found plenty of food in the abandoned homes. Striking the trail of Houston's army at the crossing of the San Bernard, they hurried on, arriving at the Texan camp at Foster's Plantation on the last day of March: a wonderful trial of speed on foot, from Goliad to the Brazos in four days and nights.

Upon reaching the army "Kentuck" and his comrades were in fearful plight; but the boys soon togged them up and, after their wounds and sores were treated, and with a little rest and sleep, they were ready for duty. They joined Houston's forces and both fought gallantly at San Jacinto. I remember seeing "Kentuck" the next day after the battle, and he was very much interested in a game of monte. I have often heard the story of his thrilling escape and flight, substantially as I have above narrated. Of the subsequent career of "Kentuck" I know nothing, and I regret that I did not learn his real name and more of his history.

Source; by kind permission of Presidio La Bahia, Goliad, Texas.

An Account of the Goliad Massacre, by Major Miller's Secretary

S.H.B. was 18 at the time of the massacre and later a merchant in Cincinnati.

The morning of the massacre was slightly foggy. Without understanding wherefore, we of Miller's command, were ordered to tie a white band around our left arms; some of us tore pieces from our shirts for that purpose. This was to distinguish us from Fannin's men, who alone were doomed. We were conducted out to a peach and fig grove, in front of the church, and in sight of two of the three parties into which Fannin's men were divided; the third being out of view behind the church, near the riverbank.

When the firing began, boy as I was, I was impressed by the varied expressions in the faces of our men, thus made unexpected witnesses of the awful tragedy. Surprise, horror, grief and revenge were depicted in the most vivid lines. At first all were startled; some became at once horror stricken; others wept in silent agony; still others laughed in their passion, swore, clinched their teeth, and looked like demons. Now, at the lapse of more than a quarter of a century, I can never think or talk of that dreadful scene with any degree of composure.

Some of the poor fellows attempted to escape, and of course outrun the Mexicans; but then the cavalry! Just as one of these men of Fannin's had gotten fairly clear of his pursuers, a mounted Mexican from close by me at once started on the chase, and catching up with him, cut him down. Never did I so want to hamstring a horse. Those not killed outright, were deliberately butchered by the Mexicans, men and women, and stripped. This over, some of them, even the women, as they passed us by on their return ladened with plunder, insulted us by the grossest vulgarities, shook their fists in our faces, swearing in taunting tones and the vilest words- 'Your turn tomorrow!

In about an hour more, the wounded left in the barracks were dragged out into the fort yard and butchered. Colonel Fannin was the last victim. When informed of his fate, be met it like a soldier. He banded his watch to the officer whose business it was to murder him, and requested that he be shot in the breast and not in the head, and likewise to see that his remains should be decently buried. With that perfidy which is so prominent a characteristic of the Mexican race, all requests were ignored. Fannin seated himself in a chair, tied the handkerchief over his eyes, and bared his bosom to receive the fire of the soldiers.

The stripped bodies of the slain were collected and placed in piles. Those of the wounded who had been massacred at the fort, Fannin among the rest were chucked stark naked into carts, like so many dead hogs, carried out and dumped on top of the others. Brush was then piled over the whole and set on fire. It took several days successive burnings to consume them. Nightly the prairie wolves gathered to feast on the half-roasted bodies, and kept up their howlings through all the long hours, and as the day dawned their execrable screams increased, in rage at being thus driven by the morning light from their horrid banquet.

Barber and Howes History of the United States 1861
Source; by kind permission of Presidio La Bahia, Goliad, Texas

FANNIN'S RETREAT AND THE BATTLE OF COLETO (AN ACCOUNT BY MS. CASH)

Kennedy says "I borrow the subsequent details from a published narrative very inartificially written, but bearing evidence of being the unvarnished account of an eye-witness." The eyewitness was Mrs. Cash, wife of Irish colonist and surveyor from the Power-Hewetson colony. The family held a land grant on the lower San Antonio River.)

Harry Ripley ... poor fellow, had his thigh broke. Mrs. Cash, at his request, helped him into her cart, and fixed a prop for him to lean against, and a rest for his rifle, while in that situation, he was seen to bring down four Mexicans before he received another wound, which broke his right arm ... "You may take me down now ... done my share, they have paid exactly two for one on account of both balls in me."

Personal safety was guaranteed to every individual; all were to have been treated as prisoners of war in civilised warfare, and private property was to have been respected ... men threw their arms on the ground, Fannin alone reserving his sword, which he handed to Urrea, who had advanced to receive it ...

"The only American non-combatant residing at Goliad (Mrs. Cash) accompanied the troops in the retreat. It was 10 o'clock ere the rear-guard had crossed the river, and they hastened towards Victoria, until Fannin, judging it time to give his oxen rest, ordered a halt, and had them ungeared for an hour; after which preparations were made for a second move, but, alas! one of the carts broke down soon after they started, thus causing another delay, to distribute the load on the other waggons, when minutes were precious to us. However, they again took up the line of march.

Fannin ordered Horton to proceed a-head with his command, and scour the timber bordering the Coleto, which lay on their left; he himself brought up the rear. Thus the troops moved on half a mile farther, when two Mexicans came out of the Coleto timber a mile behind them-not in their front, from whence Fannin supposed them likely to appear.

They halted on the edge of the prairie and reconnoitred for ten minutes, and then retired back into the woods soon after they again returned accompanied by four others, and after reconnoitring a short time, the enemy's advance guard of cavalry made their appearance, deploying into the open ground in platoons four deep.

They immediately galloped after our troops, and when within a fourth of a mile of them, they separated and passed on in double files, having the Americans between them, until their van was half a mile a-head of the battalion, in the direction of Victoria, when they wheeled from both divisions and galloped to the centre, until their ranks again met; their rear also closed in the same manner, and our friends found themselves surrounded on all sides by the enemy. For half an hour each party was preparing for the coming struggle. Our men saw at a glance that their only hope rested in their courage and decision; and they threw themselves into a hollow square, facing outwards.

'The Rise, Progress and Prospects of the Republic of Texas'
William Kennedy, 1841. *Source;* by kind permission of Presidio La Bahia, Goliad, Texas.

Sketch: Battlefield at Coleto Creek

[Kennedy's note: The effect of a piece of artillery, fired against Fannin's orders, satisfied Urrea, who had no cannon on the ground, that to win the day he must come to close quarters; he therefore ordered a charge of cavalry. They galloped in "dashing style" towards the Texan battery, when they were broken by a discharge from seven pieces.]

The shock was sudden: they halted, drew up their horses and faltered when encouraged to return to the charge; finally they turned tail and retreated to the woods, full two miles off, where they were again formed, and prepared for another attempt to carry the battery by storm. In the mean while, the enemy's infantry kept up a harmless fire, advancing on our line after each round; but they were only suffered to take ground within rifle range, where they were stopped, for the boys burnt no powder in vain.

As soon as the cavalry had retreated, our artillerists wheeled their pieces to the right about, and scattered grape and death among the infantry, who were thus in their turn driven back to the Coleto timber, about half a mile off. As soon as Urrea could restore order to his lines he again advanced.

His infantry came up in double quick time, and his cavalry in a 'slope,' until they were within range, when the artillery again checked them with a single volley, and they turned tail and scampered for the woods as before.

Their infantry advanced just near enough to lose a man at every rifle crack, but not near enough to do execution on our friends, for their ammunition was poor, and their marksmen were unaccustomed to the use of the musket; their officers were marked targets, and they suffered accordingly. As soon as the cavalry had started for the timber, our artillery wheeled about as before, and sent the infantry off also. As yet our friends' were unscathed.

In about half an hour after their second repulse, Urrea succeeded in putting his columns in order. They were reluctantly driven by their officers to the assault for the third time, for it required great exertions to induce them even to make a show as though they intended to advance to the charge; our men saw the officers beating them over the shoulders, and coaxing them on by pricking them from behind. They however succeeded in urging them to within fifty or sixty yards of our lines, on either side, but then again the grape and canister from our artillery mowed down their cavalry as if they were rushes, while their infantry suffered martyrdom under the fire of our rear and left divisions.

Their front ranks were so suddenly swept off, as almost to form a breastwork sufficient in itself to shield our friends from their assaults. The scene was now dreadful to behold; killed and maimed men and horses were strewn over the plain, the wounded were rending the air with their distressing moans, while a great number of horses without riders were rushing to and fro back upon the enemy's lines, increasing the confusion among them: they thus became so entangled, the one with the other, that their retreat resembled the headlong flight of a herd of buffaloes, rather than the retreat of a well-drilled regular army, as they were.

In the rush back, a number were overthrown and trodden under foot. The Mexican officers exerted themselves to bring them again to the charge for the fourth time, but without avail. The cavalry could not be urged within rifle range: they contented themselves with firing their carbines at such a distance as did no execution, but they met with greater success in stabbing their infantry on, for in that manner they urged them to the assault. They halted when within sixty yards of our lines, and held their ground under a murderous fire from our riflemen.

It was now about dusk; and Urrea bethought himself of a plan of attack which answered but too well. He directed the Campeachy Indians, who were better marksmen than any other of his troops, to throw themselves into the tall grass and approach, as they did, within thirty paces of our lines. They then commenced a well-directed fire upon us, which told most destructively, by wounding fifty and killing four in the space of an hour.

The darkness of the twilight and the confusion arising from the smoke prevented our men from discovering the exact position of their dangerous neighbors, who were so well concealed in the grass that they could not dislodge them. Among the wounded was Harry Ripley, a youth of eighteen or nineteen, the son of General Ripley, of Louisiana; he, poor fellow, had his thigh broken soon after the Indians first took to the grass. Mrs. Cash, at his request, helped him into her cart, and fixed a prop for him to lean against, and a rest for his rifle, while in that situation, he was seen to bring down four Mexicans before he received another wound, which broke his right arm; he immediately exclaimed to Mrs. C.,

"You may take me down now, mother; I have done my share, they have paid exactly two for one on account of both balls in me." Such was the spirit that animated every man among the assailed save two, who had run under a cart early in the engagement, and covered up their heads. One of them was familiarly called Black Hawk, as he professed to be a very fire-eater-but his courage was all in his tongue.

So soon as the darkness rendered the flashes of the Indians' guns visible, they began to pay the piper, for our boys were quick on the trigger, and at that distance took care that a second flash should not be seen from the same weapon; so they soon used them completely up, and then Urrea drew off his troops. They retired about a quarter of a mile off on each side, and rested on their arms all night. It was determined by our friends to throw up a breast-work, so the poor fellows set to work, and they dug a ditch on all sides: with the earth, their baggage, and ammunition waggons they made a very passable fortification.

The wounded suffered agonies for want of water, and by their moans and petitions for it made the situation of those who had escaped unhurt even more distressing. They, however, worked manfully, and accomplished more than could have been expected of them, wearied and thirsty as they were. During the whole night, the Mexican General caused his bugles to sound at intervals of five minutes, with the view of keeping his troops on the look-out.

[Kennedy note: Early in the morning, the Texans discovered that the labour of the night had been in vain. Urrea had received a reinforcement of 500 fresh troops, with a supply of artillery.]

The trench thrown up by the batallion offered no protection against anything but musketry, and the waggons and other baggage only served as missiles to destroy our troops, for a single discharge would have so shattered every thing that each fragment would have injured somebody.

The temporary work had not been thrown up as a defence against artillery, but against infantry and cavalry; and had the artillery not been brought against it, the probability is that Fannin would have been victorious in the end, for the previous day's engagement had cut off.600 effective men from Urrea's ranks; but such another day's work was necessary to render the relative numbers of the opposing forces so nearly equal that our friends would have merely had to turn out, and knock their opponents down with the butts of their rifles. However, as soon as Colonel Fannin saw the artillery, he knew that the fate of the engagement was determined; and he called a consultation of officers, amid the cries of our wounded that he would surrender, and thus procure water for their parched tongues.

Mrs. Cash, at the solicitation of some of the wounded, undertook to go to Urrea and ask him for water before the action again commenced. Accompanied by her little son, a lad of fourteen years of age, who had done a man's fighting during the work of the previous day, she made her way over the ground between the two contending parties, and proceeded directly to Urrea, to whom she was introduced by a Mexican officer stationed at Goliad.

She made him acquainted with her errand; he did not answer her request, but fixing his eyes upon the boy's shot-pouch and powder horn, that she had neglected to take from him before she left our lines, he exclaimed, "Woman! are you not ashamed to bring one of such tender age into such a situation?" The boy immediately answered him, "that, young as he was, he knew his rights, as did everybody in Texas; and he intended to have them or die."

The conversation was here interrupted by the raising of a white flag on the part of our troops, as a token of surrender, which was immediately torn into three pieces by the wind: it was blowing a stiff norther at the time. As soon as Mrs. Cash had left our lines, the officers consulted whether it was best to surrender, and it was so determined. Immediately after hoisting the flag,

Captain Shackelford and Captain Chadwick, the engineer and adjutant, advanced out on the prairie and met Urrea, with whom they arranged terms;-personal safety was guaranteed to every individual; all were to have been treated as prisoners of war in civilised warfare, and private property was to have been respected. Shackelford then returned to the battalion and announced the result, our men threw their arms on the ground, Fannin alone reserving his sword, which he handed to Urrea, who had advanced to receive it. One of our men threw a lighted segar among some loose cartridges, to prevent the enemy from getting the ammunition, and they exploded, together with four kegs of powder; four men were blown up by it, which served to increase the trouble and distress."

Source; by kind permission of Presidio La Bahia, Goliad, Texas.

FANNIN AND HIS COMMAND
(BY LEWIS M.H.WASHINGTON)

Comprising a brief sketch of the Organization, Military Operations and Massacre of Col. James W. Fannin and his Regiment,
in the Revolutionary Struggle of Texas - 1835 and '36.

BY ONE OF THE COMMAND
[Lewis M. H. Washington]

To the Editors of the Georgia Citizen:

As a respectable number of your readers were no doubt acquainted, not only with the late gallant, but unfortunate, Colonels Fannin and Ward—who so nobly fell as martyrs in the cause of Texas Independence—but with many of those ardent and chivalric spirits that composed their Command, and who, or nearly all of them, shared the same fate with their heroic commanders, in the memorable massacre at Goliad, in the Spring of 1836, I will endeavor, as far as my memory serves me, to give, through the medium of your valuable journal, a brief sketch of the organization, military operations, massacre, &c., of that noble and valorous Regiment, believeing [sic] that it will be read with some degree of interest. In giving this sketch to the public, I feel the necessity of asking, in advance, all due allowance, for such inaccuracies of *detail* as may, and no doubt *will* be found to exist in the body of the narrative.

As I write almost wholly from memory, no liberal-minded man will expect *every incident* connected with the history of that "*eventful and bloody drama*," (as it has been aptly termed,) to be given with unerring exactness, even to the *minutiae*. Notwithstanding the slight inaccuracies that may be discovered, it is confidently believed that in the main, this desultory narrative will be found, by those familiar with the *facts*, substantially correct:—

Many of the volunteers comprising the command of Col. Fannin, were from Georgia; the most of whom volunteered their services to battle in the cause of Texian Liberty, early in the Fall of 1835, at Macon, Milledgeville and Columbus. Those from the vicinity of the two former places were chiefly raised for the occasion, by Col. William Ward, and those from the latter, by Capt. Wm. A. O. Wadsworth; but on their route to Texas, their numbers were considerably augmented by other volunteers, who enrolled themselves at Montgomery, Ala., Mobile and New Orleans, making in all, when they debarked at Velasco, Texas, on the 20th Dec., '35, a well equipped force of about three hundred as gallant and intrepid young men as ever drew a sword or shouldered a fire-lock in the cause of human Liberty.

Within a day or two after the arrival of these voluntees at Velasco, a thorough organization of the companies was consummated, in order that they might be immediately reported for service to the Provisional Government, (Governor Henry Smith and the members of the Council,) then temporarily resident at San Felipe de Austin, the Colonal capital of Stephen F. Austin's Colony.

The number raised by Capt. Wadsworth was but little more than sufficient for one full Company, and consequently, it was determined that they should be organized as one Company, including the whole number. W. A. Wadsworth was elected Captain, and Thomas Reese and Joseph Wilson as Lieutenants. Mr. R. Rutledge, I believe, was elected Orderly.

The volunteers raised by Col. Ward were found, on landing in Texas, to number well nigh two hundred -- enough for two companies, and it was determined they should be as equally divided as practicable, and the two divisions separately organized as companies. So soon as the division was satisfactorily arranged, the companies proceeded to elect their officers respectively.

Of one of the companies, U. J. Bullock was elected Captain, Bazil Lamar and Alex. Patton, Lieutenants, and Francis M. Hunt, Orderly. The other chose a gallant young gentleman of the name of Wynn, (I think from the couty [sic] of Gwinnett,) as Captain, and Willey Hughs, as a Lieutenant. The names of the other Lieutenants and the Orderly of this company I have also forgotten.

After these companies were respectively organized, the whole three proceeded to organize as a Battalion—choosing Wm. Ward as Major, and a Mr. Chadwick as Sergeant Major.

Here it is proper that I should say something regarding the gallant and accomplished Chadwick, for I have always viewed him as decidedly one of the most thorough and effective military men, as well as one of the most polished and pleasant gentlemen, that ever battled under the bright banner of the Lone Star.

He was, I think, a native of the State of New York. At an early age he graduated at West Point Military Academy, and was breveted a 2d Lieutenant in the Draggoon Regiment, then commanded by the celebrated Dodge of "Black Hawk War" notoriety, which officer he accompanied in his first expeditions to the Rocky Mountains.

For a man of his age, (not being over thirty when he entered the service of Texas) he had seen much hard and trying service, which well fitted him to meet the hardships, privations and perils which were necessarily to be encountered in the service of his adopted country. In addition to that which he had learned from experience on the "tented field," amid the wild and inhospitable regions of the Rocky Mountains, he was a scientific and efficient tactician, as well as an accomplished Military and Topographical Engineer. Joined to these accomplishments, was a native suavity of temper and urbanity of manner, which at once made him the *pride of the Battalion*.

So soon as the "Georgia Battalion" had been properly organized and reported, Col. Jas. W. Fannin, who had but a short time before been promoted by the Provisional Government, to a Colonelcy, in consequence of his gallant and officer-like bearing as Captain of a volunteer corps, at the "Grass Fight" and the "Battle of Concepcion," during the then recent siege of San Antonio de Bexar, came down to Velasco from San Felipe, and formally mustered the troops composing it into the service of Texas, for a term of six months.

After the Battalion had been "mustered in," Col. Fannin, by proper authority, proceeded to appoint the necessary staff officers: David I. Holt was appointed Quarter Master, and a Mr. Cozart and the writer of this sketch were appointed Assistant Quarter Masters. Mr.

Green Buchanan received the appointment of Commissary of Subsistence, and Mr. James Hughes that of Issuing Commissary.

The organization of the Battalion being now completed, Col. Fannin proceeded to San Felipe for further orders; and whilst he was there, it was determined by the Provisional Government that he should as soon as practicable, repair with the Georgia Battalion to the Western frontier, and there, at such point as he might think poper [sic] to select, organize as strong a force as he could gather, with a view to crossing the Rio Grande, and making a descent upon the city of Matamoras, *provided* that after a careful reconnoissance of the city and its defences, by competent and trustworthy men - its capture should be deemed practicable.

In the event of the projected enterprize being found impracticable or too hazardous, he was to quarter the troops at some eligible point on the frontier and there fortify, for the purpose of holding any Mexican force that might attempt to penetrate the settlements, by the lower or Matamoras route, in check, until a sufficient force could be gathered, to offer successful resistance.

After everyting [sic] concerning the future operations of Col. Fannin's command had been arranged at San Felipe, and the necessary official orders delivered, the Colonel hastened to Velasco in order to make immediate preparations for the embarkation of the troops, munitions of war, etc. for Copono [sic, Copano], a point on Aransas Bay, which had long been the port of entry for all that district of country lying between the Guadalupe and Nueces rivers

On his arrival at Velasco, Col. Fannin made early application to Messrs. McKinney & Williams -- a high-minded, wealth, liberal and patriotic mercantile firm at the mouth of the Brazos river -- for vessels of sufficient capacity for the transportation of the volunteers, Commissary's stores, etc., to the aforementioned point of destination.

Those gentlemen, with their usual promptness and alacrity, soon procured such vessels as were required, and in a few days every thing was ready for embarkation.

A day or two before Col. Fannin and his command sailed from Velasco, the steamer Yellow Stone arrived from New Orleans with two more fine companies of volunteers on board—Capt. Isaac Ticknor's from Montgomery, Ala., and Capt. Duval's, (a son of the Ex-Governor Duval, of Fla.,) from Kentucky, making an acquisition to the command of something over one hundred choice and thoroughly equipped men.

Added to the American force above alluded to, was a small company of Mexicans, under command of Captain Luis Guarro—an intelligent, gallant and patriotic officer of the regular army of Mexico, who had been for years attached to the garrison at Tampico, and had participated with his company of regulars, in the disastrous revolutionary movement of the patriot General Mexia, in November, '35. After the defeat of Gen. Mexia, at Tampico, he, and those of his adherents who were not captured and shot by the Centralist troops, made their way to Velasco, on board a small schooner, bringing with them to [sic] new and elegant brass field-pieces, double-fortified long sixes.

Cap. Guarro tendered the services of himself and company, as before stated, a day or two prior to the embarkation of the troops at Velasco. Their services were accepted by Col. Fannin, more for the purpose of obtaining the beautiful field-pieces than on account of any real efficiency that was supposed to attach to the men as soldiers. It was the opinion of all; that Capt. Guarro himself was a man of unquestionable valor and fidelity to the patriotic cause in which he had so nobly embarked, and by doing which he had been compelled to flee precipitately from his own family and country, and leave a large estate subject to confiscation by the Central Despotic Government of Mexico.

Everything being in readiness, the entire command sailed from the mouth of the Brazos river on board the schooner Columbus, and another vessel, the name of which I cannot now remember, on the 23d January, 1836, and after a short, but tempestuous voyage, arrived at Copono, Jan. 30th, where we remained until February 1st.

On our arrival at Copono, we met Maj. Holt, Q. M., who had started through over land, prior to our sailing, in order to make the necessary arrangements for the transportation of our stores, baggage, etc. to the Mission de Refugio, a small but ancient village, situated on the Mission river, 15 miles above Copono, that being the point previously determined upon, at which the forces destined for the proposed descent upon Matamoras, were to concentrate. The Major had procured a number of Mexican ox-carts, and had them in readiness to "load up and go ahead." We were, consequently, detained but one night and half a day at Copono.

Late in the afternoon of the 1st February, we encamped on the *plaza*, (public square,) in front of the old stone Church, in the sequestered and antique village of Refugio, and on the following morning removed to a beautiful site for an encampment, three miles below the village, on the west side of the river, and immediately on the Matamoras road. This place had long been known as the *Lopez Rancho*, and some Mexican *peons* in the employ of Lopez, were residing there at the time.

At Copono and Refugio, we found some companies or rather small fragments of companies, awaiting our arrival. Among these, as far as my memory now serves me, were the commands of Captains Pearson, King, Wyatt, Bradford and Burke. The two former gentlemen were, I believe, of the histrionic profession, as well as some of those attached to their respective commands.Mitchell, (from Columbus, Ga.,) and a gentleman of the name of Wallace, (from Virginia,) as Majors of Battalions.

The entire force now amounting to nearly or quite enough to constitute a regiment, it was determined that a regimental organization should at once be had.The election of officers for the regiment was held at the *Rancho*, about three days after our encampment there, and resulted in the choice of James W. Fannin as Colonel[1], William Ward as Lieut. Colonel, and Doctor.

---

[1] Though Col. Fannin already bore a commission in the army of Texas, and had received *special orders* to take command of the Georgia volunteers who were at the mouth of the Brazos river at the date of his orders, and of such other companies as might thereafter arrive in the country, until it should be deemed proper by the Provisional Government to change the plan of operations, yet his commission was for the *regular* service, and he was well aware of the long established usage, (in America) of volunteers electing their own officers, consequently he resigned his commission in the regular army, in order to give the volunteers of the Regiment an opportunity of exercising their own choice in the selection of their regimental commander.

The elections being over, Col. Fannin proceeded to the appointment of his staff. Sergeant Major Chadwick was promoted to the Adjutantcy of the Regiment, and a Mr. Brooks, (a Virginian who received a thorough military education in the U.S. and in France, and had seen considerable service in the French war against the Malays,) was appointed to officiate both as Regimental Inspector and Captain of Engineers. Dr. Wm. McGee received the appointment of Surgeon, and the same Quarter Masters and Commissaries that had been previously appointed at Velasco, were re-appointed here, together with some additional Assistants in both Departments.

About the time the organization of the Regiment was completed, a spy who had been sent to observe the movements of the enemy on the Rio Grande, and especially to ascertain the strength of the defence at Matamoras, returned to our encampment and announced the arrival of a reinforcement of 3,000 of the choicest Mexican troops at that city.

The receipt of this startling intelligence resulted in a council of war being immediately held by the commissioned officers of the Regiment, all of whom, I think, (with the exception of Capt Pearson,) acquiesced in the proposition of Col. Fannin to abandon for the time being the project of taking Matamoras, and to repair forthwith to Goliad, (28 miles above Refugio, on the route to San Antonio,) and there take up quarters within the walls of the old fortified church, known as the *Mission de Espirito Santa*.

This position was deemed the most important one on the Western frontier, with perhaps the exception of San Antonio, as it was immediately on the only travelled route from Matamoras to the American settlements in Texas. Another inducement to occupy the fortress of *Espirito Santa* was that there was already a small force of Texian regulars and volunteers in possession of that post; Capt. Phil. Dimett, with a few intrepid men, having taken it by storm the preceding fall, from a considerable body of Mexican soldiers and *rancheros* that had been garrisoned there since the beginning of the revolutionary movements in the autumn of '35.

The acquisition of the troops at Goliad, together with some fine pieces of artillery that had been captured there by Dimett, was deemed of great importance, as it would add much to the strength and efficiency of the Regiment.

So soon as it was determined to occupy the fort at Goliad, preparations were made to take up the line of march for that point, and on the 11th Feb. the whole force was on their way thither—excepting Captain Pearson's company, which proceeded in the direction of Matamoras, as far as San Patricio, (an Irish settlement on the Nueces river, about 40 miles west of

Refugio,) where were Col. Frank Johnson and Capt. Grant endeavoring to concentrate a sufficient body of men to make a foray upon the Mexican settlements on the Lower Rio Grande, and, perhaps, to pay Matamoras a visit, not believing the report given by the spy relative to the arrival of a strong reinforcement at that place.

We reached Goliad on the afternoon of the 13th February, after a march of about two days over a level prairie, facing, the whole time, a furious '*Norther*,' accompanied with heavy sleet. This was the first *genuine taste* of the '*pleasures of soldiering*' that the most of the 'boys' had ever experienced; and the reader may rest assured that many were the imprecations muttered, and many the rueful grimaces of visage exhibited, whilst the draught was being swallowed. Poor fellows! there was yet reserved a far more bitter cup than that for them to drain.

On our arrival at Goliad, we found within the walls of the Fort, one company of regulars, (not full,) and several fragments of volunteer companies; the most of the men composing them having belonged to companies that had been disbanded at San Antonio after the capitulation of Cos, early in the preceding December. The garrison was under the temporary command of Capt. Westover, of the regulars, who, upon the arrival of Col. Fannin, formally resigned the command into his hands, having been instructed to that effect by official orders previously received from San Felipe.

Col. Fannin, on examining the Fort, found that much work was necessary before it could be made to answer any valuable purpose, in the event of being stormed or besieged by a strong and well-appointed force of the enemy. Consequently, he at once caused such repairs as could be made to be commenced, together with other important defences suggested by Capt. Brooks.

A number of old dilapidated stone houses were torn down and the stones of which they were built used in strengthening the walls of the fortress. A narrow ditch was dug on the inner side, four feet from the walls, and strong pickets planted firmly and closely together therein. The space between the pickets and the walls was then filled with earth and stones well beaten in, making the entire outer walls, well nigh eight feet in thickness. A deep ditch, six feet wide, was then dug, inside of these strong rampants, with draw-bridges at such places as they were thought necessary. Substantial platforms were erected at suitable points on the walls, upon which heavy pieces of artillery were mounted so as to command the different streets of the town that led to the '*military plaza*,' in the centre of which the Fort was situated. In fine, no time was lost nor pains spared to put every thing in as defensible a condition as possible.

A short time after our being quartered in the fortress *de Espirito Santa*, there was a small Artillery Corps organized, made up of men of some experience in gunnery, and commanded by a kinsman of the late gallant Stephen Decatur—Capt. Stephen D. Hurst, and who, I believe, was once also attached to the U. S. Navy. With the assistance of the accomplished Brooks of the Engineers, Capt. Hurst soon drilled his little corps into a tolerable degree of efficiency.

About a week or ten days after our arrival at Goliad, we were reinforced by Capt. Jack Shackelford, with his chivalric and finely equipped company of uniformed volunteers—the '*Red Rovers*,' from North Alabama. Our entire effective force now amounted to about 500 men; and through the indefatigable exertions of our Quarter Masters, Commissaries and Contractors, the garrison was pretty well supplied with beef, corn, flour, sugar and coffee. Our supply of ammunition was also very good; and of arms we had an abundance—good muskets, fine Harper's Ferry yagers, and close-shooting rifles—also, eleven pieces of artillery, including one mortar.

On the 25th of Feb, an express arrived from San Antonio informing Col. Fannin of the arrival at that place of a large force of Mexicans, and that the Alamo, an ancient fortified church on the east side of the river, and immediately opposite the city, was then besieged by them. This express had been dispatched by Col. Wm. B. Travis, (who was quartered in the Alamo with only about 160 men,) requesting Col. Fannin to reinforce him, if possible, from the garrison of Goliad, as his small force was wholly insufficient to withstand, successfully, any united and determined assault by a force so vastly superior in numerical strength.

On the receipt of this intelligence, Col. Fannin consulted with the commissioned officers of his command, and it was determined that the whole regiment, excepting a small force to be left in charge of the post at Goliad, should take up the line of march on the following day for San Antonio, or, rather, the Alamo. On the next day, so soon as necessary preparations could be made for the march, we were on the route to the seat of 'active operations;' but it being late in the day before we could get off, night overtook us ere we had proceeded two miles from the Fort.

After we had encamped, Col. Fannin again went into consultation with his officers, which resulted in an order for a countermarch of the whole force to the fortress de Espirito Santa. This retrograde movement was, upon due reflection, considered advisable by a majority of the commissioned officers of the regiment, for the following reasons:

1st. As the dispatch from Col. Travis had not intimated the probable number of the enemy by which the Alamo was besieged, it was highly possible that their force was of sufficient strength to prevent any reinforcement not stronger than we were, from obtaining ingress to the fortress, if not to encounter and vanquish us in an open field engagement on our approach to San Antonio.

2d. If there was really a powerful force of Mexicans then before the walls of the Alamo, (which the expressman informed us was far from being in a defensible condition even if it were strongly garrisoned,) it was probable that some decisive movement would be made ere we should reach there, as the distance was ninety miles, a march of fully four days and a half, and our subsistence and ammunition, carts, artillery, &c., were drawn by jaded and famished ox-teams that were liable to fail even before we could make half the distance.

3d. The report brought in by the spy whilst we were lying at the Lopez Rancho, was considered sufficient to warrant the supposition that it was, probably, Santa Anna's purpose to divide the Mexican forces destined for a second invasion of Texas, so that one division should penetrate the country by the upper or San Antonio road, and the other by the lower or Goliad route. In the event of that supposition being correct, the abandonment of the fortress de Espirito Santa, or the leaving of it garrisoned by so few men as were then occupying it, might be followed by the most disastrous consequences not only to the garrison but to the cause of Texian liberty.
These reasons having been duly considered, it was deemed most prudent that the regiment should return to the Fort and put themselves in readiness for its defence.

On the day after our return, a second express came in from San Antonio, bringing intelligence of the arrival at that place of a reinforcement of 5000 Mexican troops, under the immediate command of the President-General, the despot Santa Anna himself, and stating that the whole Mexican force then before the walls of the Alamo did not amount to less than 7000 men. Col. Travis had also been reinforced by the arrival of about 30 Texians, mostly from the neighborhood of Gonzales, who obtained ingress to the Fort by stealth under cover of the darkness of night.

The besieging army had made several attempts to take the Alamo by storm, but up to the time of the expressman's leaving, they had been beaten back with considerable loss on their part, and without the loss of a single man on the side of the Texians. Yet, notwithstanding the success that had so far attended the gallant defence of the Alamo, it was hardly possible that so small a force could much longer continue to repel the constant and determined assaults of a besieging army, numbering at least *twenty to their one*.

In fact, it appeared that the heroic little band foresaw the result, and were hourly expecting the last, bloody scene of the momentous tragedy in which they were engaged as prominent actors, for the expressman stated that Col. Travis and his Spartan-like command had deliberately resolved never to surrender the fortress as long as there was a living man within its walls to strike for its defence.

About this time a Mexican, Placedore, an old and respectable citizen of Texas, who had been with Johnson and Grant at San Patricio, came into Goliad at full speed and greatly exhausted, and announced that Capt. Grant's command, of which he was a member, had encountered a heavy force of the enemy near the Casa Blanco, about 15 miles west of San Patricio, and that they were all either killed or captured save himself.

Col. Fannin placed entire confidence in Placedore's statement, knowing him to be a man of veracity and of well-tested fidelity to the Texian cause.On the second day after the arrival of Placedore at Goliad, Col. Frank Johnson and Mr. John Love came in on foot and without shoes, and confirmed the intelligence brought by Placedore. They also stated that after calling up Grant's party, the Mexicans had advanced upon San Patricio at the dead of night, and had taken them, (Johnson's command,) by surprise, they only, (Johnson and Love,) effecting an escape.

It was now certain that a division of the Mexican army was approaching the coast settlements of Western Texas by the lower route, proving the correctness of the supposition which had led to the abandonment of the project of reinforcing Col. Travis, as before stated.

The company of Mexicans in the Fort were seized with great trepidation on the receipt of this, to them, appalling intelligence; and through their commander, Guarro, requested permission of Col. Fannin to repair immediately to Coxe's Point on the Bay, where they would have an early opportunity of embarking for New Orleans, urging as a reason for making the request, that if they should fall into the hands of the central troops they would instantly be shot as deserters and rebels. Col. Fannin granted their request with little hesitancy, as he had never viewed them as being efficient or reliable, though he had full confidence in both the patriotism and prowess of Capt. Guarro.

As soon as the necessary permission was given, in writing, the whole company left Goliad, as was understood, for Coxe's Point, but -- as it will hereafter appear in the course of this narrative -- all of them, except Capt. Guarro, who went to New Orleans, joined Gen. Urrea's division of the Mexican army then at San Patricio. A day or two before they asked permission to leave, a strange Mexican, representing himself to be a Ranchero, residing on the San Antonio river some ten or fifteen miles below Goliad, came into the Fort and remained with the Mexican company for several hours.

Nothing at the time was thought of this visit of the self-styled Ranchero, as Mexicans of similar appearance had often before visited their countrymen in the Fort, without having attracted special notice or excited suspicion, but it was subsequently ascertained that he was a Mexican officer in disguise, and that he was the bearer of a dispatch from Gen. Urrea to that company, offering a full pardon for their former 'traitorous and libellous conduct at Tampico,' if they would 'forsake the ignoble and criminal cause of the *pirates*,' as the Texians were termed, 'and return, like *honorable Mexicans*, to the glorious and triumphant standard of their *blessed* country.' But in the event of their obstinately refusing to accept the magnanimous offer extended to them, they would, just as certain as the *Blessed Virgin* was then looking down from Paradise upon them, be captured and executed as traitors, *without the benefit of clergy*.

On the 10th of March, a man of the name of Ayers, who had been merchandising in the town of San Patricio, arrived at Goliad apparently in the deepest trouble, stating that his family were captives in the hands of a number of Rancheros at the Mission Refugio, and expressed much fear that they would be put to death. Col. Fannin, on hearing this touching representation of what he *supposed to be facts*, immediately dispatched Capt. King's small company and Capt. Bradford's—still smaller—the whole under command of King, to effect a release of Ayers' family and such others as might have fallen into the hands of the *reputed Rancheros* subsequent to the departure of Ayers from the Mission, Ayers remaining at Goliad seemingly in the greatest mental agony.

On the 12th a young man, the son of an Irish widow who resided at the Mission, came into Goliad and reported himself to Col. Fannin as the bearer of a verbal express from Capt. King, who had upon reaching the Mission, instead of finding a mere body of Rancheros as Ayers had represented, encountered over three hundred well-drilled and well-armed Mexican soldiers. He stated that King, on finding so great an odds against him, had betaken himself to the old stone church and had, up to the time of his leaving, succeeded in defending himself from within its massive walls without any loss, but that it would be impossible for him to do so much longer, inasmuch as his men were out of provisions and had well nigh expended all their ammunition.

This intelligence reached Goliad about 10 o'clock at night, and in less than an hour from the time of its receipt, the 'Georgia Battalion'—including Capt. Ticknor's company from Montgomery, Ala.—were on a forced march for the Mission, which place they reached at an early hour on the following day. Ayers accompanied the Battalion. I will here introduce an extract from an article touching the subject matter under consideration, which appeared in the *Galveston News* in the summer of 1851, and may be relied on as substantially correct:

"Finding a great many more Mexicans there than had shown themselves at the time Capt. King forwarded the express, Col. Ward was compelled to cut his way through their lines which had been formed expressly to prevent his joining King's gallant little band, something near a mile above the Mission. In the course of this short but desperate encounter, about twelve Mexicans were killed.

"On the following morning Maj. Mitchell with a detachment, was ordered across the river to destroy the *ranchos* lying below the Mission, which order having been satisfactorily executed Col. Ward began making preparations to retreat with the families under his protection. Just at this time a difficulty arose between Col. Ward and Capt. King, the latter of whom, at the urgent request of Ayers and against the will of Ward, was about starting out for the purpose of recapturing some goods which he, Ayers, represented as having been stolen from him by the Mexicans. King with twenty-eight men, had crossed the river for the purpose already stated, when the whole of Urrea's division of the Mexican army, amounting to near sixteen hundred men, appeared in 'battle array' before them.

"Capt. King was at once cut off from all chance of escape, and he fought with furious desperation until he lost all of his men but nine, eight of whom, including himself, were taken prisoners by the Mexicans, one only making his escape."

The attention of Urrea was now turned to routing Col. Ward's command from their stronghold in the church, and the whole division, with one brass six-pounder, was brought into the fierce conflict which continued steadily until 12 o'clock at night. The loss on the part of the Mexicans was subsequently ascertained to be about 500 killed and wounded. The greater portion of them were killed on the ground.

In the meantime, intelligence had been received at Goliad of the fall of the Alamo, and Col. Fannin dispatched an express, borne by the intrepid James Murphey from Milledgeville, Geo., to Col. Ward. The brave young Murphey was discovered, ere he reached his destination, by a large body of Mexican Cavalry. They pursued, captured and killed him. The express, which gave an account of the fall of the Alamo and requested Col. Ward to retreat to Victoria, at which point Gen. Houston, the commander-in-chief, had ordered Col. Fannin to meet him, so that all the forces then in the field might be concentrated, was sent in by Urrea accompanied with a demand that Col. Ward instantly surrender, also stating that he had Capt. King and seven of his men prisoners. Ward detained the bearer of Urrea's communication and the conflict was renewed with increasing fury on both sides.

The Mexicans at one time succeeded in getting their brass piece within ten paces of the southern wall of the church, but the 'boys' poured such a deadly volley into them from their close, shooting yagers and rifles, they retreated precipitately with their 'big gun,' leaving a goodly number of their *companiaros* 'stretched' behind them.

During the whole of this sanguinary affair, Col. Ward had not a solitary man killed, and but two were wounded. One of them was a daring youth, who ran outside 'to take a crack at the d—d yellow-bellies,' to use his own language. His name was Shelton; he was not more than 15 years old, and was one of the most respectable families in the State of Mississippi. The other was a man named Winters attached to Capt. Wynn's company.

About 2 o'clock on the morning of the 14th, under cover of the darkness of night, Col. Ward began his retreat for Victoria, being compelled to leave his wounded in the church. Winters was supposed to be mortally wounded, and the tender age of young Shelton, it was thought, would elicit the compassion of the Mexican General. Vain hope! No sooner had the successful retreat of Ward and his men been discovered, than the gallant King and his seven men, together with Winters and the brave little Shelton, were shot! It was afterwards learned from some Mexicans belonging to Urrea's division, who were taken prisoners, that the Mexican General caused Capt. King to be bound firmly to a tree with *lariettas* and shot, alleging that he had *fought like a beast*, and it was proper that he should *die like a beast.*

In consequence of the total ignorance of all the command of the geography of the country, Victoria was not reached until the 20th, making the time occupied on the march six days, when if there had been any competent guide along the distance could have been accomplished in less time than three days.

During those six days the greatest suffering was experienced from fatigue, hunger and thirst, being out of provisions and so closely approximated to the bay that no fresh water could be obtained.

On arriving at Victoria, instead of finding Col. Fannin and Gen. Houston there as was expected, Col. Ward was met by an overwhelming Mexican force. At this place, Lieut. Wilson of Captain Wardsworth's [sic] company, on venturing up to a house in search of something to eat, was killed.

It was there that the first tidings were had of Col. Fannin's defeat and surrender, a desultory account of which will be given hereinafter.

Col. Ward now commenced a retreat to the timber of the Guadalupe bottom, in the dense fastnesses of which his command lay for several hours. They then resumed their march, bearing to the East, and on the next day reached Placedore's creek running between Victoria and a noted landing on Lavaca bay, known at that time as Linn's Landing, since as Linnville. Whilst lying on this creek, Col. Ward was overtaken by a heavy force of Mexicans that had been dispatched by Gen. Urrea in pursuit of him.

A flag of truce was sent in, borne by a Polish officer of the Mexican army, Col. Holzinger, who spoke English tolerably well. He demanded a surrender upon the same terms characterizing the surrender of Col. Fannin, viz:"That they should lay down their arms and agree, upon the honor of soldiers, never again to take up arms against the Republic of Mexico. After doing this, they were to be safely conveyed, at the expense of the Mexican Government, to the city of New Orleans, and there set at liberty."

Col. Ward, with characteristic contempt of death, at once ordered his men to prepare to "*fight to the knife*," telling them at the same time that *he* would "rather die than surrender." His men, or rather the most of them, being worn down and famished, and without ammunition, indicated a willingnes [sic] to "*come into terms*." This enraged the Colonel, but his good sense pointed out the course he should pursue. He immediately ordered them, they being already formed in line, to come to an "*order*."

This done, he explained the terms upon which a surrender was demanded, and expressed an entire lack of confidence in *Mexican Honor*, in any form watsoever [sic] that it could be pledged. He then ordered those who were "in favor of surrendering, to *stand at an order*, and those who were disposed to *do or to die*, to *come to a shoulder*." But few comparatively came to a *shoulder*, consequently the required surrender was at once made.

The Command was then marched back to Goliad, with the exception of about 23, who were sent to Victoria as mechanics, where Col. Fannin and his command were already confined in the Fort. They had separated but a short time before, the bosom of each filled with "high-bounding pride" and glorious aspirations. They now met, worn down and emaciated; and what was the most humiliating of all, they were subject to the insults, and other harsher usage of a people far inferior, in every essential particular, to the meanest of their own negro servants.

Col. Fannin evacuated the Fort at Goliad, I think, on the 17th of March, for Victoria. The *van-guard* was composed of mounted men, the most of whom had been raised in Matagorda county, by Col. A. C. Horton, formerly of Georgia. The command had not proceeded over eight miles, before Col. Fannin discovered himself surrounded by about 6,000

Mexicans, including a very large force of cavalry. The whole cavalry force dashed in between the *van-guard* and the main body of the Texian army, and effectually cut of [sic] Col. Horton's command.

Col. Fannin entrenched himself as speedily and as well as possible, and soon the engagement became general. Charge after charge was made by the enemy's cavalry, but the fine pieces of artillery which were dexterously and judiciously formed into batteries by the experienced Brooks, mowed them down so dreadfully that they soon began to display more caution and less impetuosity.The conflict continued the better part of two days, and the slaughter was great, especially on the side of the enemy.

General Fillisola, who was in the engagement with his Division, acknowledged afterwards, in his report of the battle published in Matamoras, that the entire loss of the Mexican army, killed and wounded, did not fall short of 750.On the side of the Texians, the loss was about 60 killed and wounded. Col. Fannin himself was severely wounded, also the gallant and accomplished Brooks. The chief command, had the engagment [sic] continued longer, would neccessarily [sic] have devolved upon Maj. Chadwick, whose thorough efficiency has already been alluded to.

On the second day, a flag of truce was sent in by the Mexican General, borne by the afore mentioned Col. Holzinger, demanding a surrender of the Texians. After Col. F. had chsulted [sic] with his officers, a *parley* was agreed upon. This was acquiesced in by Urea [sic], and the parley resulted in a capitulation, the substance of the terms of which has already been given.

Col. Fannin could not retreat himself as near sixty of his men being wounded, and there being no possible means of conveying the wounded from the field.When Santa Anna, then at San Antonio, heard that the Texians had surrendered, he dispatched an order to Urrea to cause every one of them to be instantly shot. Urrea positively refused to carry the savage order into execution, it "being in direct violation of the terms upon which the capitulation was based." He received a second order, couched in the most peremptory language; this he also refused to carry into execution, protesting in the strongest terms against such "*wholesale butchery*." Fillisola was then ordered to cause the prisoners to be executed; and on the eighth day after the surrender the brutal massacre of those brave Americans was perpetrated.

Colonels Fannin and Ward, with all the wounded of their commands, were shot within the walls of the church. Their warm life-blood flowed freely through the gloomy aisles of the *sanctuary*, and curled and clotted around the *holy alter* where stood the high-reared *crucifix*, and where was placed the *beauteous image of the "Blessed Virgin."*

All the ballance [sic], with the exception of the three, Dr. Joseph H. Barnard, then Post Surgeon, Captain Jack Shackleford, who was a physician, and a German of the name of Voss, who was kept to assist Doctors Barnard and Shackleford in dressing the wounds, amputating the limbs, &c., of the wounded Mexicans, marched out of the Fort, in three separate divisions, and in three different directions, under the pretext of procuring a supply of wood and water, and then *deliberately fired upon, from the rear*, by over a thousand Mexican soldiers, they having been sent *ostensibly* as guards, but *really* as *cold-blooded executioners*.

Not exceeding twenty escaped. Some were shot fully two miles distant, having been overtaken in their flight and despatched without being shown the least quarters.Those who fell at the first fire, were collected together by the Mexicans, and burnt. The ballance [sic] remained just where they fell until the latter part of the following May, when Gen. Rusk, then commander-in-chief of the Texian army, caused their bones to be collected and buried in one common grave, with military honors.
The zephyrs that softly sigh o'er their grave,
Are requiems fit for the rest of the brave.

A literal and complete copy from: Washington, Lewis M. H., *narrative, reprinted from the Georgia Citizen, Texas State Gazette, Austin: William H. Cushney, June 18, 1853.Source;* by kind permission of Presidio La Bahia, Goliad, Texas.

THE ALAMO

The Mexican Army arrived in San Antonio on February 23. The Texian garrison was completely unprepared for the arrival of the Mexican army and had to quickly gather food from the town to supply the Alamo. By late afternoon Bexar was occupied by about 1500 Mexican troops, who quickly raised a blood-red flag signifying no quarter. For the next thirteen days, the Mexican army besieged the Alamo. Although there were several small skirmishes, they had little real impact In the early hours of March 6, the Mexican army attacked the fort in what became known as the Battle of the Alamo.

Almost all of the Texian defenders, estimated at 182–257 men, were killed, including James Bowie, Davy Crockett and William B. Travis. Most Alamo historians agree that 400–600 Mexicans were killed or wounded This would represent about one-third of the Mexican soldiers involved in the final assault, which Todish remarks is "a tremendous casualty rate by any standards".Soon, Santa Anna divided his army and sent flying columns across Texas. The objective was to force a decisive battle over the Texian Army, now led by General Sam Houston.

Col. Juan Seguin in his native Castillian language at the burial of the ashes of the Alamo Defenders 25 Feb 1837

> *"Compañeros de armas: Estos restos que hemos tenido el honor de conducir en nuestros hombros son los de los valientes héroes que murieron en el Alamo. Sí mis amigos, ellos prefirieron morir mil veces a servir el yugo del tirano. Que ejemplo tan brillante, digno de anotarse en las páginas de la historia. El genio de la libertad parece estar viendo en su elevado trono de donde con semblante halagueño nos señala diciendo: "Ahí tenéis a vuestros hermanos, Travis, Bowie, Crockett y otros varios a quienes su valor coloca en el número de mis héroes.--Yo os pido a que poniendo por testigo a los venerables restos de nuestros dignos compañeros digamos al mundo entero. Texas será libre, independiente o pereceremos con gloria en los combates."*
>
> "Comrades in arms: These remains which we have had the honor of carrying on our shoulders are the ones of the brave heroes who died in the

> Alamo. Yes, my friends, they preferred a thousand deaths rather than surrender or serve the yoke of the tyrant. What a brilliant example.
> Worthy indeed of being recorded in the pages of history. The genius of liberty seems to be witnessing from its high throne, from whence with praising look points out the deed saying: "Here you have your brothers, Travis, Bowie, Crockett and a few others whose valor, places them in the number of my heroes. The worthy remains of our venerable companions bearing witness, I ask you to tell the world, Texas shall be free and independent or we shall perish with glory in battle."

By kind permission of ; Presidio La Bahia, Goliad, Texas.

The Apache Wars

The first conflicts between Apaches (known by themselves as *T`Inde, Inde, N`de, N`ne* = "people") and Americans began in 1847 during the Taos Revolt of the Mexican-American War. The Apaches were fighting in defense of Mexico with their New Mexican allies. The first campaigns specifically against the Apache came in 1851 and would end with the surrender of Geronimo in 1886. However, Apache attacks on white and Mexican settlers would continue as late as 1900.

Apaches were not new to warfare, they had fought the Spanish and Mexicans for decades before conflict with Americans. The major campaigns of the period occurred all around the present day Tucson, Arizona. The Apaches failed to drive the Spanish and Mexicans off their homeland and other lands conquered by Apaches from other native American tribes. This led to the later Apache conflicts. The various Apache groups ruled a vast area of land stretching from southern California to western Texas, northern Arizona to Mexico and a region in Oklahoma.

In the early years of the wars, the Apache war battles were often the result of stolen property and massacres of whites and Mexicans. This period; roughly from 1851 to 1875. The latter period from 1875 to 1886, the United States engaged Apaches in order to settle them on reservations or to keep them from escaping the reservations. From 1886 to 1900, small battles occurred between United States Cavalry expeditionary forces, white settlers, and little groups of Apaches who evaded the U.S. Army's reservations. A lot of the time, individual Apache warriors were reported to have made attacks.

Sometimes the Native Americans were provoked by white and Mexican settlers, speculators of the federal Indian Reservation policy. Apache leaders like Mangas Coloradas of the Bedonkohe; Cochise of the Chokonen; Victorio of the Chihenne band; Juh of the Nednhi band; Delshay of the Tonto; and Geronimo of the Bedonkohe led war or raiding parties against non-Apaches and resisted the military's attempts, by force and persuasion, to relocate their people to various reservations.and subsequent Comanche Wars left northern Mexico,(then including present day New Mexico, Texas, Cochiulla,) in a state of perpetual instability.

Consequently, present day Juarez remained the northern most fortified city with the Río Bravo del Norte, (Rio Grande) and Apache attacks dissuading settlement and development north. The water of the river, the city walls and citadels of Jurez provided a natural defense against further raids although in some decades several thousand strong Apache armies made raids deep into Mexico, slaughtering the male population and enslaving the women and children. As a result, the Rio Grande proved a boundary line of actual Spanish presence.

Nonetheless, in the early years of Spanish power, several attempts were made which successfully colonized areas north for some decades. Being a grassland then, the few Spanish grandees and their mixed-race Mexican servants, and the small community of Spanish friars and their Mexican-Indian wards ranched the area and developed small scale but successful agriculture consisting of vineyards and fruits. However, in 1680, after the successful Pueblo Revolt

El Paso – Local History, 1540 -1821

Spanish Lay Claim to El Paso (El Paso del Norte)

Inhabited for centuries by various Indian groups, El Paso saw its first Europeans when Spaniards passed through in the mid-1500s. During 1540 to 1542, an expedition under Francisco Vázquez de Coronado explored the area now known as the American Southwest. These earliest Spanish explorers saw on their approach from the Rio Grande two mountain ranges rising from the desert, with a deep chasm between. They named the site "El Paso del Norte, " or "the Pass of the North."

The Rodríguez-Sánchez expedition in 1581 was the first party of Spaniards to explore the Pass of the North, bringing about the beginning of El Paso's modern history. Further expeditions followed, culminating in an April 30, 1598 ceremony near the site of present-day San Elizario in which expedition leader Juan de Oñate took formal possession of the territory drained by the Rio del Norte (now the Rio Grande). Called "La Toma," (the claiming) this act brought Spanish civilization to the Pass of the North, laying the foundation for more than two centuries of Spanish rule.

Population of the area grew when the Pueblo Indian Revolt of 1680 sent Spanish colonists and *Tigua Indians of New Mexico* southward in search of safety. By 1682, five settlements were thriving on the south bank of the Rio Grande. By the middle of the eighteenth century, approximately 5,000 people populated the El Paso area; among them were Spaniards, *mestizos,* and Indians. The region became known for its vineyards, with residents producing wine and brandy. In 1789, the presidio of San Elizario was founded to defend the El Paso settlements against encroaching Apaches.

Spanish Rule Ends, Tensions Begin

With Mexico's independence from Spain in 1821, the entire El Paso area became part of Mexico. Agriculture and commerce flourished, but the unpredictable levels of the Rio Grande made for difficulties with crops, fields, and structures frequently damaged by the rising water levels. In the

1830s, the river flooded much of the lower Rio Grande valley, creating a new channel and displacing several towns.

May 1846 saw more difficulties as hostilities erupted between the United States and Mexico. During the Mexican War, Col. Alexander Doniphan and a force of American volunteers defeated Mexican fighters at the battle of Brazito, entering El Paso del Norte. The Treaty of Guadalupe Hidalgo on February 2, 1848, ended the dispute and again changed the boundary between the two nations, bringing El Paso territory under the blanket of the United States.

El Paso's settlements grew in 1849 as easterners rushed west in search of gold. Lines between Mexico and the United States were revised yet again; this time the three Mexican towns of Ysleta, Socorro, and San Elizario ended up on the United States side of the line. The military post of Fort Bliss was established in 1858; one year later pioneer Anson Mills completed his plat of the town of El Paso. The name "El Paso" brought about confusion with the Mexican town across the Rio Grande, El Paso del Norte, so the Mexican town's name was changed to Ciudad Juárez in 1888.

During the Civil War, El Paso's alliance was to the South, though the Union presided and local Southern sympathizers eventually received pardons. In 1877, Texans and Mexicans became embroiled in a bitter civil war, the Salt War of San Elizario, which lasted six months.

MY EARLY TIMES in TEXAS
T E X A S

by, J .C DUVAL.

Overview of 'Early Times in Texas'

'Early Times in Texas' was first published in Burke's Weekly of Macon. Georgis in 1867.

John Duval grew up in Tallahassee when his father was appointed to a federal judgeship in what was then Florida Territory. Duval returned to Bardstown in 1831 with his mother to continue his education at St. Joseph College.

Written by John C. Duval (a pseudonym of *John Crittenden Duval*, (1816 – 1897) born in Bardstown, Kentucky, March 14, 1816 The book is about his own experiences as " Jack Dobell" in the ill-fated expedition by James W. Fannin's army. He was the last survivor of the Goliad Massacre in 1836.Late in 1835 he left the college to join a small company organized by his brother Capt. Burr H. Duval to *fight with the Texans against Mexico.* The brothers *were with* James W. Fannin's *army* when it surrendered to the Mexican forces under José de Urrea. In the Goliad Massacre on Palm Sunday, 1836.

Burr Duval was killed, *but John escaped.* Later he entered the University of Virginia to study engineering. He returned to Texas by 1840 and became a land surveyor. In 1845 he was, alongside William A. A. (Bigfoot) Wallace, a member of John C. (Jack) Hays's company of Texas Rangers. Duval did not favor secession, but he joined the Confederate Army as a private, declining a commission. He was a captain by the war's end. He returned to Texas by 1840 and became a land surveyor. In 1845 he was, alongside William A. A. (Bigfoot) Wallace, a member of John C. (Jack) Hays's company of Texas Rangers. Duval did not favor secession, but he joined the Confederate Army as a private, declining a commission. He was a captain by the war's end. He died in Fort Worth on January 15, 1897.

CONTENTS.

EARLY TIMES IN TEXAS.

(Note; Page numbers refer to the original book)

Early Times in Texas, or the Adventures of Jack Dobell

CHAPTER I

INTRODUCTORY; THE VOLUNTEERS FOR TEXAS DOWN THE MISSISSIPPI , NEW ORLEANS , OUT ON THE GULF.

In 1835 the people of Texas, or rather the settlers from the "States," determined to throw off the Mexican yoke, and resist to the last extremity any further encroachment upon their liberties.

At that time, with the exception of Mexicans and Indians, there was probably not more than twenty thousand people in the colonies, and although the Mexican government for several years previously had shown a disposition to ignore the rights and privileges guaranteed them under the Constitution of 1824, I hardly think the colonists with their limited means and numbers would have ventured to rebel against its authority, if they had not counted largely on getting all the aid they should need to carry out the revolution successfully, from their friends and brethren in the United States. In this expectation they were not disappointed. Many young men, from almost every State in the Union, armed and equipped at their own expense, hastened to the assistance of the colonists, as soon as the standard of rebellion was raised.

A volunteer company was organized for this purpose in my native village, and although I was scarcely old enough to bear arms, I resolved to join it. But it was no aspiration for "military fame" that induced me to do so. One of the frequent visitors at ray father's house was an old friend of his, who had been in Texas and traveled over a considerable portion of it, and who subsequently held a position in the cabinet of the first president.

He was enthusiastic in his praise of the country, and insensibly an ardent longing sprang up in my bosom to see for myself the " broad prairies," the beautiful streams and vast herds of buffalo and wild horses of which he had so often given me glowing descriptions. By joining this company I thought an opportunity would be afforded me of gratifying it which perhaps might never again offer itself and so, in spite of the opposition of relatives and friends, my name was added to the muster roll.

I purchased a good Kentucky rifle (with the use of which was already well acquainted), shot pouch, powder horn, tomahawk, and butcher knife, and thus equipped, with my knapsack on my shoulders, I fell into ranks, and amid the waving of handkerchiefs and the cheering of bystanders I bid adieu to my native village and started for the " promised land " of Texas.

It was the latter part of November when we left Bardstown and though not very cold, the snow was some three or four inches deep on the ground, which retarded our march so much that we only made about twenty miles by sunset, when we halted for the night in a grove near the margin of a stream that empties into the famous " Salt River." We cleared away the snow from under the trees, built up log heaps for fires, and after eating our supper of hot coffee, "hard tack" and fried middling, for which our tramp had given us excellent appetites, we spread our blankets upon the fallen leaves and turned in for the night.

The next morning we were on the road by sunrise, and about dusk, after a toilsome and fatiguing march through the slushed mud (for a thaw had set in) we reached the city of Louisville and took up our quarters at the Gait House. The next day we purchased a supply of provisions enough to last us for the voyage and went on board of a steamer bound for New Orleans.

Nothing worthy of note occurred on the passage. Occasionally, by way of varying the monotony of our daily life,we would go ashore when the boat landed for wood or freight, and get up an, impromptu " shooting match," in which the skill of our Kentucky riflemen was exhibited, greatly to the astonishment of the "natives." It was no unusual thing for many of them to put three balls out of five, at the distance of one hundred yards, into a paper not larger than a silver dollar.

The second day of the voyage we left the snow and ice behind us, and on the fourth we came to the region of "Spanish moss." The trees on both banks of the river were draped in its long funereal folds, which waving slowly back and forth in the breeze, was too suggestive of any but cheerful thoughts.

The next day we came to the "coast," a strip of country so called, extending along the river for more than a hundred miles above the city of New Orleans. It is protected from overflow (though not entirely) by what are termed "levees," or embankments, thrown up on each side of the river, a few paces back from the margin, but these are sometimes broken through in very high stages of water.

The river was unusually full at the time we passed, and in one place we noticed where the water had made a breach in the embankment more than a hundred feet in width, through which it was rushing with the velocity of a mill race, and had already inundated the coast country on that side as far as the eye could extend.

From the time we struck the "coast" we experienced no more cold weather. Everywhere the forests were still green, and the orange and pomegranate were bending down with the weight of their ripened fruit. Here, too, we first observed extensive fields of cotton and sugar cane, in the former of which gangs of negroes were seen, bearing huge baskets filled with the "snowy fleece" upon their woolly heads.

The fifth day, we reached New Orleans, fortunately just in time to secure a passage on a schooner that was to sail the next day for Velasco, a small port at the mouth of the Brazos river. The following day, before the schooner was ready to sail, I had an opportunity to see the city, of which I was glad to avail myself.

The great number of vessels moored in along line to the wharves, the puffing of steamboats, the clatter of drays and carts, the noise and bustle on the levee, and the jargon of foreign tongues were all calculated to fill with astonishment and wonder the mind of a youth who had never before been beyond the precincts of his native village. In the evening we embarked with all our goods and chattels on the schooner, and having made fast to a tow-boat, in company with two ships and a bark, we were soon under way, and bade farewell to the "Crescent City," and its forests of masts and tapering spires quickly faded away in the distance.

From New Orleans to the mouth of the Mississippi the scenery along the river is monotonous and dreary. Low swampy lands extended back in an unbroken level as far as we could see, in some places entirely covered with water and in others with a rank, luxuriant growth of reeds and coarse grass, among which cranes and many other aquatic birds could be seen silently standing in rows, or stalking solemnly about in search of the reptiles with which these marshes abounded.

Along the shores immense piles of drift wood were heaped up, amongst which, and scarcely to be distinguished from the decayed logs composing them, the black scaly sides of an alligator could now and then be seen, to be saluted whenever within range, by a shower of bullets from our rifles. The Mississippi empties into the gulf by three mouths and about 10 o'clock the day after we had left New Orleans, we entered the one called the "Southwest Pass" and an hour or so afterwards we had crossed the "bar" and were rolling and tossing upon the blue waves of the gulf of Mexico.

The line was cast off from the tow-boat, sails hoisted and soon we were scudding along before a fair wind in the direction of the distant shores of Texas. For a long way out we noticed that the blue waters of the gulf refused to "fraternize" with the vast muddy stream continually pouring in from the mouths of the Mississippi.

In a few hours we lost sight of the low shores of Louisiana, and nothing was to be seen but the sky and the apparently interminable waste of blue water. Our schooner was a small one, and with more than fifty passengers on board, it can easily be imagined we were packed rather too closely together for comfort. For my share of the sleeping accommodations, I appropriated a large coil of chain cable, in the hollow of which by doubling up after the fashion of a jack knife, I managed to snooze pretty comfortably at night.

CHAPTER II

CATCHING A NORTHER FLYING FISH LAND HO! THE CITY of VELASCO
DANGEROUS BAR FIRST VIEW OF THE "PROMISED LAND"
NOT VERY PROMISING CAMP AT THE RIVAL CITY OK QUINTANA
MUSTERED INTO SERVICE FALSE ALARM ORDERED ON BOARD OF SLOOP OF WAR INVINCIBLE GALVESTON ISLAND, LOSS OF ONE OF OUR BOAT'S CREW UNSUCCESSFUL CRUISE.

The second day of our voyage about sunset, we observed a black cloud towards the north, which spreading rapidly soon obscured the whole heavens. Sails were hauled down and reefed, the hatches secured, and every precaution taken for the safety of the vessel in the approaching "norther" one of those fierce winds that frequently occur during the winter season in the gulf of Mexico, as well as on the prairies of Texas.

We had scarcely made "all snug" when the norther struck the schooner with unusual violence, carrying away our maintop mast, and forcing the vessel almost upon her beam ends. She soon righted however, and away we flew before the blast that whistled and shrieked through the cordage in a way not at all pleasant and enlivening to the ears of a landsman.

In a little while the waves began to rise and the vessel to toss and pitch like an unbroken mustang, and feeling some of the premonitory symptoms of sea sickness, such as a frantic effort to throw up my boots, I retired to my coil of cable below; but the tossing of the schooner, the rushing of the waves along side and the trampling of sailors en deck effectually drove away sleep.

The next morning the storm had abated, the sun shone out clear and warm, and from that time until we reached Velasco we had no more bad weather. Whilst the storm lasted, a number of flying fish fell upon the deck of the schooner, which the sailors secured, and we found them to be an agreeable addition to our ordinary fare of sea biscuit and "salt junk." They are a delicate little fish, from six to eight inches in length with two long fins resembling wings projecting from the upper portion of the body.

When chased by the dolphin or other large fish, they may be seen rising in schools from the tops of the waves, and flying forty or fifty yards in the direction of the wind; then dipping again into the crests of tbe billows, from which they quickly rise for another flight, should their enemies still continue to pursue them. Their flight rarely exceeds forty or fifty yards, for the reason that their fins cannot serve the purpose of wings unless frequently moistened by contact with the water.

On the morning of the seventh day after leaving Southwest Pass, the shores of Texas were dimly discernible from the masthead, looking like a long low cloud on the western horizon. The wind was "dead ahead" and we were nearly the whole day beating up within sight of the beach and the few miserable little shanties that then constituted the city of Velasco. Finding it was impossible to cross the bar with the wind ahead, we cast anchor in the roadstead, hoping it would be more favorable the next day.

But the next morning it was still from the same quarter, and tired out with our confinement on board of the vessel a dozen of us manned the long boat, resolved to make a landing in spite of "wind and weather." But in this we "reckoned without our host," for we missed the channel, got into the breakers which came very near swamping our boat, and we were glad to make our escape from them back to the schooner again.

Two years subsequently I saw a boat capsize amongst those same breakers, and although in full view of many people on shore, everyone on board of her was drowned before any assistance could be given them.

Not long after our return to the schooner, to our great joy the wind hauled around to the east, which enabled us to cross the bar, and soon we were safely anchored in the mouth of the Brazos River. The country in the immediate vicinity of Velasco is low, and back of it a dead level prairie extended as far as the eye could reach ; consequently I must confess I was not much pleased with the first view of the "promised land." Velasco was a miserable little village consisting of two stores and a hotel, so called, and five or six grog shops, dignified with the name of saloons." Opposite to it, on the south bank of the river was the rival city of Quintana, containing about the same number of shanties and a mixed population of Yankees, Mexicans and Indians.

We landed upon the Quintana side and pitched our camp upon the beach, adjoining the camps of several other companies that had arrived a few days previously. Here we remained two weeks or more, and as we were liberally supplied with rations by the patriotic firm of McKinney & Williams, and game and fish were to be had in abundance, we "fared sumptuously" every day. In hunting and fishing, making tents, cleaning our guns, and preparing in other ways for our anticipated campaign, our time passed pleasantly enough.

Whilst at this place our company was formally mustered into service of the embryo Republic of Texas. It was left optional with us to enlist for twelve months or for "during the war," and we unanimously chose the latter upon the principle of "in for a penny, in for a pound," or as Davy Crockett would have said, we resolved to "go the whole hog or none."

One day whilst we were encamped at Quintana we had quite an exciting scene, which bade fair for a time to initiate us into the realities of actual warfare. Two vessels were seen in the offing, one of them evidently in hot pursuit of the other.

As soon as they had approached near enough to be distinctly seen through a glass, it was asserted by several who claimed to know, that the smaller vessel was the Invincible, a schooner recently purchased by Texas, and the larger one in pursuit was the Bravo, a noted Mexican privateer. In this opinion we were confirmed, as a sharp cannonading began between the two vessels.

Our company was at once ordered on board of a small steamer lying in the mouth of the Brazos, the instructions to hasten to the assistance of the Invincible with as little delay as possible. We quickly got up steam and notwithstanding the violence of the breakers on the bar, which on two occasions broke entirely over our little steamer, we were soon alongside of the foremost vessel, which proved to be as we had supposed, the Texas schooner, Invincible.

By this time the other vessel had approached near enough to be recognized as the Brutus, lately purchased also for the Texas navy, and after the interchange of some signals the firing ceased. Each vessel, it seems, had mistaken the other for the Bravo, and hence the pursuit of the Brutus, and the attempt of the Invincible to escape, as she had only a sailing crew on board; and we were compelled to return to camp without having had an opportunity of "fleshing our maiden swords."

A few days afterwards, our company was ordered to take up our quarters on board the Invincible, to serve as a kind of Marine Corps for her protection until a regular crew could be enlisted. Whilst on board of her, in the hope of meeting the Bravo, we took a cruise along the coast as far as the east end of Galveston island. Here an incident occurred, which as being indicative of the great changes that have taken place since the times of which I write, may be worth mentioning.

We were lying at anchor off the point of the island, and as we were running short of wood and water, a boat was sent ashore for a supply. The former could be had in any quantity along the beach, and the latter, though slightly brackish, by digging shallow wells at the base of the sand hills.

When the boat was ready to return, it was found that one of the crew, who had wandered off from the well whilst the others were filling the casks, was missing. Search was made for him, but he was nowhere to be seen, and as there was every appearance of a "norther" coming up, the officer in command of the boat thought it most prudent to hasten back to the vessel, leaving the missing man on the island. In a few minutes after the boat reached the vessel the norther struck us, and we were compelled to hoist anchor and run before it.

Three days elapsed in beating back to our anchorage, and a boat was immediately sent ashore with a crew of half a dozen men, to look for our lost comrade. At length he was found , five or six miles below the place where he had been left, wandering on the beach, searching for oysters and clams, upon which he had subsisted since leaving the vessel. His mind was considerably affected by exposure to the norther, his fear of wild beasts and savages, and the apprehension of our failing to return.

For several days he talked in a wild and incoherent manner, and he did not entirely regain his mind for two or three weeks. For three days this man had wandered about the island without seeing a living soul, and yet it is probable he was at no time more than four or five miles from where Galveston, a city of forty thousand inhabitants, now stands!

After an unsuccessful cruise in search of the Bravo, we returned to Quintana, and pitched our tents again upon our old camping ground.

CHAPTER III.

LEAVE FOR COPANO LAND,
AT MATAGORDA ISLAND
RENDEZVOUS OF LAFITTE THE PIRATE,
ARRIVE AT COPANO, TEXAS RANGERS MARCH TO REFUGIO ,
OLD CHURCH AT REFUGIO, PRIMITIVE METHOD OF SKIMMING MILK
MARCH TO GOLIAD, CARANCHUA INDIANS, DEER
WILD HORSES, NORTHERS ARRIVAL AT GOLIAD.

A day or so after our return to Quintana, the officer in command of the Invincible was instructed to take our company on board and to sail immediately for Copano, on Aransas bay, where we were to disembark and march from thence to Goliad. It was rumored that a considerable force had already been concentrated at that point, under the command of Col. J. W. Fannin, destined for the invasion of the border States of Mexico, and of course we surmised that our company would form a part of the invading army.

We set sail about dark, and a brisk norther springing up, by daylight the next morning we were in sight of Aransas Pass, which we shortly entered without difficulty, and cast anchor in a secure harbor behind the southwest point of Matagorda island. This harbor had been, in times past, a rendezvous for the vessels of the famous pirate, Lafitte.

On the island the embankments around his old camping grounds or fortifications were still visible, and along the beach were many posts yet standing with iron rings affixed to them, which undoubtedly had been used for securing the small boats that plied between the vessels and the shore. "The pass" was known then only to Lafitte and his followers, and here in security they could repair their vessels, supply them with wood and water, and divide among themselves the spoils of their piratical expeditions. On the east end of Galveston island they had a similar place of rendezvous, near where the city now stands, and the remains of their fortifications could be plainly seen when I first landed on the island, in 1837.

A few years ago, while excavating sand near these old fortifications, some workmen found a considerable amount of old Spanish coin, buried there no doubt, by some pirate on the eve of his departure upon some marauding expedition, from which probably he never returned.

We remained on the island several days, passing the time very pleasantly hunting and fishing, and gathering oysters* which were abundant in the bay, and then we embarked on board of a small vessel for Copano, which at that time was the principal port of Southwest Texas. In a few hours we reached the pert, and landing, we pitched our tents on the bluff just back of it. Here we found a company of Texas Rangers who had been on frontier service for six months, during all of which time they had not seen a morsel of bread.

They had subsisted solely upon beef and the game they killed. We gave them a part of the "hard tack" we had brought with us, and though wormeaten and musty, they devoured it with as keen a relish as if it had been the greatest delicacy. Although they had had no bread for so long a time, they were healthy and in "good order," which convinces me that Byron was right in saying that man was a carniverous animal, and would bear vegetables "only in a grumbling way" especially beans.

From Copano (which consisted mainly of a warehouse and a large tank of fresh water) we took up the line of march lor Refugio, distant about twenty miles. It is situated on a little stream called Mission river, near the bank of which we pitched our tents, just before sunset. Refugio at that time contained about two dozen adobe huts (inhabited by a mixed population of Irish and Mexicans), and an old, dilapidated church, built, I was told, the same year that Philadelphia was founded.

A few months subsequently Refugio was the scene of a hard fought battle between thirty-five Americans under Capt. King, and a large body of Mexican cavalry. The old church, where King and his men defended themselves for some time against a host of Mexicans, when I last saw it, still showed evidence of the severity of the conflict in its battered walls and its roof perforated with shot from the Mexican artillery.

Observing a number of fat cows in the vicinity of the village, I concluded to go out and forage for a little of the "lacteal fluid," of which we had not had a drop since leaving Kentucky. So taking a camp kettle in my hand I went to the nearest house and inquired of a woman standing at the door, if she had any milk for sale. "Faith, and I have," said she, "any kind you may want, swate milk, butter milk, clabber milk and blue Johns." I told her I would take some of the "swate," whereupon she led me to a small out-house,in which were a number of pans filled with milk.

Selecting one containing the "swate," she rolled up her sleeve and deliberately proceeded to skim it with her open hand, which looked to me to have been unacquainted with soap and water for some time past. When she had finished skimming the milk in this primitive fashion, she poured the contents of the pan into my camp kettle, at the same time saying : "There,my little mon, there's a pan of milk for yez that's fit for the Pope of Room, Heaven protect His Holiness."

I said nothing, though like the owl I did a good deal of thinking, paid for the milk and returned to camp, where my hungry mess-mates speedily emptied the kettle, wondering that I took coffee in preference to such nice new milk. I told them of the skimming process I had witnessed, but men in camp are not usually very "squeamish," and they merely said " that what would not poison would fatten;" that they had to " eat their peck of dirt anyhow," and the sooner they got through with the job the better.

The next morning we continued our march for Goliad, about thirty miles distant, but as we got a late start, we only made twenty miles or so by sunset, and pitched our camp near a pool of fresh water, under the shelter of some spreading live oak trees. Here we found encamped a band of the Caranchua tribe of Indian?, at that time professing to be friendly to the Americans

We were told that these Indians were cannibals that they always devoured the prisoners they took in their conflicts with their enemies. They were the largest Indians I have ever seen, scarcely a man among them being less than six feet high, and many of them over six feet. The men were entirely naked, saving a breech cloth fastened around the waist, and being hideously painted, one can readily imagine that they presented a most ferocious and savage appearance.

Their language was the most peculiar jargon of gutteral sounds I ever heard, the words seeming to be articulated by some spasmodic action of the throat without any aid from the tongue or lips. They were armed with long lances, bows and arrows, and a few with old flint-lock muskets.

These Indians some time afterwards captured several Americans and killed and "barbecued" them, which so enraged the settlers that they organized an expedition against them and succeeded in exterminating the whole tribe with the exception of a small remnant that fled to Mexico. These Caranchuas, I believe, were the only Indians known to be cannibals, on the North American continent.

Along the whole route from Copano to where we were encamped, we had seen great numbers of deer, sometimes as many as two or three hundred in a drove, and so unused to being hunted or disturbed by man, that even when we approached within a few yards of them they showed no signs of fear. Of course we had no difficulty in getting fresh meat whenever we wanted it.

Once, too, at the distance of half a mile we saw a large drove of mustangs, but they were much wilder than the deer, for when several of us attempted to approach them, they circled around us out of range of our rifles,every now and then stopping a moment, stamping and snorting, until at last one of them that seemed to be the leader of the drove, started off at full speed, and the rest following, in a short time nothing but a cloud of dust indicated the direction they had taken.

Some years subsequent to this, a company of rangers to which I belonged, when in pursuit of Indians in the country between the Nueces and Rio Grande rivers, met with a drove of mustangs so large that it took us fully an hour to pass it, although they were traveling at a rapid rate in a direction nearly opposite to the one we were going.

As far as the eye could extend on a dead level prairie, nothing was visible except a dense mass of horses, and the trampling of their hoofs sounded like the roar of the surf on a rocky coast. Most persons probably would be inclined to doubt this 'horse story," and to consider it one to be told to the "horse marines" alone; yet it is literally true, and many are still living who were with me at the time, who can testify that my statement is in no manner exaggerated.

During the night a norther came up, but as we were well protected by thick timber, which afforded plenty of fuel for our fires, we managed to keep pretty comfortable. These "northers," as they are called in Texas, are winds that spring up suddenly from the North, during the winter season, sometimes "dry," at other times accompanied with rain or sleet.At first they blow with considerable violence, but gradually subside in the course of one, two or three days, and are followed usually by a week or so of clear, pleasant weather. To travelers unprepared for them they are very disagreeable visitants, and instances have been known of persons freezing to death in them when caught out in the open prairies where there was no shelter from the wind nor means of making a fire.

Early the next morning we took the road for Goliad again,and in the course of three or four hours we came in sight of the dome of the old Mission. Not long afterwards we entered the town and took up our quarters in an empty stone building near the old church. Here we found about four hundred men under the command of Colonel J. W. Fannin, the force with which it was designed to invade the border States of Mexico.

CHAPTER IV.

DESCRIPTION OF GOLIAD DAILY DRILLS AND "DRESS PARADES"
OUR COMPANY CALLED THE "MUSTANGS" AND THE REASON WHY!
MARCH TO SAN PATRICIO

CARLO'S RANCH CAPTURE OF THE REV. PADRE AND HIS COURIERS
AN ALARM THAT ENDED VERY PLEASANTLY INDEBTED TO THE REV. PADRE FOR A "GOOD SQUARE MEAL"
TIRED OF TRAMPING THE BOYS MOUNT MUSTANGS BUT DISMOUNT VERY EXPEDITIOUSLY,RETURN ON FOOT TO GOLIAD.

Goliad, at the time we arrived there, contained a population of about two thousand Mexicans who were professedly friendly to the Texans, but who afterwards, when Santa Anna invaded the country, proved to be their most vindictive foes.I must, however, make an exception in favor of the "Senoritas,"who generally preferred the blue-eyed, fair complexioned young Saxons to their copper-colored beaux.

Goliad is situated on the south side of the San Antonio river about forty miles above its mouth, and ninety-five miles below the city of San Antonio. The American town of Goliad, built up since the war, is situated nearly opposite the old town, on the north side of the river. After the defeat of Santa Anna, the great majority of the inhabitants of the old town abandoned the place and went to Mexico.

When I last saw it, in 1877, its population had dwindled down to one or two hundred miserable "peons" and most of the "jacals" or huts were gone. The Old Mission, with its dilapidated walls half a dozen stone tenements and a few adobe houses alone remained to designate the spot where once had stood the old town of Goliad.

The lands around the place are rich and productive, and the locality (though we did not find it so) is a healthy one. Thousands of fat beeves roamed the prairies in its vicinity j and as corn could be had in abundance upon the neighboring ranches, we were well supplied with provisions. Besides,when theTexans took possession of the place,several months previous to our arrival, a large amount of sugar and coffee was found in the Mexican commissary department, which, of course, we did not scruple to appropriate to our own use.

In order to render his little force as effective as possible,when the time for action should come, Colonel Fannin ordered daily drills, which were my detestation and from which I invariably absented myself whenever I had a pretext for doing so. I greatly preferred hunting deer in the prairies and attending the "fandangos" or dances that took place daily and nightly in one part of the town or the other.

Not long after our arrival at Goliad the soubriquet of Mustangs or Wild Horses was acquired by our company from the following incident: M., our second lieutenant, was a man of great physical powers, but withal one of the most peaceful and most genial men when not under the influence of liquor. But occasionally he would get on a "spree" and then he was as wild as a "March hare" and perfectly uncontrollable. The Mexicans seemed to know him and to fear him, also, and when he was on one of his "benders" they would retreat into their houses as soon as they saw him and shut their doors. This proceeding, of course, was calculated to irritate M., and he would forthwith kick the door from its hinges. On a certain occasion he battered down the doors of half a dozen houses in one street, and from that time the Mexicans called him the "Mustang," and finally the name was applied to the company.

Few events occurred to vary the daily routine of our life at Goliad. The following, however, I will mention: Our company was detailed on one occasion to go to San Patricio, an Irish settlement about fifty miles southwest from Goliad, for the purpose of securing a couple of field pieces left there by the Mexicans. This we accomplished without difficulty, and without any opposition, although our scouts had informed Colonel Fannin that a considerable force of Mexican guerillas was in the vicinity of the place.

On another occasion our company was detailed to march to Carlos Ranch, a Mexican village about twenty miles below Goliad, with instructions to arrest certain of the inhabitants, who, it was ascertained, were constantly transmitting intelligence of our movements to Santa Anna, and among the number was the old padre or priest of the village. In order that the Mexicans might not suspect our object and frustrate our plans by giving the padre and his friends timely warning of our intentions, we left the town quietly after dark in the opposite direction to the one we designed taking. When safe beyond observation, we turned our course down the river, and making a forced march, we reached the village a little before daylight and surrounded it without alarming any of the inhabitants.

A detachment then entered the padre's house, and caught the bird in his nest, together with five or six other suspicious characters (supposed to be his couriers, as in fact they were), and the whole of them were "bagged" without alarming any of the people in the village. Having thus accomplished our object we marched to a point on the river about a quarter of a mile above, where we halted in a grove to rest and prepare something for breakfast. Placing a guard over the padre and his couriers, we stacked our guns and soon every one was busily engaged in cooking such "grub" as we had in our knapsacks.

By this time the sun had risen, and we were just seating ourselves on the grass around the scanty fare we had prepared for our breakfast (consisting of hard tack, jerked beef and the inevitable coffee), when our attention was drawn to shrieks and doleful cries in the direction of the village, and seeing a crowd of people coming from it towards us, we hastily sprang to our guns, thinking the Mexicans were about to make an attempt to rescue the prisoners, but as the crowd drew nearer, we saw that it was composed mostly of women and children. It seems that they had just found out we had captured their Reverend padre and they were coming to bid him farewell and obtain his parting blessing.

I had heard that the Mexicans were completely under the control of their priests, but I had but a faint conception of the fact until I witnessed the scene that ensued. As the came up the women knelt at his feet, weeping and mourning, and kissed his hand and even the hem of his priestly robes. Presently another crowd of women came from the village, bringing with them plates filled with hot "tortillas," pots of coffee, "dulces," etc., intended for the padre's breakfast, and that of the other prisoners, and when they deposited them on the grass before them we took possession of them as the "legitimate spoils of war" and found they were much better than our course of hard tack and dried beef. Such conduct on our part.

I admit, bordered closely on the "sacrilegious,'" but then you must remember we had been marching all night and of course were very hungry and as the Mexicans said themselves, "what better could you expect from 'Gringos'and heretics! " Seeing that the Rev. padre would have but little chance to get his breakfast until we had ours, the women continued to bring in fresh supplies of eatables as fast as we disposed of them. Finally however, when our hunger was appeased, the Rev. padre and his couriers had a show at what was left.

In the vicinity of the place where we had halted, we noticed a large "corral" in which several hundred head of mustangs were penned. We were all tired of "trudging" on foot, and concluded we would "press" into the service (a military term for appropriating property belonging to others) a sufficient number of these mustangs to mount the whole company.

Accordingly we compelled the Mexicans to rope and equip with saddles and bridles about fifty of them. We were all I suppose pretty good horsemen, as the term is understood in the "old States," but we knew that these mustangs were only partially broken to the saddle, 'and we anticipated having some "fun " when we mounted them though nothing like as much as we really got, for at the time, we were totally ignorant of that peculiar trick of mustangs called " pitching, " by which they manage almost invariably to get rid of a "green" rider.

When the mustangs with considerable difficulty, after roping them closely to trees, had all been saddled and bridled,at the word of command, we mounted (except five or six who failed to do so) and the next instant a scene of horses kicking, reai ing and plunging ensued, of which only a confused recollection remained upon my mind, and in less time than it takes me to tell of it, we were all put " hors de combat, " (no pun intended.)

As for the part I took individually in this equestrian performance, I have only to say that I had hardly seated myself in the saddle, when my unruly steed humped his back like a mad cat, reared up, and then came down on his stiffened fore legs with such force, that if " next week " had been lying on the ground ten or fifteen feet ahead of me, I would certainly have knocked out the middle.

I was partially stunned by the fall but soon rose to my feet, and was much relieved and consoled looking round, to find that all the rest had been served in the same way, except one rider who managed to stick upon his horse in spite of all the animal's efforts to get rid of him the Mexicans no doubt had purposely selected the wildest horses in the corral, and it is probable the most of them had never been backed half a dozen times even by the rancheros themselves, who are unsurpassed by any people in horsemanship. I am confident that the padre and his flock enjoyed this equestrian performance much more than the actors, but as heretofore the laugh had been all on our side, we did not blame them for the pleasure they took in our discomfiture.

However, we concluded to dispense with our unmanageable steeds, "unpressed " them by restoring them to their lawful owners, and resulted our march on foot for Goliad. The Mexican padre was sent to San Felipe on the Brazos, where he was securely caged until Santa Anna and his army were defeated and driven from Texas. He had the reputation of being a great scoundrel and an inveterate gambler, and his sinister countenance did not belie "the soft impeachment" I will do him the justice however, to say that we were indebted to him for the best breakfast we had eaten since landing in Texas. Peace to his ashes.

CHAPTER V

RUMORS OF THE ADVANCE OF GEN. SANTA ANNA INTENDED INVASION OF MEXICO
ABANDONED PREPARATIONS FOR THE DEFENSE OF GOLIAD SANTA ANNA CROSSES THE RIO GRANDE, CAPT. KING WITH THIRTY-FIVE MEN SENT TO REFUGIO BATTLE
AT REFUGIO AND MASSACRE OF CAPT. KING AND HIS MEN DISPATCH FROM GENERAL HOUSTON COL. FANNIN DETERMINES TO RETREAT PREPARATIONS FOR RETREATING THE RETREAT - AND FATAL "HALT" IN THE PRAIRIE.

Some time after our arrival at Goliad, information was obtained from some friendly Mexicans that General Santa Anna was preparing to enter Texas at the head of a large army ;consequently all idea of invading Mexico, was abandoned* and we set to work to render the fortifications around the old missions as defensible as possible. We strengthened the walls in many places, built several new bastions on which artillery was placed in such a way as to command all the roads leading into the town

Every day we were drilled by our officers for three hours in the morning and two in the afternoon, which, as I have said before was a great bore to me, as I would have preferred passing the time in hunting and fishing. We also deepened the trenches around the walls, and dug a ditch from the fort to the river, and covered it with plank and earth, so that we might obtain a supply of water, if besieged, without being exposed to the fire of the enemy. We were well supplied with artillery and ammunition for the same, and also with small arms, and had beef, sugar and coffee enough to last us for two months but very little bread.

Some time in February, a Mexican from the Rio Grande arrived at Goliad who informed Col. Fannin that Santa Anna had already or would shortly cross the river into Texas with a large army which would advance in two divisions, one towards Goliad and the other towards the city of San Antonio.

Some days afterwards, two or three Texans came in from San Patricio, bringing the news that Capt. Grant and some twenty-five or thirty men stationed at that place, had been surprised by a force of Mexican guerillas and all of them massacred. About this time also a courier from Refugio came in who stated to Col. Fannin that he had been sent by the people of that place, to ask for a detachment of men to escort them to Goliad, as they were daily expecting an attack from the guerillas.

In compliance with this request, Col. Fannin sent Capt.King and his company (about thirty-five men) to act as escort for those families who desired to leave. When Capt. King and his men reached Refugio, they were attacked on the outskirts of the town by a large force of Mexican cavalry, and being hard pressed they retreated into the old mission, a strong stone building, at that time encompassed by walls.

There they defended themselves successfully, and kept the Mexicans at bay until their artillery came up, when they opened fire upon it with two field pieces which soon breached the walls,and the place was then taken by storm. Capt. King and some seven or eight of his men (the only survivors of the bloody conflict), were captured and led out to a post oak grove north of town, where they were tied to trees and shot. Their bones were found still tied to the trees, when the Texan forces re- occupied the place in the summer of '36.

About this time a courier arrived bringing a dispatch from Gen. Houston to Col. Fannin, and it was rumored in camp that the purport of this dispatch was "that Col. Fannin should evacuate Goliad and fall back without delay towards the settlements on the Colorado." But as to the truth of this I cannot speak positively.

At any rate Col. Fannin showed no disposition to obey the order if he received it on the contrary, hearing nothing from Capt. King, although he had sent out three scouts at different times to obtain information of his movements, all of whom were captured and killed, he despatched Maj. Ward with the Georgia Battalion (about one hundred and fifty strong) to his assistance.

They were attacked before they reached Refugio by a large force of Mexican cavalry. They made a gallant defense for some time against the vastly superior numbers of the enemy, but at length their ammunition was exhausted and they were compelled to retreat to the timber on the river, where they were surrounded by the Mexican cavalry, and most of them finally captured. This division of our small force in the face of an enemy so greatly our superior in numbers, was, in my opinion, a fatal error on the part of Col. Fannin.

Hearing nothing either from Capt. King or Major Ward, and satisfied from information obtained by our scouts that a large force of Mexicans was in the vicinity of Goliad, Col. Fannin and his officers held a council of war in which it was determined to evacuate the place and fall back as rapidly as possible towards Victoria on the Guadalupe river. The same day, I believe, or the next after this council of war was held,a courier came in from San Antonio bringing a dispatch, as I was informed, from Col. Travis, to the effect "that he was surrounded in the Alamo by Santa Anna's army, and requesting Col. Fannin to come to his relief without delay."

Rations for five days and as much ammunition as each man could conveniently carry were immediately issued, and our whole force, including a small artillery company with two or three field pieces, started for San Antonio, crossing the river at the ford a half mile or so above town. After crossing the river and marching a short distance on the San Antonio road, a halt was made and our officers held a consultation, the result of which (I suppose) was the conclusion that we could not reach San Antonio in time to be of any assistance to Col. Travis. At any rate we were marched back to Goliad, re-crossing the river at the lower ford.

A few hours after we had got back to our old quarters, a detachment of Mexican cavalry, probably eighty or a hundred strong, showed themselves at a short distance from the fort apparently bantering us to come out and give them a fight.Col. Horton, who had joined us a few days previously with twenty-five mounted men, went out to meet them, but when he charged them they fled precipitately, and we saw them no more that day. That evening preparations were made to abandon the place; to that end we spiked our heaviest pieces of artillery, buried some in trenches, reserving several field pieces, two or three howitzers and a mortar to take with us on our retreat.

We also dismantled the fort as much as possible, burnt the wooden buildings in its immediate vicinity and destroyed all the ammunition and provisions for which we had no means of transportation. The next morning we bade a final farewell, as we supposed ,to Goliad, and marched out on the road to Victoria. We had nine small pieces of ordnance and one mortar, all drawn by oxen as were our baggage wagons. Our whole force comprised about two hundred and fifty men, besides a small company of artillery and twenty-five mounted men under Col. Horton.

We crossed the San Antonio river at the ford below town,and a short distance beyond Menahecil. a creek we entered the large prairie extending to the timber on the Coletto, a distance of eight or nine miles. When we had approached within two and a half or three miles of the point where the road we were traveling entered the timber (though it was somewhat nearer to the left a halt was ordered and the oxen were unyoked from guns and wagons, and turned out to graze.

What induced Col. Fannin to halt at this place in the open prairie, I cannot say, for by going two and a half miles further, we would have reached the Coletto creek, where there was an abundance of water and where we would have had the protection of timber in the event of being attacked. I understood at the time that several of Col. Fannin's officers urged him stiongly to continue the march until we reached the creek, as it was certain that a large body of Mexican troops were somewhere in the vicinity; but however this may be, Col. Fannin was not to be turned from his purpose, and the halt was made.

Possibly he may have thought that two hundred and fifty well armed Americans under any circumstances would be able to defend themselves against any force the Mexicans had within striking distance, but as the sequel will show the halt at this place was a most fatal one for us. Up to this time we had seen no Mexicans, with the exception of two mounted men, who made their appearance from some timber a long way to our right and who no doubt were spies watching our movements.

CHAPTER VI

MEXICAN CAVALRY SURROUND US ON THE PRAIRIE
COL. HORTON AND HIS MEN CUT OFF FROM US AND FORCED TO RETREAT
THE BATTLE OF COLETTO

IN A "TIGHT PLACE" I LET OFF MY SCOPET AND COME VERY NEAR KILLING ALL "BEHIND IT."
CARISE INDIAN SHARPSHOOTERS
CAPT. D. GIVES FOUR OF THEM THEIR "QUIETUS"
MEXICAN PRISONERS "HOLE" THEMSELVES, RETREAT OF THE ENEMY.

At length after a halt of perhaps an hour and a half on the prairie, and just as we were about to resume our march for the Coletto, a long dark line was seen to detach itself from the timber behind us, and another at the same time from the timber to our left. Some one near me exclaimed, "Here come the Mexicans!" and in fact, in a little while, we perceived that these dark lines were men on horseback, moving rapidly towards us.

As they continued to approach, they lengthened out their columns, evidently for the purpose of surrounding us,and in doing so displayed their numbers to the greatest advantage. I thought there were at least ten thousand (having never before seen a large cavalry force), but in reality there were about a thousand besides several hundred infantry (mostly Carise Indians). In the meantime we were formed into a "hollow square" with lines three deep, in order to repel the charge of the cavalry, which we expected would soon be made upon us.

Our artillery was placed at the four angles of the square, and our wagons and oxen inside. Our vanguard under Col. Horton, had gone a mile or so ahead of us, and the first intimation they had of the approach of the enemy was hearing the fire of our artillery when the fight began.

They galloped back as rapidly as possible to regain our lines, but the Mexicans had occupied the road before they came up and they were compelled to retreat. The Mexicans pursued them beyond the Coletto, but as they were well mounted they made their escape. The loss of these mounted men was a most unfortunate one for us. Had they been with us that night after we had driven off the Mexicans, we would have had means of transportation for our wounded, and could easily have made our retreat to the Coletto.

When the Mexicans had approached to within half a mile of our lines they formed into three columns, one remaining stationary, the other two moving to our right and left, but still keeping at about the same distance from us. Whilst they w'ere carrying out this maneuver, our artillery opened upon them with some effect, for now and then we could see a round shot plough through their dense ranks.

When the two moving columns, the one on the right and the one on the left were opposite to each other, they suddenly changed front and the three columns with trumpets braying and pennons flying, charged upon us simultaneously from three directions. When within three or four hundred yards of our lines our artillery opened upon them with grape and cannister shot, with deadly effect, but still their advance was unchecked, until their foremost ranks were in actual contact in some places with the bayonets of our men. But the fire at close quarters from our muskets and rifles was so rapid and destructive, that before long they fell back in confusion, leaving the ground covered in places with horses and dead men.

Capt. D's company of Kentucky riflemen and one or two small detachments from other companies formed one side of our "square," and in addition to our rifles, each man in the front rank was furnished with a musket and bayonet to repel the charge of cavalry. Besides my rifle and musket I had slung across my shoulders an "escopeta," a short light blunderbuss" used by the Mexican cavalry, which I had carried all day in expectation of a fight, and which was heavily charged with forty "blue whistlers" and powder in proportion. It was my intention only to fire it when in a very "tight place," for I was well aware it was nearly as dangerous behind it as before.

In the charge made by the Mexican cavalry they nearly succeeded in breaking our lines at several places, and certainly they would have done so had we not taken the precaution of arming all in the front rank with the bayonet and musket. At one time it was almost a hand to hand fight between the cavalry and our front rank, but the two files in the rear poured such a continuous fire upon the advancing columns, that, as I have said, they were finally driven back in disorder. It was during this charge and when the Mexican cavalry on our side of the square were in a few feet of us, that I concluded that I had got into that "tight place" and that it was time to let off the "scopet" I carried. I did so, and immediately I went heels over head through both ranks behind me.

One or two came to my assistance supposing no doubt I was shot (and in truth I thought for a moment myself that a two ounce bullet had struck me) but I soon rose to my feet and took my place in the line again just as the cavalry began to fall back. Now, I don't assert that it was the forty "blue whistlers" I had sent among them from my "scopet" that caused them to retreat in confusion. I merely mention the fact that they did fall back very soon after I had let off the blunderbuss among them. My shoulder was black and blue trom the recoil for a month afterwards. When I took my place in the line again, I never looked for my "scopet," but contented myself while the fight lasted with my rifle.

The Mexicans had no doubt supposed they would be able to break our lines at the first charge, and were evidently much disconcerted by their failure to do so; for although they reformed their broken columns and made two more attempts to charge us, they were driven back as soon as they came within close range of our small arms.

When they were satisfied that it was impossible for them to break our lines, the cavalry dismounted and surrounding us in open order, they commenced a "fusilade" upon us with their muskets and escopetas, but being very poor marksmen,most of their bullets passed harmlessly over our heads. Besides, this was a game at which we could play also, and for every man killed or wounded on our side I am confident that two or three Mexicans fell before the deadly fire from our rifles.

But there were with the Mexicans probably a hundred or so Carise Indians, who were much more daring, and withal better marksmen. They boldly advanced to the front, and taking advantage of every little inequality of the ground and every bunch of grass that could afford them particular cover, they would crawl up closely and fire upon us, and now and then the discharge of their long single barrel shot guns was followed by the fall of some one in our ranks.

Four of them had crawled up behind some bunches of tall grass within eighty yards of us, from whence they delivered their fire with telling effect. Capt. D who was using a heavy Kentucky rifle, and was known to be one of the best marksmen in his company, was requested to silence these Indians.He took a position near a gun carriage, and whenever one of the Indians showed his head above the tall grass it was perforated with an ounce rifle ball, and after four shots they were seen no more. At the moment he fired the last shot Capt. D had one of the fingers of his right hand taken off by a musket ball. When the Mexicans quit the field, we examined the locality where these Indians had secreted themselves, and found the four lying closely together, each one with a bullet hole through his head.

At the commencement of the fight a little incident of a somewhat ludicrous character occurred. We had some five or six Mexican prisoners (the couriers of the old padre, captured at Carlos' Ranch). These we had placed within the square, when the fight began, for safe keeping, and in an incredibly short time, with picks and shovels, they dug a trench deep enough to "hole" themselves, where they lay "perdue" and completely protected from bullets.

I for one, however, didn't blame them, as they were non-combatants, and besides to tell the truth when bullets were singing like mad hornets around me, and men were struck down near me, I had a great inclination to "hole up" myself and draw it in after me. The fight continued in a desultory kind of way, until near sunset, when we made a sortie upon the dismounted cavalry, and they hastily remounted and fell back to the timber to our left, where, as soon as it was dark, a long line of fires indicated the position of their encampment.

That night was anything but rest for us, for anticipating a renewal of the fight the next morning, all hands were set to work digging entrenchments and throwing up embankments, and at this we labored unceasingly till nearly daylight. We dug four trenches enclosing a square large enough to contain our whole force, throwing the earth on the outside, on which we placed our baggage and everything else available that might help to protect us from the bullets of the enemy.

Before we began this work, however, Col. Fannin made a short speech to the men, in which he told them "that in his opinion, the only way of extricating themselves from the difficulty they were in, was to retreat after dark to the timber on the Coletto, and cut their way through the enemy's lines should they attempt to oppose the movement." He told them there was no doubt they would be able to do this, as the enemy had evidently been greatly demoralized by the complete failure of the attack they had made upon 'us. He said, moreover, that the necessity for a speedy retreat was the more urgent, as it was more than probable that the Mexicans would be heavily reinforced during the night.

He concluded by saying that if a majority were in favor of retreating, preparations would be made to leave as soon as it was dark enough to conceal our movements from the enemy. But we had about seventy men wounded (most of them badly) and as almost every one had some friend or relative among them, after a short consultation upon the subject, it was unanimously determined not to abandon our wounded men, but to remain with them and share their fate, whatever it might be.

CHAPTER VII

OUR LOSS IN THE COLETTO FIGHT
DISMAL NIGHT ON THE PRAIRIE
THREE DESERTERS SHOT BY THE MEXICAN PATROL .
PREPARATIONS FOR RENEWING THE FIGHT
HOISTING THE "WHITE FLAG"
COL. FANNIN REFUSES TO SURRENDER AT DISCRETION:,
"MEXICANS FIRE UPON us WITH ARTILLERY;
WHITE FLAG AGAIN PARLEY-
GEN.URREA COMES INTO OUR LINKS CAPITULATION AGREED UPON AND TERMS REDUCED TO WRITING MEXICAN LOSS IN THE FIGHT,
MARTINI : THE TREACHEROUS FRIEND.

Our loss in the Coletto fight was ten killed and about seventy wounded (Col. Fannin among the latter), and most of them badly, owing to the size of the balls thrown by the Mexican escopetas, and the shotguns of the Indians. The number of our casualities was extremely small considering the force of the enemy, and the duration of the fight, which began about three o'clock and lasted till nearly sunset.

I can only account for it by the fact that the Mexicans were very poor marksmen, and that their powder was of a very inferior quality. There was scarcely a man in the whole command who had not been struck by one or more spent balls, which, in place of mere bruises would have inflicted dangerous or fatal wounds if the powder used by the Mexicans had been better.

I can never forget how slowly the hours of that dismal night passed by. The distressing cries of our wounded men begging for water when there was not a drop to give them were continually ringing in my ears. Even those who were not wounded, but were compelled to work all night in the trenches, suffered exceedingly with thirst.

Even after we had fortified our position as well as we could, we had but littie hopes of being able to defend ourselves, should the Mexicans as we apprehended, receive reinforcements during the night, for we had but one or two rounds of ammunition left for the cannon, and what remained for the small arms was not sufficient for a protracted struggle.

Some time during the night it was ascertained that three of our men (whose names I have forgotten) had deserted, and shortly afterwards as a volley of musketry was heard between us and the timber on the Coletto, they were no doubt discovered and shot by the Mexican patrol.

Daylight at last appeared, and before the sun had risen we saw that the Mexican forces were all in motion, and evidently preparing to make another attack upon us. When fairly out of the timber, we soon discovered that they had been heavily reinforced during the night. In fact, as we subsequently learned from the Mexicans themselves, a detachment of seven hundred and fifty cavalry and an artillery company had joined them shortly after their retreat to the timber. In the fight of the previous day they had no cannon.

They moved down upon us in four divisions, and when within five or six hundred yards, they unlimbered their field pieces (two brass nine pounders) and opened fire upon us. We did not return their fire, because as I have said, we had only one or two rounds of ammunition left for our cannon and the distance was too great for small arms. Their shot, however, all went over us, and besides, the breast works we had thrown up would have protected us, even if their guns had been better aimed. We expected momentarily that the cavalry would charge us, but after firing several rounds from their nine pounders, an officer accompanied by a soldier bearing a white flag, rode out towards us, and by signs gave us to understand that he desired a ' 'parley."

Major Wallace and several other officers went out and met him about half way between our "fort" and the Mexican lines. The substance of the Mexican officer's communication (as I understood at the time) was to the effect "that Gen. Urrea, the commander of the Mexican forces, being anxious to avoid the useless shedding of blood (seeing we were now completely in his power), would guarantee to Col. Fannin and his men, on his word of honor as an officer and gentleman, that we would be leniently dealt with, provided we surrendered at discretion, without further attempt at hopeless resistance."

When this message was delivered to Col. Fannin, he sent word back to the officer "to say to Gen. Urrea, it was a waste of time to discuss the subject of surrendering at discretion that he would fight as long as there was a man left to fire a gun before he would surrender on such terms." A little while afterwards the Mexicans again made a show of attacking us, but just as we were expecting them to charge, Gen. Urrea himself rode out in front of his lines accompanied by several of his officers and the soldier with the "white flag."

Col. Fannin and Major Wallace went out to meet them, and the terms of capitulation were finally agreed upon, the most important of which was, that we should be held as prisoners of war until exchanged, or liberated on our parole of honor not to engage in the war again at the option of the Mexican commander in chief. There were minor articles included in it, such as that our side arms should be retained, etc.

When the terms of capitulation had been fully decided upon, Gen. Urrea and his secretary and interpreter came into our lines with Col. Fannin, where it was reduced to writing, and an English translation given to Col. Fannin which was read to our men. I am thus particular in stating what I know to be the facts in regard to this capitulation, because

I have seen it stated that Gen. Santa Anna always asserted there was no capitulation, and that Col. Fannin surrendered at discretion to Gen. Urrea. This assertion I have no doubt was made to justify as far as possible his order for the cold blooded murder of disarmed prisoners. Gen. Urrea,I believe, never denied the fact of the capitulation, and I have been informed, when the order was sent him by Santa Anna to execute the prisoners, he refused to carry it into effect, and turned over the command to a subaltern.

I have always believed myself that Gen. Urrea entered into the capitulation with Col. Fannin in good faith, and that the massacre of the prisoners, which took place some days afterwards, was by the express order of Santa Anna, and against the remonstrances of Gen. Urrea. If Gen. Urrea had intended to act treacherously, the massacre, in my opinion,would have taken place as soon as we had delivered up our arms, when we were upon an open prairie, surrounded by a large force of cavalry, where it would have been utterly impossible for a single soul to have escaped, and consequently he could then have given to the world his own version of the affair without fear of contradiction.

I have said nothing as yet of the Mexican loss in the fight and I cannot do so with any certainty, of my own knowledge; but there is no doubt it was much greater than ours. They told us after we had surrendered that we had killed and wounded several hundred. Dr. Joseph Barnard, our assistant surgeon, who was saved from the massacre to attend their wounded, told me afterwards that he was confident we had killed and wounded between three and four hundred, and his opportunities for forming a correct estimate of the number were certainly better than those of any one else. After our surrender we were marched back to Goliad, escorted by a large detachment of cavalry, and there confined within the walls surrounding the old mission.

Among the Mexican officers there was a lieutenant by the name of Martinez, who had been educated at a Catholic college in Kentucky, where he had been a room-mate of a member of Capt. D 's company, by the name of B, everyday whilst we were prisoners he used to come and talk with B, and professed his great regret to find him in such a situation, but he never gave him the slightest intimation of the treacherous designs of the Mexicans, nor, as far as I know, made the least effort to save his college room-mate.

CHAPTER VII1

MAJOR WARD AND HIS BATTALION BROUGHT TO GOLIAD,
CAPT. MILLER AND HIS MEN CONFINED IN THE OLD MISSION
MEXICAN SOLDIER ASTONISHED TYRANICAL OFFICER
AN OFFICER TRYS TO MAKE A GOOD CATHOLIC OF ME,
BUT FAILS
ORDERED TO COPANO MARCHED OUT IN THREE DIVISIONS
THE MASSACRE - I NARROWLY ESCAPE THE BAYONET,
SWIM THE RIVER AND HALT UPON THE OPPOSITE SIDE.

A day or so after our return as prisoners to Goliad, Maj.Ward and his battalion, or rather those who survived the engagement they had with the Mexicans, near Refugio, were brought in and confined with us, within the walls enclosing the old mission; and also a company of about eighty men under the command of Maj. Miller, who had been surprised and captured at Copano just after they had landed from their vessel. These men were also confined with us, but kept separate from the rest, and to distinguish them, each had a white cloth tied around one of his arms. At the time, I had no idea why this was done, but subsequently I learned the reason.

The morning of the sixth day after our return to Goliad, whether the Mexicans suspected we intended to rise upon the guard, or whether they merely wished to render our situation as uncomfortable as possible, I know not, but at any rate from that time we were confined in the old mission, where we were so crowded we had hardly room to lie down at night. Our rations too, about that time, had been reduced to five ounces of fresh beef a day, which we had to cook in the best way we could and eat without salt.

Although, thus closely confined and half starved, no personal indignity was ever offered to us to my knowledge, except on two occasions

Once a Mexican soldier pricked one of our men with his bayonet, because he did not walk quite fast enough to suit him, whereupon he turned and knocked the Mexican down with his fist. I fully expected to see him roughly handled for this "overt act," but the officer in command of the guard, who saw the affair, came up to him and patting him on the shoulder, told him he was "muy bravo," and that he had served the soldier exactly right. At another time one of our men was complaining to the officer of the guard of the ration issued to him, who ordered one of the soldiers to collect a quantity of bones and other offal lying around, and throwing them on the ground before, the man, said, "There, eat as much as you want good enough for Gringoes and heretics."

One day an officer who was passing, asked me some question in Spanish, and when I answered him in Spanish, he took a seat by me, and talked with me for some time. He asked me a great many questions about the United States, our form of government, the number of our regular army, what State I came from and what induced me to come to Texas, etc., to all of which I frankly answered. He expressed much astonishment at the correctness of my pronunciation, and asked where I had learned to speak Spanish, saying he was sure

I had not learnt the language among the Mexicans. I told him I had studied Spanish under a teacher of modern languages at a Catholic institution in Kentucky. He then asked if I was a Catholic myself, and when I told him I was not, he seemed disappointed, and tried in various ways to get some sort of admission from me that I had more faith in the Catholic religion than any other.

The talk I had with this officer made but little impression upon me at the time, but I have since thought on account of my youth, or because I had in some way gained his favor, he was desirous of an excuse or pretext to save me from the fate he probably knew was in store for us. I know that several of our men were saved from the massacre, for no other reason that I am aware of, than that they professed to be members of the Catholic church.

Several times afterwards the officer above mentioned came to talk with me, and he insisted I was a Catholic if I would but own it; but I strenuously denied "the soft impeachment" to the last. If I had suspected his object in getting me to admit that I was a Catholic, it is probable I might have sought temporal as well as eternal safety the bosom of the church. It would have been very easy for me to have passed for a "good Catholic," for Catholicism (at least among the lower class of Mexicans) consists mainly in knowing how to make the sign of the cross, together with unbounded reverence first, for the Virgin Mary, and secondly, for the saints generally and the priests. But I did not suspect the object this officer had in view when he tried to make a convert of me to the true faith, and I am afraid I have lost the only chance I shall ever have of becoming a "good Catholic."

On the morning of the 25th of March, a Mexican officer came to us and ordered us to get ready for a march. He told us we were to be liberated on "parole," and that arrangements had been made to send us to New Orleans on board of vessels then at Copano. This, you may be sure, was joyful news to us, and we lost no time in making preparations to leave our uncomfortable quarters. When all was ready we were formed into three divisions and marched out under a strong guard. As we passed by some Mexican women who were standing near the main entrance to the fort, I heard them say "pobrecitos" (poor fellows), but the incident at the time made but little impression on my mind.

One of our divisions was taken down the road leading to the lower ford of the river, one upon the road to San Patricio, and the division to which my company was attached, along the road leading to San Antonio. A strong guard accompanied us, marching in double files on both sides of our column.

It occurred to me that this division of our men into three groups, and marching us off in three directions, was rather a singular maneuver, but still I had no suspicion of the foul play intended us. When about half a mile above town, a halt was made and the guard on the side next the river filed around to the opposite side. Hardly had this maneuver been executed, when I heard a heavy firing of musketry in the directions taken by the other two divisions.

Some one near me exclaimed "Boys! they are going to shoot us!" and at the same instant I heard the clicking of musket locks all along the Mexican line. I turned to look, and as I did so, the Mexicans fired upon us, killing probably one hundred out of the one hundred and fifty men in the division.

We were in double file and I was in the rear rank. *The man in front of me was shot dead*, and in falling he knocked me down. I did not get up for a moment, and when I rose to my feet, I found that the whole Mexican line had charged over me, and were in hot pursuit of those who had not been shot and who were fleeing towards the river about five hundred yards distant.

I followed on after them, for I knew that escape in any other direction (all open prairie) would be impossible, and I had nearly reached the river before it became necessary to make my way through the Mexican line ahead. As I did so, one of the soldiers charged upon me with his bayonet (his gun I suppose being empty). As he drew his musket back to make a lunge at me, one of our men coming from another direction, ran between us, and the bayonet was driven through his body..

The blow was given with such force, that in falling, the man probably wrenched or twisted the bayonet in such a way as to prevent the Mexican from withdrawing it immediately. I saw him put his foot upon the man, and make an ineffectual attempt to extricate ^the bayonet from his body, but one look satisfied me, as I was somewhat in a hurry just then, and I hastened to the bank of the river and plunged in.

The river at that point was deep and swift, but not wide, and being a good swimmer, I soon gained the opposite bank, untouched by any of the bullets that were pattering in the water around my head. But here I met with an unexpected difficulty. The bank on that side was so steep I found it was impossible to climb,it, and I continued to swim down the river until I came to where a grape vine hung from the bough of a leaning tree nearly to the surface of the water.

This I caught hold of and was climbing up it hand over hand, sailor fashion, when a Mexican on the opposite bank fired at me with his escopeta, and with so true an aim, that he cut the vine in two just above my head, and down I came into the water again. I then swam on about a hundred yards further, when I came to a place where the bank was not quite so steep, and with some difficulty I managed to clamber up.

CHAPTER IX

HEADED OFF BY A PARTY or LANCERS
WITNESS THE MURDER OF FIVE OR SIX OF OUR MEN
DODGE THE LANCERS AND FINALLY ESCAPE,
HIDE IN GROVE OF TIMBER TILL NIGHT
EXECUTION OF COL. FANNIN,
WONDERFUL ESCAPE OF WM. HUNTER

The river on the north side was bordered by timber several hundred yards in width, through which I quickly passed and I was just about to leave it and strike out into the open prairie, when I discovered a party of lancers nearly in front of me, sitting on their horses, and evidently stationed there to intercept any one who should attempt to escape in that direction. I halted at once under cover of the timber, through which I could see the lancers in the open prairie, but which hid me entirely from their view.

Whilst I was thus waiting and undecided as to the best course to pursue under the circumstances, I saw a young man by the name of Holliday, one of my own messmates, passing through the timber above me in a course that would have taken him out at the point directly opposite to which the lancers were stationed.

I called to him as loudly as I dared and fortunately, being on the "qui vive," he heard me, and stopped far enough within the timber to prevent the lancers from discovering him. I then pulled off a fur cap I had on, and beckoned to him with it. This finally drew his attention to me, and as soon as he saw me he came to where I was standing, from whence, without being visible to them, the lancers could be plainly seen.

A few moments afterwards we were joined by a young man by the name of Brown, from Georgia, who had just swam the river, and had accidentally stumbled on the place where Holliday and I were holding a "council of war" as to what was the best course to pursue. Holliday, although a brave man, was very much excited, and had lost to some extent his presence of mind, for he proposed we should leave the timber at once and take the chances of evading the lancers we saw on the prairie.

I reasoned with him on the folly of such a proceeding, and told him it would be impossible for us to escape in the open prairie from a dozen men on horseback. "But," said Holliday, "the Mexicans are crossing the river behind us, and they will soon be here." "That may be," I replied, "but they are not here yet, and in the mean time something may turn up to favor our escape." Brown took the same view of the case I did, and Holliday's wild proposition to banter a dozen mounted men for a race on the open prairie was "laid upon the table."

Whilst we were debating this (to us) momentous question, some four or five of our men passed out of the timber before we saw them, into the open prairie, and when they discovered the lancers it was to late. The lancers charged upon them at once, speared them to death, and then dismounting robbed them of such things as they had upon their persons. From where we stood the whole proceeding was plainly visible to us, and as may be imagined, it was not calculated to encourage any hopes we might have had of making our escape.

However, after the lancers had plundered the men they had just murdered, they remounted, and in a few moments set off in a rapid gallop down the river to where it is probable they had discovered other fugitives coming out of the timber.

We at once seized the opportunity thus afforded us to leave the strip of timber which we knew could give us shelter but for a few moments longer, and started out, taking advantage of a shallow ravine which partially hid us from view. We had scarcly gone two hundred yards from the timber, when we saw the lancers gallop back and take up their position at the same place they had previously occupied. Strange to say, however, they never "observed us, although we were in plain view of them for more than a quarter of a mile, without a single brush or tree to screen us.

We traveled about five or six miles and stopped in a thick grove to rest ourselves, where we staid until night. All day long we heard at intervals irregular discharges of musketry in the distance, indicating, as we supposed, where fugitives from the massacre were overtaken and shot by the pursuing parties of Mexicans.

As the undergrowth was pretty dense in the grove where we had stopped, we concluded the chances of being picked up by one of these pursuing parties would be greater if we traveled on than if we remained where we were, and we determined to "lie by" until night. In talking the matter over and reflecting upon the many narrow risks we had run in making our escape, we came to the conclusion that in all probability we were the only survivors of the hundreds who had that morning been led out to slaughter; although in fact as we subsequently learned, twenty-five or thirty of our men eventually reached the settlements on the Brazos.

Drs. Shackleford and Barnard, our surgeons, were saved from the massacre to attend upon Mexicans wounded in the fight on the Coletto, and when their forces retreated from Goliad after the battle of San Jacinto these were taken to San Antonio,where they were ultimately liberated. Our own wounded men, or rather those of them that survived up to the time of the massacre, were carried out into the open square of the fort, and there cruelly butchered by the guard. Captain Miller and his men were saved, because, as I was subsequently informed, they had been captured soon after they landed from their vessel, without any arms, and of course without making any resistance.

Col. Fannin, who was confined to his quarters by a wound he had received at the fight on the Coletto, soo'n after *the massacre of his men*, was notified to prepare for immediate execution. He merely observed that he was ready then,,as he had no desire to live after the cold-blooded, cowardly murder of his men. He was thereupon taken out to the square by a guard, where he was seated on a bench, and his eyes blindfolded.

A moment before the order to u fire" was given, I was told (though I cannot vouch for the truth of the statement) he drew a fine gold watch from his pocket, and handing it to the officer in command of the guard, requested him as a last favor to order his men to shoot *him in the breast and not in the head*. The officer took the watch, and immediately ordered the guard to fire at his head. Col. Fannin fell dead and his body was *thrown into one of the ravines near the fort.* Thus died as brave a son of Georgia as ever came from that noble old State.

The escape of Wm. Hunter was most wonderful. At the first fire he fell pierced by a musket ball. A Mexican soldier thinking he was not quite dead, cut his throat with a butcher knife, but not deep enough to sever the jugular vein, stabbed him with his bayonet, and then beat him over the head with the breech of his musket, until he was satisfied his bloody work was accomplished. He then stripped him of his clothing and left him for dead upon the ground where he had fallen.

Hunter laid there in a perfectly unconscious state until dark, but after night came on, the cool air and dew revived him, and by degrees he regained his senses. For a time all that had occurred since morning appeared like a troubled dream to him, but gradually the reality of the events that had taken place forced itself upon his mind, and he cautiously raised his head to reconnoitre. All was still around and not a moving, living creature was visible, nothing but the pallid upturned faces of his murdered comrades dimly seen in the waning light of day.

He found himself extremely weak from loss of blood, and his limbs were sore and stiffened; but he was suffering intensely from thirst, and he resolved, if possible, to drag himself to the river. With much pain and difficulty, he succeeded in reaching the water, and after quenching his thirst, he bound up his wounds as well as he could with strips of cloth torn from his shirt.

Before daylight he had recovered his strength so far that he was able to swim the river, and took his way to a Mexican ranch on the Manahuila creek, with the people of which he had had some previous acquaintance, thinking it was better to trust himself to their tender mercies than to attempt to travel through a wilderness in his wounded and weakened condition.

When near the ranch he met a Mexican woman, who told him he would certainly be killed if he went there. She advised him to secrete himself in a thicket she designated, and told him as soon as it was dark she would come out to him and bring him some food and clothing.

Hunter had his suspicions that she intended to betray him, yet there was no alternative but to trust her, and he hid himself in the thicket she had pointed out to him, and anxiously awaited her reappearance. True to her promise, a little while after dark, she returned, bringing some provisions and water, together with a suit of Mexican clothes.

For nearly a week this Mexican woman came to his place of concealment every night, fed him and dressed his wounds until he was sufficiently restored to travel. She then supplied him with as much provisions as he could carry and also a flint and steel for making fire, and bidding him "adios" she returned to the ranch. Thus recruited and supplied with clothing and provisions, Hunter took his course through the wilderness, and having a pretty good idea of the "lay of the land," after many narrow escapes he eventually made his way to the Texan army under General Houston.

CHAPTER X.

TAKE THE ROAD AGAIN INDIAN SIGN NARROW ESCAPE FROM LANCERS
SUFFERING FROM HUNGER LOST AND TAKE THE BACK TRAIL
RETURN NEARLY TO GOLIAD
MAKE A FRESH START AND HOLLIDAY TAKES THE LEAD AND AGAIN GOES WRONG
REBEL DISAGREEMENT IN CAMP
COMPROMISE, AND I FINALLY LEAD THE WAY

As soon as it was dark we left our hiding place and set out in a northeasterly direction, as nearly as we could determine, and traveled until daylight, when we stopped an hour or so in a grove to rest. We then proceeded on our course again till near sunset, when we encamped in a thick "mot" of timber without water.

An unusually cold norther for the season of the year was blowing, and a steady drizzling rain was falling when we stopped. Brown, who had pulled off his coat and shoes before he swam the San Antonio river, suffered severely, and I was apprehensive, should we be exposed all night to such weather without a fire, that he would freeze to death. I had a little tinder box in my pocket containing a flint and steel, but all the tinder there was in it was a small piece not much larger than a pin head.

This I carefully placed on a batch of cotton taken from the lining of my fur cap, and after many unsuccessful efforts I managed at last to ignite it. With this we started a fire, and then the first thing I did was to tear off a portion from my drawers, which I partially burned, thus securing a good supply of tinder for future use. Before going to sleep we collected fuel enough to last until daylight, with which we occasionally replenished the fire so that we passed the night in tolerable comfort.

The next morning Brown, who as I have previously stated, had pulled off his coat and shoes and thrown them away when he swam the river, found himself so sore and crippled he was unable to travel. The prairie we ad passed over the day before, had been recently burned off and the sharp points of the stubble had lacerated his naked feet dreadfully. It was evident he could not go on without some sort of covering for his feet.

I cut off the legs of my boots, and with a pair of scissors which he happened to have in his pocket, and some twine, I contrived to make him a pair of sandals,such as I had seen worn by Mexican soldiers. After thus shoeing him (by way of remuneration, I suppose,)Brown separated the two blades of the scissors and gave me one of them, which was of great service to me, for by whetting it on stones I gave it an edge, and it answered pretty well in place of a knife.

The grove of timber in which we had passed the night,covered perhaps an acre of ground, and just outside of it there was a strip of sandy soil almost bare of grass. In the morning when we left the grove we observed a good many fresh mocassin tracks which must have been made during the night by a party of Indians, who probably had been drawn to the locality by the light from our fire. Why they did not attack us I cannot imagine, unless it was because they were ignorant of our number and that we were without arms. At any rate, but for their tracks in the sand we would not have known they had been around our camp during the night.

The next morning we set out, as we supposed, in the direction we had traveled the day before, and in about one hour we came to some timber, bordering upon what I thought was one of the branches of the Coletto creek. Here we laid ourselves down on the grass to rest for a few moments, and scarcely had we done so when a party of ten Mexican lancers made their appearance, riding along a trail that ran within fifty yards of where we were lying. As luck would have it, just as they came opposite to where we were, they met another soldier and stopped to have a talk with him.

For nearly an hour, it seemed to me, but in fact, I suppose, for only a few minutes, they sat on their horses conversing together within a few paces of where we were lying, and without a single bush or tree intervening to hide us from their view, but fortunately they never looked toward us or we would inevitably have been discovered. At length they rode on, and we were very glad when we lost sight of them behind a point of timber.

The weather still continued cloudy and drizzly, and not being able to see the sun we had nothing to guide us, and in consequence were doubtful as to whether or not we were pursuing the right course. However, we traveled on until night, and again encamped in a thick grove of timber. Having eaten nothing since we left Goliad, and only a small piece of beef for two days previously, we had begun to suffer severely from the pangs of hunger. Game we had everywhere seen in the greatest abundance, but having no guns, the sight of herds of deer and flocks of wild turkeys, suggestive as they were to our minds of juicy steaks and roasts, only served to aggravate the cravings of our appetite.

It was at a season of the year, too, when no berries or wild fruits were to be found, and the pecans and other nuts had fallen and been destroyed by wild hogs, deer and other animals. But in spite of our hunger we slept pretty well on our beds of dry leaves, except that we were occasionally aroused from our slumbers by the howling of wolves, which were sometimes so impudent 1 as to approach within a few paces of the fire about which we were lying.

In the morning the weather was still cloudy and cold, and we set out again upon our travels. Holliday being by several years the oldest of our party, had heretofore taken the lead to which Brown and I had made no opposition, but after an hour or so I was convinced he was leading us in the wrong direction, and in this opinion I was confirmed when in a little while we came to a creek I was pretty sure was the Manahuila, the same we had crossed the day after leaving Goliad.

I told Holliday I was confident he was taking the back track, but he thought not, and so we kept on until toward evening, when we came to several groves of live oak timber which I remembered having seen when hunting in the vicinity of Goliad. Holliday, however, had but little faith in my recollections of the locality, and proposed that Brown and myself should wait in one of these groves until he reconnoitered the country ahead, and we consented to do so.

In about an hour he returned and told us that he had been in sight of Goliad, and that he had distinctly heard the beating of drums and the bugle calls of cavalry in the town. We felt very much discouraged, as may well be supposed, to find ourselves, after traveling so long, almost at the same point we had started from; but it was useless to repine, and we set out again in the right direction, Holliday, as usual, leading the way.

After an hour or so I found that Holliday was gradually turning his course toward Goliad again. Time with us was too precious to be wasted. I came to a halt and told Holliday I would follow him no farther. He insisted he was going the right direction, and I as positively that he was going directly contrary to the course we ought to pursue. He was obstinate, and so was I. Holliday, I knew, had been born and raised in a city whilst I had lived the greater part of my life on the frontier, and had been accustomed to the woods ever since I was old enough to carry a gun.

Besides, I knew that I possessed to a considerable degree what frontiersmen call "hog-knowledge," by which is meant a kind of instinctive knowledge that enables some people to steer their way through pathless woods and prairies without a compass or any landmarks to guide them. I therefore told Holliday, if he persisted in traveling in the direction he was going, we would certainly have to part company, although I was very loath to do so under the circumstances.

Thereupon and without further parley I turned and took the opposite course to the one we had been traveling. Brown, who made no pretensions to being a woodman, followed me, for the reason, I suppose, that he had lost confidence in Holliday as a guide, and thought possibly I might do better. Holliday remained standing where we had left him, apparently undetermined what to do, until we had gone perhaps a hundred yards, when he turned and followed us.

As he came up he merely said that he would rather go wrong than part company, and no allusion afterwards was made to the subject, but from that time on I always took the lead as a matter of course, recrossing the Manahuila creek, and night coming on shortly afterwards, we encamped by the side of a pool of water in a thick "island" of timber.

By this time we were suffering greatly with hunger, nevertheless I slept soundly through the night, although in my slumbers I was constantly tantalized by dreams of juicy steaks, hot biscuits and butter, etc., which always mysteriously disappeared when I attempted to "grab" them.

CHAPTER X1

SUFFERING FROM HUNGER "TURKS HEADS, A SUBSTITUTE FOR WATER
REACH THE GUADALUPE RIVER
UNSUCCESSFUL ATTEMPT TO GET VEAL STEAKS, BUT CAUGHT THE PIGS AND ATE FIVE FOR SUPPER,
INDIANS, MEXICAN LION.

The next morning we again took our course across the prairie, but owing to the rank growth of grass with which in many parts it was covered, and our increasing weakness, our progress was slow and painful. On the way, Holliday found about a dozen wild onions, which he divided with Brown and myself; but the quantity for each was so small that it seemed only to aggravate the pangs of hunger. During the day, we saw in the distance several parties of Mexicans or Indians,we could not tell which, as they only came near enough for us to see that they were men on horseback.

That night we again encamped in a strip of woods bordering a small creek, but we slept very little on account of our sufferings from hunger, which had now become excruciating. The next morning Brown was so weak he could scarcely walk two hundred yards without stopping to rest, nevertheless we went on as fast as we could travel. A part of the way was over high rolling prairie, on which no water could be found, and the pangs of thirst were added to those of hunger,until alleviated by the juice of some "Turks heads" which we found growing on the top of a pebbly knoll.

These plants are, I believe, a species of the cactus, about the size of a large turnip, grow on top of the ground, and are protected on the outside by a number of tough, horny prickles. The inside is filled with a spongy substance, which when pressed yields a quantity of tasteless juice that answers as a tolerable substitute for water.

The evening of the fifth day after leaving Goliad, we descried a long line of timber ahead of us, and just before sunset we came to a large stream, which from my knowledge of the geography of the country I was sure must be the Guadalupe. At the point where we struck it, the prairie extended up to the bank, which was high and very steep. A few hundred yards above us we saw a cow and her calf grazing near the edge of the bluff, and approaching them cautiously we attempted to drive them over it, hoping that one or the other would be disabled or killed by the fall, but after several ineffectual efforts to force them to take the leap, they finally broke through our line and made their way to the prairie, taking with them some steaks we stood very much in need of.

Completely exhausted by our exertions, and suffering extremely from hunger, we looked around for a suitable place to camp, as it was now nearly night, and coming to a pit or sink twelve or fourteen feet deep, which would protect us from the cold wind blowing at the time, we built a fire at the bottom, laid down upon the leaves, and in a little while we all went to sleep.

How long I had slept I do not know, but I was at length aroused from my slumbers by a rattling among the sticks and dry leaves above me, and looking up I discovered a wild sow with her litter of pigs coming down the almost perpendicular bank of the sink. I silently grasped a billet of wood lying near me, and awaited their approach. The old sow came on, totally unsuspicious that three ravenous chaps were occupying her bed at the bottom (for by this time our fire had burnt out), and when she and her pigs were in striking distance I suddenly sprang up and began a vigorous assault upon the pigs.

The noise aroused Brown and Holliday, and comprehending at once the state of affairs they sprang to my assistance, and before the sow and her pigs could make their escape up the steep sides of the pit we had "bagged five of the latter. We made a desperate attack on the old sow also, but weak as we were from starvation, and with our inefficient weapons, she routed us completely, leaving us however in possession of the field and the "spoils of war."

We immediately started our fire again, and with no other preparation than a slight roasting on the coals, enough to singe off their hair, we very expediously disposed of the five pigs we had killed nearly a pig and a half for each one, but then you must remember that they were small sucking pigs, and that we had not had a mouthful to eat for five days except a handful of wild onions. Greatly refreshed by our supper of scorched pig, we laid down again upon the leaves at the bottom of the sink, and slept soundly until the sun was an hour or so high.

As soon as we awoke, we left the sink and went out to make a reconnoisance of the river, to see what the chances were for crossing it. Though not very wide at that point, we soon perceived we had a difficult job to undertake, for the river was much swollen by recent rains, and its turbid waters were rushing along at a rapid rate. Holliday and I were both good swimmers, and I felt sure we could reach the opposite bank safely; but I had my doubts about Brown. He was a poor swimmer, and consequently was timid in water.

However, there was no alternative but to make the attempt,and we therefore stripped off our clothes, tied them in a bundle on our heads to keep them as dry as possible, and plunged in the turbid flood. Holliday and I soon reached the opposite bank, but hardly had we done so when I heard Brown cry out for help, and' looking back I saw that he was still some distance from the shore, and evidently just on the eve of going under.

At the very point where I landed there happened to be a slab of dry timber lying near the water, which I instantly seized, and swimming with it to the place where Brown was struggling to keep his head above the surface, I pushed the end of the slab to him, which he grasped and to which he held on with the usual tenacity of a drowning man, and with the assistance of Holliday I at last got him to the shore and dragged him out of the water. It was very fortunate for Brown that Holliday and I, between us, had taken his clothes, as otherwise no doubt he would have lost them all.

Continuing our course, we passed through a heavily timbered bottom more than a mile wide, and then came to a large prairie in which we saw many herds of deer and some antelopes. The antelope is a beautiful animal about the size of a deer, but much more fleet. They do not run as deer do, by springs or bounds, but evenly, like the horse.

Their horns consist of two curved shafts, with a single prong to each. A man on a good saddle horse can easily overtake a fat deer on the prairie, but it would require a thorough bred racer with a light rider to come up with an antelope. We also saw to-day a party of Indians on horses, but we eluded them by concealing ourselves in some tall grass that grew in the bottom of a ravine. About dusk we came to the timber on the farther side of the prairie, in which we encamped under the spreading branches of a live oak tree.

Next morning we continued our route, and after passing through some open post oak woods, we came to a small stream not more than knee deep, and of course easily forded. Crossing this stream, we went through more post oak woods, and then entered another large prairie, and it was late in the evening, owing to the difficulty of making our way through tall and tangled grass, before we reached the timber on the opposite side, where we encamped in a little open space surrounded by a dense growth of underwood. Here we made a fire, and slept soundly till morning.

As soon as daylight appeared we were off again, and passing through a skirt of woods we came to another small stream, which was also fordable. Crossing it, we entered a large prairie, on the opposite side of which a long line of timber was dimly visible in the distance. All day long, stopping occasionally to rest, we toiled through the matted grass with which this prairie was covered, and just at sunset we came to the woods we had seen, where we encamped near a pool of water.

Whilst collecting a supply of fuel for the night, I came upon a heap of brush and leaves, and scraping off the top to see what was beneath, I discovered about half the carcass of a deer which apparently had been recently killed and partly eaten by a panther or Mexican lion, and the remainder "cached" in this heap for future use.

Of course, under the circumstances, I had no scruples about appropriating the venison, and calling Brown and Holliday to my assistance we carried it to camp, where, after cutting off the ragged and torn portions of the meat, we soon had the balance spitted before a blazing fire. After making a hearty supper on our stolen venison, we raked a quantity of dry leaves close to the fire and "turned into bed."

During the night, at various times, we heard the roaring of a Mexican lion (very probably the lawful owner of the larder that had supplied us with supper), and for fear he might be disposed to make a meal of one of us in place of venison, we took good care not to let our fire burn down too low. There is no animal, I believe, on the American continent, with the exception of the grizzly bear, that has ever been known to attack a man sleeping near a fire.

The Mexican lion is, I think, described in books of natural history under the name of puma or South American lion. They are of a tawny or dun color, about the size of the East Indian tiger, have a large round head and a short mane upon the neck. Their nails are very long, sharp and crooked coming to an edge on the inner side as keen as that of a knife. Their roar is very similar to that of the African lion. They are fierce and strong, but cowardly; although when pressed by hunger, they have been known to attack men in open daylight.

One instance of this comes within my own knowledge. Several teamsters, with their wagons, were traveling the road from San Antonio to Victoria, and a teamster needing a staff for his ox whip, went to a thicket eighty or a hundred yards from the road to cut one; whilst occupied in cutting down a small sapling with his pocket knife, a Mexican lion stealthily crawled up behind him and sprang upon him before he was aware of its presence. The man's cries for help were heard by one of the teamsters, who hurried to his assistance.

The only thing he had in the shape of a weapon was his ox whip, but with that he boldly attacked the lion, which, frightened by his approach and the loud popping of the whip, let go its prey and made a rapid retreat, but the poor fellow he had caught was dreadfully bitten and torn, and it was a long time before his wounds were healed. The Mexican lion is now rarely seen in Texas except among the dense chapparals between the Nueces and Rio Grande rivers.

CHAPTER XII

THE LAVACA RIVER, MEXICAN MARAUDERS,
BROWN AND MY SELF TEMPORARILY CAPTURED,
ATTEMPT TO ESCAPE I SUCCEED BUT BROWN IS RECAPTURED,
I PASS A LONELY NIGHT IN THE LAVACA BOTTOM,
TAKE THE PRAIRIE AGAIN SOLITARY AND ALONE
SLEEP ONCE MORE UNDER A ROOT.

As soon as it was fairly light we again started, and passing through a heavily timbered bottom, came to the Lavaca or Cow river, a small stream about thirty yards wide where we struck it. In going through the bottom we noticed several piles of rails and some clapboards, the first indications we had seen of settlements since we left Goliad. We also saw a drove of hogs in the bottom, which confirmed us in the opinion that there had been an American settlement somewhere in the vicinity.

These hogs were of the genuine "razorback" species, and as wild and fleet as deer; consequently, although our hunger was almost as pressing as ever, we did not care to exhaust our strength in what we knew would be a hopeless attempt to capture one of them.

We swam the river without difficulty, and stopped an hour or so on the bank to rest ourselves and dry our clothes. We then went on, but as the bottom on that side was very wide, and the day being cloudy, we lost our way and it was nearly sunset when we reached the open prairie. A few hundred yards below where we came out of the timber we observed ten or a dozen horses "staked," and, on approaching them, we heard people talking in the woods near by.

I advised an immediate retreat from the locality, but for some reason Holliday came to the conclusion that the horses belonged to a company of Texan scouts, and proposed that we conceal ourselves in a clump onrushes from whence we could see any one who might come to look after them and thus satisfy ourselves without running any risk as to whether the owners were Americans or Mexicans. Holliday's counsel prevailed, and Brown and I hid ourselves in a small bunch of bushes and Holliday in another.

A dog which was at the camp, all this time kept up an incessant barking, and probably it aroused the suspicions of the owners that some one was trying to steal their horses; at any rate, in a few moments after we had hidden ourselves, a strapping "ranchero" came out of the timber, and when he had looked to see if the horses had been disturbed in any way, he came as straight as he could walk to the bunch of bushes in which Brown and myself had taken our position and was just on the eve of entering it when he saw us.

He instantly sprang back exclaiming, "Hey! Americanos! What are you doing here? Do you want to steal our horses?" He then made signs for us to follow him, which we did, knowing that resistance, weak as we were and without arms, would be useless, and that one shout from the ranchero would bring those in camp to his assistance. Holliday, as I have said, was concealed in a separate clump of bushes, and, keeping quiet, the ranchero did not discover him. Brown and I got up and followed him, but I was fully determined from the start not to follow him as far as his camp.

I saw that his course would take him very close at one point to the timbered bottom, and as we went along Brown and I agreed upon a plan to escape from our captor, which was to follow him quietly until near the timber, and then suddenly "break ranks" and get under cover as speedily as possible. Then we were to take different directions and meet at the same place the next morning. The ranchero, although he could plainly see that Brown and I were unarmed, kept some paces ahead of us all the time, every now and then looking back to see if we were following.

Before Brown and I separated I told him I would meet him at the Mexican camp the next morning, as it was probable they would leave it before we could return there. In pursuance of our plan, as soon as we came very close to the edge of the timber, we suddenly left our ranchero with- out even saying "adios," and in a moment we were hidden from his sight by the dense undergrowth. When we thus so unceremoniously left our new acquaintance we were so near the camp that we could plainly hear the rancheros conversing with each other, and the moment we made a "break" our escort shouted to his companions to hasten to his assistance.

"Here are Americans, come quick and bring your guns," but just at this juncture Brown and I had some little matters of our own that required immediate despatch, and we did not wait for the Mexicans "to come and bring their guns with them." Brown went one way and I another as soon as we entered the timber, and I never saw him again until several weeks afterwards when he came to the army on the Brazos.

The sun had just set when we entered the timber, and night soon set in dark and cloudy. After going perhaps a mile, I concluded it would be impossible for the Mexicans to find me and I pitched my camp, which was speedily done by pitching myself on the ground at the foot; of a tree on which there was a thick growth of Spanish moss, that served to protect me in a measure from a drizzling rain that commenced falling. I did not dare to start a fire for fear the light from it might bring the Mexicans to the locality, should they be in pursuit.

I had never felt so despondent since making my escape from Goliad as I did that night. My separation from my companions, my uncertainty as to their fate, the thought of my helpless situation, without arms of any kind to protect myself against the attacks of wild beasts and still more merciless Mexicans and Indians, together with the mournful howling of wolves in the distance, all conspired to fill my mind with gloomy forebodings and anticipations. However, notwithstanding such unpleasant thoughts and surroundings, I soon fell asleep and slept soundly until morning.

When I awoke day was beginning to break, birds were singing and squirrels chattering in the trees. The rain had ceased, and after brushing off the damp leaves that adhered to my clothes, my toilet was made, and I started back in the direction of the place where Brown and I had separated. I came out of the bottom very near the place where I had entered it the evening before, but no living thing was visible on the prairie as far as I could see, except some herds of deer and a flock of wild turkeys.

I proceeded cautiously along the edge of the timber until I came to where the Mexicans had staked their horses. They were gone, and hearing no sounds from the woods in which they had camped, I ventured in to reconnoitre. Their fires were still burning, but the camp was deserted and there was nothing left to indicate the probable fate of my companions.

I was exceedingly hungry, and I searched the camp closely, hoping the Mexicans might have forgotten some remnant of their provisions when they went off, but I found no vestiges of eatables of any kind except a few egg shells. Leaving the camp, I returned to the prairie and traveled up and down the timber above and below it for several miles hoping to meet with one or the other of my companions. I continued my search for them until late in the evening, when having abandoned all hopes of finding them, 1 struck out across the prairie in the direction I intended going.

Before I had gone a quarter of a mile I happened to look back towards the river and saw a house that had been hidden from my view, when searching for my companions, by a strip of timber. As I was suffering much from hunger, I concluded to return and make an examination of this house and premises, hoping I might find something to eat. I approached the house cautiously for fear it might be occupied by a marauding party of Mexicans, but seeing nothing to excite my suspicions, I ventured up. Everything about the house furniture broken and strewn in fragments over the floor, beds ripped open and their contents scattered around, plainly indicated that it had been visited by some plundering band of rancheros or Indians.

However, in an outhouse near the main building, I found a piece of bacon and four or five ears of corn. The corn, I ground upon a steel mill that was fastened to a post in the yard, and starting a fire in one of the chimneys of the main building, I very soon prepared a substantial meal of "ask cake" and broiled bacon, to which I paid my sincere respects. By this time night had set in, and, spreading some tattered bed clothes left in the house upon the floor, I slept comfortably until morning.

CHAPTER XIII

UNSUCCESSFUL SEARCH FOR MY COMPANIONS
BROWN RECAPTURED FINALLY RELEASED AND MAKES HIS WAY SAFELY TO THE ARMY
HOLLIDAY PURSUED BY A PARTY OF MEXICANS CAPTURED BY TWO RUNAWAY NEGROES THEY DETERMINE TO KILL HIM, BUT HE TALKS THEM OUT OF THE NOTION ARRIVES SAFELY AT COLUMBIA ,HIS SUBSEQUENT FATE,

I SET OUT AGAIN "SOLITARY AND ALONE" TRAILED BY INDIANS, BUT I GIVE THEM THE DODGE.

The next morning, after making a hearty breakfast on ash cake and bacon, as there was no urgent necessity for hurrying forward, I concluded I would make another attempt to find my companions, and I again traveled for several miles above and below, along the edge of the timber, but seeing nothing of them I at length became satisfied that they had been captured by the Mexicans, or had gone on without waiting for me. The facts were, however, as I afterwards learned from both of them when I subsequently met them on the Brazos, about as follows:

After Brown and I broke away from the ranchero and went off in different directions, he pursued Brown, came up with him and took him back to the camp. There they tied him securely to a tree, and then proceeded leisurely to cook and eat their supper. Brown, who could speak a little Spanish, told them he was starving and begged them to give him something to eat, but they said it was useless to do so as they intended to shoot him in the morning.

He then told them if such was their intention to shoot him at once and not keep him tied up to a tree like a dog all night, but the Mexicans paid no attention to his request and when they had finished their supper, they laid down upon their blankets and went to sleep. Brown tried his best to untie himself, but the ranchero had fastened him so securely to the tree that he found it impossible to get loose, and was compelled to remain in a standing position all night.

The next morning, as soon as it was fairly light, one of the rancheros walked up to Brown and pinned a piece of white cloth to his breast, telling him it was a mark for them to shoot at. Four or five rancheros then stationed themselves a few paces in front of him, cocked their guns and presented them as if about to shoot. All this time, Brown, who had been rendered perfectly desperate by pain and hunger, was cursing the Mexicans as much as his imperfect knowledge of the language would permit.

He told them they were a set of cowardly scoundrels, and that the bravest feat they had ever performed was the murder of unarmed and helpless prisoners, and so on. Brown said he was suffering and had suffered so excruciatingly from pain and hunger all night that he really wanted the

Mexicans to shoot him and put him out of his misery, but they seemed much astonished at his boldness and sang froid, and the one in command of the party came to where he was tied, cut the ropes and told him to go, that he was "muy bravo" (very brave), and that in place of shooting him they would leave him to perish of hunger.

Then they saddled their horses and mounting them rode off. Some days afterwards Brown was again captured by a party of Mexicans, but in some way he managed to escape from them, and finally, more by good luck than anything else, for he was a poor woodman, he made his way to the army on the Brazos.

Holliday, as I have before stated, was not seen when the ranchero captured Brown and myself, and as soon as it was dark he left his hiding place and took his course across the prairie. Subsequently he had many narrow escapes from marauding parties of Mexicans and Indians. On one occasion a party of Mexicans pursued him so closely that he took refuge in a lake. He waded on until the water was up to his neck, when the Mexicans amused themselves for some time by firing off their scopets at his head, but fortunately for Holliday night came on, and under cover of the darkness he skipped out and dodged his pursuers.

Another time, two runaway negro men caught him in a house to which he had gone in search of something to eat. They asked him if he was a Texan, and upon his replying in the affirmative they told him they intended to kill him. Whereupon they tied him securely in the room and went out, but in a few moments returned, each one with a heavy club in his hand, and they told him to say his prayers speedily, as they were going to beat out his brains.

Holliday, however, "reasoned" the matter with them, telling them it wasn't fair to kill him for what other white men might have done to them that he had never injured them in any way, etc. His talk seemed to produce some effect upon one of the negroes but the other still insisted on killing him. Finally, however, the one who was inclined to favor him prevailed upon the other to abandon his intention of beating out his brains, and they said they would not kill him but would take him to the camp of some Mexican guerrillas near by.

Holliday thought that this would be worse than "jumping out of the frying pan into the fire;" that such a proceeding would not be better than having his brains knocked out, and he urged all the arguments he could think of against it. At length, much to Holliday's relief, they agreed to let him go, and before they left they not only gave him provisions, but directions that enabled him to make his way through an unknown country to the Texan army under General Houston.

He came into Columbia, on the Brazos, about ten days after I did. Holliday was subsequently appointed to a captaincy in the Texas regular army, was again taken prisoner in the unfortunate Santa Fe expedition, carried to the City of Mexico, and, after his liberation, died of yellow fever on the voyage from Vera Cruz to New Orleans, and was buried at sea.

Giving up all hopes of finding my companions, I started out across the large prairie that extended in the direction I was going as far as my eye could reach. The game on this prairie was more abundant than I had seen it elsewhere. I am sure that frequently there were a thousand deer in sight at a time. Here, too, I first saw the pinnated grouse, or prairie hen. At first I supposed the call of the cock was the distant lowing of wild cattle, some of which were grazing on the prairie. Wild turkeys were also numerous, and so unused to the sight of man, that they permitted me at times to approach within a few paces of them.

During the day I saw several parties of Mexicans or Indians on horses, but they did not come near me. About three o'clock in the evening I reached the timber on the Navidad, where I stopped to rest a while and lunch on some of the ash cake and bacon I had brought along with me. I then proceeded on my course through the bottom, and after going probably half a mile I came to the Navidad river, at that place thirty or forty yards wide. It was swollen by recent rains and not fordable, so I was compelled to swim it, which I did easily, stripping of my clothes and tying them on a piece of dry wood, and pushing it before me as I swam.

As soon as I reached the bank I dressed myself and continued my course through the bottom, which was much wider on that side. I had gone perhaps half a mile, when my attention was drawn to the continuous barking of a dog in the direction from which I had come. At first I did not notice it particularly, supposing it was some dog left behind by the settlers on the Navidad when they fled from the invading Mexican army. But at length I observed that although I was traveling at a pretty rapid walk the barking of this dog seemed to be nearer and nearer to me, and I supected he was trailing me and that probably there was some one with him. I therefore hurried on as fast as possible, and in an hour or so came to the open prairie on the north side of the river.

All this time, I could hear the baying of the dog at apparently about the same distance behind me as when I first noticed it. I was sure then he was trailing me, and never halted for a moment, but continued on my course into the prairie for several hundred yards and then turned short round and retraced my steps to the edge of the timber, where I sprang as far as I could to one side and went down the edge of the timber about a hundred yards to a fallen tree, among the limbs of which I concealed myself, and from whence I could have a distinct view of anything coming out of the bottom at the point I left it.

After I had thus "holed" myself, the barking of the dog grew louder and nearer every moment, and in a little while I saw the dog, followed by three Indians, emerge from the timber, precisely at the point where I had left it. One of the Indians held the dog by a leash, and was armed with a gun the other two had their bows and lances. If I had been armed with the poorest pot-metal, muzzle-loading shot-gun that was ever manufactured at Birmingham, I would not have feared them, but as I had no weapon more formidable than the scissor blade given me by Brown, I "laid low" and watched them from my hiding place.

When the Indians following the dog came to the place in the prairie from whence I had turned back on my trail, the dog lost it of course, but the Indians (taking it for granted, I suppose, that I had gone on in the same direction) urged and led the dog that way until finally they went out of sight. If I had not thrown them off my trail in the manner described, there is no doubt I would have lost my scalp on that occasion, and I took considerable credit to myself for having beaten them at their own game.

I remained but a little while in the hiding place after the Indians left. But the course I wished to travel was the one they had taken, and for that reason, and because my provsions were nearly exhausted, I determined to keep up along the edge of the timber, hoping to find gome settlement and replenish my larder. I followed up the margin of the timber for several miles, and at length came to a "clearing," on the opposite side of which I saw a house.

I cautiously advanced towards the house until I was satisfied it was not occupied, and that I could venture up with safety. On entering it I found that a marauding party of Mexicans had lately been there and appropriated to their own use whatever there might have been eatable on the premises. I searched the house thoroughly, but could find nothing in the way of "provender."

CHAPTER XIV

SLEEP LUXURIOUSLY IN A BED ROUSED UP BY HOGS. SUCCEED IN CAPTURING ONE MURDER HIM IN COLD BLOOD WITH A MAULBUTCHER HIM WITH A PIECE OF DRAWING KNIFE MAKE ANOTHER START UNSUCCESSFUL ATTEMPT TO STEAL A MEXICAN SOLDIER'S GUN.

By the time I had finished my fruitless search for something to eat the sun was about setting, and as there was a bed in the house, which looked very inviting to me after sleeping so long on the ground, I concluded to accept the invitation and pass the night in it. After a very frugal and unsatisfying repast upon the small remnant of ash cake and bacon in my knapsack, I turned into my bed and was soon fast asleep.

It must have been near midnight when I was aroused by some noise. I listened attentively and soon ascertained that the noise was nothing but the grunting of several hogs that had taken up their quarters under the house whilst I was asleep. The house was set upon blocks, a foot or so above the ground and the space beneath the floor was therefore sufficiently roomy for their accommodation.

The floor was made of puncheons or slabs, which were held in their places solely by their weight. Hunger as well as necessity is the mother of invention, and it occurred to me that I might bag one of these porkers by quietly lifting a puncheon immediately above the spot where they were lying and then quickly grabbing the first one I could get hold of.

I therefore got up from my comfortable bed , and after listening awhile to their grunting so as to ascertain what part of the floor they were under, I slowly and noiselessly lifted a slab above them and laid it aside. Thrusting my arm down through the opening I had made, I felt around until my hand came in contact with the leg of a hog, when I suddenly seized it, and the row began. I had got hold of a hog much too large for me to manage well, and found it no easy matter to induce him to come up into my comfortable quarters.

He struggled vigorously to get loose, squealing all the while in the most ear-piercing manner, and for some time I thought it very doubtful how the contest would end whether I would succeed in hauling the hog up into the room, or the hog in dragging me under the floor. But I knew if I "let go" there would be no pork steaks for breakfast, as the other hogs had been frightened by the squealing and struggling, and had left for parts unknown. But the idea of having no steak for breakfast gave me more than my usual strength, and at last,but not until he had cut me severely with his hard hoofs and rasped a good deal of the skin off my knuckles against the sharp edges of the puncheons, I drew him by main "strength and brutality" into the room and replaced the puncheon.

I had secured my hog, but how to kill and butcher him was the next question. I had nothing to do it with except one of the blades of the little pair of scissors given me by Brown, and that I knew was totally inadequate for the purpose. I could find nothing in the room that would do, so I slipped out, carefully lastening the door after me, to see if there was anything about the premises with which I could dispatch the porker.

The moon was shining brightly, and I looked all around for something that would enable me to convert my hog into pork, but could find nothing better than a large maul that had been used for splitting rails, and with this I re-entered the room and made a determined assault upon the hog.

The maul, however, was so heavy and unwieldy I could not handle it with sufficient celerity to inflict a stunning blow. Round and round the room we went for a quarter of an hour or more, the hog squealing all the while and his hoofs clattering and rattling on the puncheons and making altogether such a "racket" as might have been heard at the distance of half a mile. At last, however, I got a fair lick at his cranium, which brought him to the floor, where I finished him by continuous "mauling."

When the bloody deed had been committed, I was so completely exhausted that I tumbled back on the bed, was asleep in a few moments, and did not awake until the sun was high in the heavens. I got up, and the first thing I did was to drag my hog to a spring near the house, where I butchered him after a fashion, with a piece of broken drawing knife I picked up in the yard.

After finishing this job I started a fire, and roasted four or five pounds of the pork for breakfast. When I had breakfasted, I packed as much of the pork as I could carry in my knapsack, and started up the bottom again, keeping close to the edge of the timber so that I might readily take shelter in the event that I should meet with a party of Mexicans or Indians. I had come to the conclusion by this time that previously I had been steering my course too low down the country, and I thought it best to keep up the river some distance before I resumed it again, in order to avoid the lagoons and swamps which I supposed abounded in the vicinity of the coast.

I traveled five or six miles without seeing anything worthy of note, and at noon stopped an hour or so at a pool of water to rest and cook some of my pork, and to "barbacue" the remainder -so as to prevent it from spoiling. It was late in the evening before I started again, and about sunset, not finding another house, I concluded to encamp in a point of timber near a pool of water.

Just after I had turned into a bed of dry grass for the night, I saw a light spring up, apparently five or six hundred yards above, on the edge of the bottom, and I concluded to get up and see what caused it. The moon had not as yet made her appearance, and I thought I could reconnoiter the locality with safety, even if the light should prove to be from the camp fire of Mexicans or Indians. Guided by the light, which continued to shine steadily, I went perhaps a quarter of a mile, when I saw that it came from the chinks of a small log cabin.

I approached it silently, and when near it, I saw there were several other cabins near it, but no lights were visible in them. The chinks between the logs of the cabin in which the light was shining were all open, and I carefully crept to the side nearest me and peeped through one of them.I had heard for some time a queer kind of rasping sound proceeding from with the cabin, for which I could not account until I looked through the chink, and then I saw a Mexican soldier sitting on the floor, shelling corn into a tub, which he did by rasping the ears on the edge.

He had on his shot pouch and powder horn, but his gun I noticed was leaning against the wall next to me, and as there was an opening between two of the logs it was leaning against wide enough to shove my arm through, it occurred to me that possibly I might be able to draw the gun through this opening before the Mexican was aware that any one was in the vicinity, as his back was turned towards me. So I reached in, seized the gun cautiously, near the muzzle, and began to draw it slowly through the chink between the logs.

There is no doubt I would have succeeded in my attempt to get the gun, but when the barrel was fairly outside and I felt sure I had secured the prize, to my great disappointment the breech was so large that it stuck hard and fast between the logs. In my effort to pull the gun through, I unavoidably made some noise that attracted the attention of the soldier, and he turned and uttered an exclamation of fear and astonishment when he saw his gun thus mysteriously disappearing through the chink in the cabin, and he instantly sprang forward and clutched it by the breech.

The noise aroused three or four dogs sleeping near the cabin, and they began to bay me furiously. I was sure there were more Mexican soldiers in the adjoining houses, and thinking I might find a "healthier" location than the one where I was, I made off at "double quick" for the bottom, closely pursued by the dogs.

When I reached the timber, I picked up a club, turned upon the dogs and drove them back. I heard a good deal of shouting and " carahooing" at the cabins, but as the night was quite dark I had no fear of being pursued, and leisurely took my way along the edge of the timber. When I had got I suppose a mile from the cabins,I went into the timber and encamped in a secure place.

My failure to get the soldier's gun was a great disappointment to me. Every house I had visited since I struck the settlements, I had searched closely for a gun, hoping that one might have been left by the occupants when they hurriedly fled before the invading army, but my search was always fruitless. People had abandoned a great deal of valuable property, but whatever arms they had they carried off.

I had an abundance of ammunition, for at one of the houses I had searched I found powder and shot, which I secured, and all I lacked was a gun. I would willingly have given all the money I had in the world (amounting to seventy-five cents in specie) for the poorest pot-metal gun that was ever manufactured, and taken the chances of its bursting whenever I fired it.

CHAPTER XV

FOLLOWED BY WOLVES MEXICANS AND INDIANS, INDIAN CAMP PASS THE NIGHT THERE SIGNAL SMOKES
LOST ON THE PRAIRIE, TRES PALACIOS, TRY MY HAND AT MANUFACTURING A BOW.

Just at daylight I was aroused from ray slumbers by the clucking and gobbling of wild turkeys. I had encamped very near a large "roost," and as I made no fire I had not disturbed them. Many of the trees in the vicinity were literally filled with them, and they were so tame I could easily have killed one with a bow and arrow if I had had them, and I determined I would try my hand at manufacturing these primitive weapons, if I could find some suitable tool to work with.

After I had reconnoitered from the edge of the timber and ascertained that there were no Mexicans in sight, I went on up the bottom three or four miles, and then struck across the prairie in the direction I had been traveling. My route was through an open prairie interspersed with groves of timber. In one of these I stopped about noon, and broiled a piece of my pork for dinner.

After resting an hour or so I continued on my way, an ' about sunset came to some timber bordering a small stream. I had scarcely entered this timber, which was open and free from undergrowth, when I noticed several large wolves -trotting along behind me. Every now and then they set up a howl, which was answered by others in the distance, and before long numbers of them had gathered around me, attracted, I suppose, by the howling of those I had first seen, or by the smell of the fresh meat I had with me.

I had no fear of an immediate attack from them, nevertheless, I hurried on as fast as I could until I came to the small stream I have mentioned, on the bank of which I pitched camp, near a large fallen tree that would afford sufficient fuel to keep a fire burning all night. I am confident if I had not had a fire that night, the wolves would have torn me to pieces; as it was, they sometimes ventured up to within a few feet of the fire, howling and snarling, and evidently inclined to make a dash at me at all hazards. It was impossible to sleep, so I took my spite out of them by occasionally throwing a fire brand amongst the crowd. This would silence them for a moment, but they would soon begin their howlings again. Towards daylight they raised the siege and departed, and I got a little nap before sunrise.

To-day, while crossing another large prairie, I saw in the distance a considerable body of Mexicans or Indians, I could not tell which, who were traveling at a rapid rate, and I soon lost sight of them. In this prairie I passed many herds of deer, generally fifty to a hundred in a herd, which were so gentle they frequently permitted me to approach within a few paces of them before they noticed me at all. I also saw several droves of mustangs, which were much wilder than the deer, and invariably whenever I got within five or six hundred yards of them they would raise their heads, gaze at me for a few moments, and then with much snorting and " cavorting " they would go off like the wind, and never slacken their speed as long as they were in sight.

In a small grove of timber where I had halted to rest awhile, I saw for the first time a horned frog. I had heard of the tarantula and centipede of Texas, and supposing the harmless frog was one or the other I picked up a stick about ten feet long (not venturing to approach nearer such a poisonous reptile) and mashed him as flat as a pan-cake. Continuing my course, about sunset I came to a belt of timber bordering another small stream.

On the bank of this stream there was an Indian encampment that appeared to have been occupied a day or so previously. Several of their fires were still smoking, and from their number I supposed there were thirty or forty in the party. Around these fires was scattered a great quantity of bones, mostly those of deer, though the head of a mustang here and there showed that they varied their diet by an occasional feast on horse flesh.

A cold misting rain had begun to fall just before I came to this camp, and seeing it was likely to continue through the night, I took possession of a shanty built of small poles and covered with slips of bark. In this I stowed myself and baggage and made myself perfectly "at home." With a large fire in front of it and plenty of hog, but no hominy, I passed a very comfortable night, serenaded as usual by wolves. Next morning the rain had ceased, and the sun was shining brightly when I woke up. Cooking a piece of my pork, I made a hasty breakfast for fear the owner of the shanty might return and ask me to pay for my night's lodging, and again started on my journey.

During the day I saw several "signal smokes," made I suppose by Indians, but they were a long way off. These signal smokes are curious things. Often when traveling over the plains of Western Texas, I have seen a column of smoke rise perpendicularly into the air (no matter how strong the breeze might be blowing) to a great height, when it would spread out at the top like an umbrella, and after remaining stationary for a moment "puff" it would suddenly disappear, to be answered perhaps by another, twenty or thirty miles away.

They are no doubt intended for signals to warn others of the proximity of foes, and to indicate their own position. I have asked many old Frontiersman how it was the Indians made smokes, but none of them could ever explain the matter satisfactorily to me. I have occasionally seen four or five of these signal smokes rising up in various directions at the same time.

To-day, for the first time, I saw what I know now was a tarantula, a very large and exceedingly venomous spider? that haunts the dry and elevated prairies of Western Texas. They are not often seen in the timbered lands or in the immediate vicinity of settlements. The body of a full grown one is as large as a hen's egg, and is covered with scattering hairs or bristles. They have two curved fangs protruding from the mouth, about as long and very similar in appearance to those of the rattlesnake.

When provoked they are very pugnacious, rising upon their hind legs and springing towards the assailant five or six inches at a time in successive leaps. The Mexicans say their bite is certain death, and one can readily credit the assertion after seeing them. I made but little if any progress to-day, for not long after I had started it clouded up and commenced misting again, so that I lost sight of the timber towards which I was steering my course. Finally I became completely bewildered and after wandering about all day I came to a belt of timber I had good reason to suppose was the same I had started from in the morning. At any rate the sun just then showed itself for a few moments, and I found I was traveling in the direction directly opposite the one I should have pursued.

It was too late to take the prairie again, and I picked out a suitable place for camp, started a fire and cooked some of my pork for supper, which for want of salt was getting to be rather too much tainted to suit the taste of any one but a Frenchman. During the night the wolves favored me with another concert of howlings, but they were much less impudent than upon a former occasion, and did not approach near enough to enable me to salute them with fire brands.

In the morning I rose betimes, and unpacking all the pork I had left, I spitted it on sticks stuck up before a blazing fire. I thought by roasting it in this way to keep it from spoiling entirely. The clouds had blown off and the sun shone out warm and pleasant, and having eaten some of my roasted pork which had decidedly too much of the "gout," I started out again across the open prairie. This time I made the trip without difficulty, and about mid-day I came to a small stream which I afterwards learned was called the Tres Palacios or Three Palaces. How it acquired the name I cannot say, but I am sure I saw no palaces in its vicinity.

Where I crossed it, I noticed a few small cedar trees growing near the bank, and I determined to cut one of them down and make a bow. This was no small job, as you may suppose, considering I had nothing to cut it with except a small piece of the blade of a drawing knife the same I had found at the house where I killed the hog, and which I had carried in my knapsack ever since. By the time I cut the sappling down, I was both tired and hungry, so I knocked off work to rest a while and cook some pork.

I then resumed my task, and chopping off about six feet from the butt end of the sapling I split it into four pieces with a wooden wedge and maul. From these I selected the one that was freest from knots and other defects, out of which, by patience and perseverance and with the aid of my piece of drawing knife I manufactured a very good bow. Arrows I knew I could easily get anywhere in the bottoms among the thickets of swamp dog wood or young cane. By the time my bow was finished night came on, and I pitched my camp near the creek in a little open space completely surrounded by a thick growth of underwood. Here I built my fire, warmed over some of my roasted pork, and after supper '-turned in" to a bed of Spanish moss which I had gathered from a tree near by.

The next morning I gave the finishing touches to my bow and then for the first time it occurred to me that I had nothing that would answer for a string. I tried to make one of the bark of several shrubs, and of the leaves of bear grass, but although I taxed my ingenuity to the utmost, I failed to make a cord strong enough for the bow, and I had at last to abandon the attempt altogether.

This was a great disappointment to me as I had calculated largely on supplying myself with an abundance of small game by means of my bow. I had heard of people having "two strings to their bows," and yet under the most pressing necessity I was unable to get one for mine which convinces me that things are very unequally divided in this world.

The day was so far gone when I had finished my unsuccessful attempt at cord making that I thought it best to remain where I was for the night and make a fresh start in the morning. It must have been twelve or one o'clock, when something awoke me, and finding that my fire had pretty well gone out, I was just in the act of getting up to throw some sticks on it, when I heard the stealthy but heavy tread of some large animal near by.

I laid still and listened attentively, and was convinced there was some heavy animal cautiously approaching the spot where I was lying. Just then, fortunately probably for me, a chunk rolled off a log I had placed behind the fire, and blazed up brightly.

By the light thus made, I saw distinctly either a large panther or Mexican lion, not twenty feet distant, crouching down as if about to spring upon me. I instantly jumped, and seizing my "bed clothes" (the dry Spanish moss I had gathered) I threw it on the fire and it blazed up at once as high as my head.

This must have frightened the animal, whatever it was, for when I turned to look it was gone. Possibly it did not intend to attack me, but the way in which it had approached me, was to say the least of it very suspicious. The loss of my "bed clothes" did not discommode me much, as I sat up the balance of the night to keep my fire supplied with fresh fuel, although the night was quite warm.

CHAPTER XYI

NARROW ESCAPE FROM INDIANS, REACH THE COLORADO RIVER AND SWIM IT OLD CANEY CREEK IMMENSE CANE BRAKE, ENCOUNTER WITH A WILD CAT GET INTO COMFORTABLE QUARTERS

As soon as the sun rose, I made haste to leave the locality where I had passed such an unpleasant night. Late in the evening I came to an extensive body of timber, in which I supposed I would find a considerable stream. On the edge of this timber I saw a house, and as by this time what remained of my pork was so strong of the "gout" that I don't think even a Frenchman would have relished it, I determined to go to the house and search for something to eat. I entered the woods some distance below it, and kept under cover until I was near enough to see there was no one about, when I ventured up. On entering I soon saw that it had been ransacked by the Mexicans, who had consumed or taken away whatever there might have been in it in the way of eatables.

In the vicinity, however, as I was leaving, I came across a half-grown hog, which evidently had very recently been shot by some one, who had taken only a small part of it, and I appropriated as much of what was left as I could conveniently carry. As the sun was about setting, I went some distance into the timber, so that the light from my fire would not be visible to any one passing along the prairie, where I "bivouackecl" for the night at the foot of a tree.

By sunrise I was up and on my way again, crossing in a mile or so a considerable creek. To-day I passed over a country mostly prairie, but interspersed here and there with groves of live oaks, hackberry, etc., which gave it a park -like appearance. In one of these groves, thickly settled with underbrush, I stopped to rest, and was just in the act of leaving it, when I heard the tramping of horses' hoofs and the jangling of spurs and other accoutrements.

Looking through the bushes I saw about twenty Indians slowly jogging along in single file upon their horses. They had no guns and were armed only with bows and lances. They rode within thirty paces of where I was lying low, but did not halt, and in a few moments they were hid from my view by another grove. I remained where I was half an hour longer than I would have done otherwise, in order to give these Indians full time to get out of my way, and then proceeded on my course.

A little before sunset I came to a clear running creek, on the farther side of which I encamped. (At that time, all the creeks and small water courses, and even the ponds in Western Texas were clear and pure, but now many of them have lost that character to a greater or less extent, owing to the cultivation of adjacent lands and the tramping of stock.) I had made my camp beneath some low spreading live oaks, which appeared to be a favorite roosting place for wild turkeys. Just at dusk they came flocking into them from every direction, and they were so unused to being hunted, I could easily have killed one with a pocket pistol but as I didn't have the pistol I had to content myself with roast pork instead of. roast turkey.

I had noticed before dark that a very extensive prairie lay to the north and east, and I was up and on my way the next morning before daylight, in order that I might reach the timber on the opposite side as speedily as possible. I ran but little risk comparatively when traveling in timber, but on the open prairie I was in constant danger of being picked up by parties of Mexicans or Indians.

I pushed on as fast as I could until noon, when I stopped to rest in a grove near a small lagoon that seemed to be well stocked with fish, for I saw numbers of bass and perch swimming in the shallow water near shore. On the margin of this lake I found some wild onions growing, which I dug up and ate raw, and which were a great treat to me, as I had not had anything in the vegetable line, fresh and green, for a long time.

In the evening I continued on my way across the prairie on the farther side of which I could see a long line of unbroken timber stretching from northeast to southwest, as far as my eye could reach. It was nearly night when I came to this timber, and I had gone but a little way in it, when I saw a large river before me, which I knew must be the Colorado.

The river was very high and rapid, and I thought it best to encamp for the night and wait until morning before I attempted to swim it. Where I struck it, it was about two hundred yards wide and much swollen by recent heavy rains, and although I was a good swimmer, I felt some hesitation the next morning in "taking water." However, I looked around and found a suitable piece of dead timber, to which I tied my boots and clothes, and launched forth with it on the turbulent stream, pushing it before me as I swam. Finally I made a landing safely on the north bank of the river, but was carried by the strength of the current a considerable distance below the point where I had entered the water.

After resting myself a while and drying my clothes, I took up the line of march again through a heavily timbered bottom about a mile and a half wide, from which I at length emerged into the open prairie. Without halting I continued on my course until late in the evening, when I came to the timber on old Caney Creek. Along this creek, which apparently in times gone by was the bed of the Colorado River, from its head to its mouth, a distance of sixty or seventy miles, there was a continuous cane brake.

Where I struck the timber on old Caney, there had been a considerable settlement, as some four or five houses were in sight, but on examination, I found that all of them had been plundered by Mexicans, who had taken everything of any value left on the premises. At one of these houses whilst searching the rooms to see if anything in the way of provisions had been overlooked by the Mexicans.

I heard a hen "squawking" as if some "varmint" was in pursuit of her. I stepped to the door to look out, and saw a hen racing around the yard and a very large wild cat following her closely. Having seen nothing eatable anywhere, except this hen, I determined to put in a "bid" for her myself, and picking up a billet of wood, I stepped out boldly towards the cat. When he saw me coming, he quit his pursuit of the hen, but showed not the slightest disposition to abandon the field. I advanced to within a few paces of where he stood humping his back and showing his teeth, and threw the stick I had in my hand at his head. I missed my aim, but struck him a severe blow on the side, and instantly he gave a scream and sprang furiously towards me.

I retreated precipitately and ingloriously for the house, which I reached just in time to rush into the door and slam it to in the face of the infuriated cat. If I had had a few feet further to go, he would have nabbed me to a certainty. The cat stopped some time in front of the door, as if he intended to besiege me in the house, or was bantering me to come out and give him a fair fight, which, under the circumstances, I declined doing, but after a while he went off leisurely towards the woods and I saw him no more.

In the mean time "the bone of contention," the hen, had gone to roost in tree near by. She undoubtedly owed her life to me, but for a very little while, for after dark I climbed up to her roost, grabbed her by the leg, and wrung her neck. With my prize, I retreated as speedily as possible to the house, for fear the wild cat might return and assert his claim to it again, and as I had no weapon I was very sure he would get the better of the contest and the hen too.

I remained all night at this house, and after breakfasting on the hen I had saved from the wild cat, I started off down the bottom to reconnoitre the country in that direction. When I had gone a mile or two I came to a small prairie connected with the main one by a very narrow neck and surrounded everywhere else by thick woods and cane brakes. This I concluded to explore, and after proceeding some distance in it, I saw there was a house at the farther end.

When I had approached within a hundred yards of the house, a half a dozen dogs came rushing out of it, seemingly with the intention of tearing me to pieces. I picked up a stick to defend myself, but when the dogs got near enough to see that I was an American, instead of attacking me they began to leap and jump around me as dogs do when they see their masters after a long absence. How they found out so quickly I was an American, I do not know, for exposure to sun and weather had tanned my complexion, until it was as dark as that of a Mexican or Indian.

With my escort of dogs I went to the house, and entering it, saw at once that the Mexicans had never been there, for everything remained, evidently, just as it had been left by the occupants furniture untouched, cases filled with books and articles of wearing apparel, cribs with corn and smoke house containing at least a thousand pounds of bacon. In a kind of shed room I also found a barrel of brown sugar and half a sack of coffee, and in the crib, besides corn, a quantity of potatoes and pumpkins.

There were a great many chickens and ducks in the yard, which no doubt, had been protected from "varmints" by the pack of dogs that still continued to escort me about the premises. In the smoke house as I have said, there was a large quantity of bacon, and the first thing I did was to take a "middling" and cut it up for the dogs. I then built a fire in one of the chimneys and in a little while had cooked for myself a first rate dinner together with a cup of coffee, the first I had tasted since leaving Goliad. After dinner I turned into one of the beds in the house and had a comfortable snooze.

When I awoke I got up and continued my investigations. In a back room I found quite a library, a rare thing at that time in Texas. I found also many articles of clothing in a closet, some of which fitted me tolerably well, and from which without any tear of being arrested for "petit larceny," I replenished my scanty wardrobe. Among other things I found in this house something I wished for exceedingly was a gun, but unfortunately it was without a lock, and consequently useless. Not far from the main building there was a row of log cabins, that evidently had served as negro "quarters," which induced me to believe that the place belonged to some well to do cotton planter.

As I had been much weakened by starvation and fatigue and the exposure I had undergone in my route through the wilderness, I concluded I would "stop over" a day or two at this house and recuperate my strength a little before I set out on my journey again. There were beds in several of the rooms, in one of which I slept at night, while my pack of dogs kept watch outside. These dogs were not mongrels or "curs of low degree," neither were they of the "suck egg" breed, as was evident from the fact that although they were in a starving condition when I came, and that the chickens had laid their eggs almost everywhere in the house and yard, not one had been touched by them for which I was thankful, being particularly fond of eggs myself.

CHAPTER XVII

IN CLOVER, TOO MUCH FRIED CHICKEN
PITCHING A TENT OVER A "DEN" OF RATTLESNAKES
FOLLOWED BY MY PACK OF DOGS AND COMPELLED TO RETURN, GAVE THEM THE DODGE IN THE NIGHT,
BUT ONE TRAILS ME AND PERSISTS IN KEEPING ME COMPANY
I NAME HIM "SCOUT."

I remained several days in my comfortable quarters, feasting on the good things I found in them, and reading books I selected from the library. On the evening of the third day of my sojourn at the house, feeling a little unwell (I rather think I had been indulging somewhat too freely in "fried chicken"), I concluded I would take a short stroll around my domains by way of exercise. After going a few hundred yards I turned to take a bird's eye view of my surroundings, and I exclaimed as Crusoe did on his island: *"I am monarch of all I survey, My right there is none to dispute,"* except, I mentally added, a marauding party of Mexicans or Indians, and now and then a wild cat.

Whilst passing through some tall grass, I came very near treading on a rattlesnake, the first I had seen in Texas, although some portions of the country I had passed over was much infested with them; but the season then was hardly far enough advanced to bring them out of the dens or holes in which they take up their winter quarters.

Often since, when passing over some of the uninhabited plains between the Nueces and Rio Grande rivers, I have found them so numerous in particular localities, that I was scarcely ever out of hearing of the sound of their rattles. They're not, however, nearly so vicious in Texas as they are in some other countries, and seldom attempt to strike, unless attacked. I have slept with them, ridden and walked over them frequently, and instead of trying to bite me they always did their best to get out of the way except on one occasion.

I was stalking some deer one day on the prairie, when I stepped upon a rattlesnake lying coiled up in the grass. I knew even before I saw it, by the peculiar soft squirmy feel under my foot that I had put it on a snake, and I promptly "lit out" without waiting for orders. As I did not wish to shoot him for fear of alarming the deer, and as they are easily stunned by a very slight tap on the head, I drew the ramrod from my rifle and gave his head a smart blow with it. I then mashed his head by repeated blows with the breech of my gun, and thinking of course I had killed him, I went on after the deer.

Two days subsequently when passing the place again, that same snake came very near biting me. I knew it was the same, for one of his eyes was out, and his whole head bruised and bloody from the blows I had given it with the breech of my rifle. I really believe he recognized me as the "author of all his ills," for when I attempted to go near him he would raise his head a foot or more from the ground, and with his rattles going incessantly, would glare at me with his one eye in the most vindictive way. I determined to make sure of him this time, and leveling my rifle at his head, I took good aim and fired. The bullet knocked his head into fragments, and one of the pieces struck me on the forehead, making a slight wound.

The idea immediately occurred to me that I had been struck by one of his fangs, and that I was fated to be killed by this particular snake. However, after bathing the scratch in a pool of water, and finding that my head had not swelled up as big as a bushel, I went on my way, congratulating myself upon my second escape from my vindictive foe.

On the morning of the fourth day of my sojourn at this house, I concluded I had regained my strength sufficiently to take the road once more, or rather the woods and prairies. Preparatory to leaving, I packed up as much sugar, coffee and bacon as I could carry, together with five or six pounds of meal, which I had ground upon a steel mill. I also put a tin cup in my knapsack, and several other articles which I thought would be useful to me. When ready to start I stuck a couple of carving knives (which I had also found* at this house) in my belt, and, bidding adieu to my dogs, after I had given them middlings enough to last them for a month.

I set out on my travels again. But, to my great dismay, when I had got a few hundred yards from the house, I found I had not consulted the wishes of the dogs about leaving them, and that the whole pack was following close at my heels suspecting I suppose, from the preparations they had seen me making, that I was going "for good." I tried to drive them back by throwing sticks and other things at them, but it was all to no purpose. They would stop whenever I did, but the minute I started they followed on. I knew it would be impossible for me to travel safely through a country in which I would be liable at any time to meet marauding parties of Mexicans and Indians with a half dozen dogs at my heels, and finding I could not get rid of them, I determined to go back to the house, wait there until night, and then quietly leave them.

So I returned, and passed another day very pleasantly at my house, looking over the books in my library, and cooking and eating at short intervals. Before I retired to my apartment, I noticed particularly where the dogs were sleeping, and about midnight I got up, quietly shouldered my pack of provisions, and left the house. I had gone perhaps half a mile down the edge of the cane brake when I heard the pattering of feet behind me, and in a few moments one of the dogs came up. I beat him severely with a stick, but he only whined and crouched down at my feet.

Finally, I determined to kill him with one of my butcher knives, but as I grasped him by the neck, and drew my carving knife, he looked up at me so piteously that I hadn't the heart to use it, and abandoned my murderous intention. I thought I could manage to keep one dog under control, and that the risk I ran of being killed or captured would not be increased to any great extent by having a dog with me; besides, I came to the conclusion that the company of a dog was better than none. Like the Frenchman, I think that solitude is very pleasant at times, provided there is some one with you to whom you can say "how delightful is solitude."

The dog that followed me was a very large "and powerful one a cross, I think, between the English bull and the Newfoundland. I found him to be tractable and, at the same time, as courageous as a lion, In a few days I had him perfectly under control; could make him lie down at a word, and remain at camp to guard it when I went off foraging or reconnoitering. I named him Scout.

CHAPTER XVIII

CAMP IN CANE BRAKE, TRY TO CUT A ROAD THROUGH THE BRAKE - FRIGHTENED BY A BEAR, FIND COMFORTABLR QUARTERS AGAIN SPLENDID BREAKFAST, VAIN SEARCH FOR A ROAD, CROSSING BRAKE NARROW ESCAPE FROM A PARTY OF LANCERS, GIVE MY DOG A LESSON.

After traveling a mile or two down the brake, I thought I had gone far enough to getaway from the other dogs, and I encamped for the balance of the night near a lagoon. I heard no wolves at this camp, but several times during the night I was roused by the noise made by some large animal forcing its way through the cane. I suppose it was a bear, as I noticed next morning a great many tracks in the soft ooze near the margin of the lagoon.

Whilst lying awake the next morning, upon my bed of dry leaves, my attention was drawn to a rustling among them, and turning them over, I found an ugly reptile about six inches long, which I thought then, and know now, was a centipede. Not fancying such a bed-fellow, I quickly dispatched him with a stick. They resemble somewhat the reptile called the "thousand leg worm," but they are much larger and flatter, and although they are well provided with legs, they have not quite a thousand.

They are of a dark brown color on the back, and the under side a dirty white. Their tail is forked, and has a long sting in the end of each prong, besides smaller stings on each foot, and, to complete their means of inflicting wounds, the mouth is furnished with fangs. They are a. disgusting looking "varmint," and are said to be very venomous. An old Texan speaking about them, said: "When they wound you with their feet alone, it hurts considerable; when they sting you with their forked tail it's a great deal worse, but when they pop you with all their stings, and bite you too say your prayers."

As soon as I had cooked and eaten breakfast and Scout had cleaned the dishes by licking them, I began to search again for a road that would lead me across the brake. Failing to find one after searching for several hours along the edge of the brake, I determined, if possible, to cut my way through it. I therefore attacked the cane, green briers and bushes with a carving knife, and after working faithfully till late in the day, I found I had gone about three hundred yards.

Such slow progress was exceedingly discouraging, for at that rate, if the brake was as wide as I thought it to be, I would be several weeks getting through it. There were a few scattering trees among the cane, and in order that I might be able to form some idea of the width of the brake, I climbed one of the tallest, from whence I could see an ocean of cane, extending at least four miles in the direction I wished to go, and beyond the scope of vision to the Northwest and Southeast.

The length of time and the amount of labor that I knew would necessarily be required to cut my way for so long a distance through this dense mass of vegetation, induced me to give over the attempt, and, descending from the tree, I took the path I had cut back to the prairie. Feeling considerably fatigued by my labors, when I got to the edge of the brake, I sat down at the root of a large tree to rest awhile.

Gradually I fell into a doze, from which I was suddenly aroused by the growling of Scout, and a scuffling, scratching noise overhead, and looking up, I caught a glimpse of some huge black animal sliding down the tree a few feet above my head. I sprang off quickly to one side, and at the same instant a bear struck the ground and took his way into the cane, which struck the ground and took his way into the cane, which popped and cracked as if a wagon was going through it. It would be hard to say which was the most frightened, I or the bear, and even Scout was so demoralized by his unexpected appearance that he made no attempt to pursue him.

The bear, of course, was up the tree when I took my seat at the foot of it, and as the tree was densely covered with Spanish moss, I had not noticed him. From my protracted stay at the foot of the tree, I suppose bruin had come to the conclusion that I was laying siege to him regularly, and getting desperate, he had charged down upon me in the manner I have related. Had I known it was a bear when I first caught a glimpse of him, I should not have been alarmed, as I had never heard tell of their attacking any one except when wounded and brought to bay. Several days afterwards, however, two of them exhibited such evident signs of hostile intentions towards me that I was induced to believe that they were not so non-combative as generally supposed.

After this little adventure, I continued on along the edge of the brake, hoping I might find some road or trail leading across it. I examined every nook and indentation, and finally came to quite a large trail leading from the open prairie towards the brake. Along this trail the old traces of wagon wheels were distinctly visible. I followed it for some distance running almost parallel with the brake, and at length came to where it abruptly turned and entered it. After crossing a strip of cane about two hundred yards wide, with a small lagoon near the center of it, spanned by a rude bridge of logs.

I came to a small prairie perhaps a mile in length and half a mile wide, a considerable part of which had been in cultivation. At the farther end of this prairie I saw a house, to which the trail I was following seemed to lead. When I had approached to within three or four hundred yards of the house, I halted for a few moments to make sure whether or not there was any one about the premises.

I heard the crowing of chicken cocks and the squealing of pigs, but as I saw no smoke issuing from any of the chimneys or any other signs to indicate that the house was occupied, I ventured up.There were a great many chickens, ducks and pigs in the yard, but no dogs came to welcome us. The house was a comfortable log building, consisting of four rooms with a wide passage between them and a broad piazza. in front, and was sheltered by some large live oak and pecan trees.

Everything in the house remained just as it was when abandoned by the occupants, which convinced me that it never had been discovered by the Mexicans. Indeed so secluded was the locality and so completely hidden from view by the strip of tall cane on the lagoon before mentioned, that no one passing along the main prairie outside would have suspected there was a settlement in the vicinity. This house was furnished even in better style than the one I stopped at last, which, together with the number of out-houses and negro quarters, convinced me it had been the residence of a wealthy planter. In the barns and cribs I found a large quantity of corn, potatoes, etc., and plenty of sugar and coffee in a store-room.

By the time I had made a thorough examination of the premises, the day was pretty well spent, and I determined to take up my quarters for the night in the house. Besides, it had clouded up and a cold, misting rain had begun to fall. I therefore proceeded to make myself at home without the least ceremony. I lolled upon the sofa, read the books, smoked a pipe (which the proprietor of the premises had left behind in the hurry of departure, with a box of tobacco), and after I had supped sumptuously on boiled eggs and peach preserves, I turned into a large double bed that looked as if it had just been spread for my special accommodation, and with Scout keeping watch at the door I slept like a prince until the sun was an hour high.

For my breakfast I had fried chicken, ash cake, boiled eggs, coffee and honey. After breakfast, I filled my knapsack with fresh provisions, and bidding adieu as I thought forever to these pleasant quarters, I set out again to search for a road that would lead me across the brake. Little did I think that five days would pass before I bade a final farewell to these quarters yet such was the fact.

All that day I searched for a road that would lead me across the interminable cane brake that barred my further progress. Occasionally I would fall into a cattle or deer trail leading into it, but they either gave out entirely after penetrating it a short distance, or else split up into half a dozen blind paths that did not seem to lead anywhere or in any particular direction. Wearied and disheartened by my failure to find a road, I returned to my domicile, feasted again on fried chicken, eggs, honey, etc., and again took possession of my double bed for the night.

The next day this same program was gone through with and the next, and the next, with the same results, and I almost began to despair of ever finding a way through this apparently endless wilderness of cane, briers and brush. However, it was some consolation to me to know that after the fatigues and disappointments of the day, I had such comfortable quarters to fall back upon at night.

Nevertheless, as I was very anxious to get on as speedily as possible, I left my domicile one morning with the determination that I would follow the brake up to the head of old Caney, providing I could find no road crossing it. I went on up the brake, examining closely every nook and indentation without success, until I had traveled, as I suppose, five or six miles. Here I struck out into the open prairie, to avoid a deep lagoon that lay in the way, and ere long I came to a well beaten road, running almost parallel with the brake.

This road had evidently been traveled a day or so previously by a large body of cavalry. I concluded I would follow it a short distance, and was going along leisurely, when I heard the clattering of horses' hoofs behind me, and turning to look, I saw a troop of Mexican lancers advancing rapidly, not more than four or five hundred yards distant. There was not a tree or bush to screen me, nearer than the brake, at least hah a mile to my right, and I knew it would be impossible for me to reach it before I was overtaken by the lancers.

For a moment I gave myself up for lost, but fortunately on one side of the road there was a patch of rank dead grass, and as there was no time for consideration, I seized Scout by the neck, dragged him twenty or thirty paces into the grass, threw him down and laid myself by his side, holding him tightly by the muzzle to prevent him from growling or barking at the lancers as they passed.

In a few moments they came up and when opposite the place where Scout and I were hidden, they halted. I could see them plainly through the grass, and could hear them talking, but not with sufficient distinctness to understand what was said. Scout, too, was aware of their proximity, and when they halted he gave a low growl, and tried to get up, but I choked him severely until he lay quiet. The lancers had evidently caught a glimpse of us before we left the road, for after they halted, several dismounted and examined the road for tracks, but luckily at that place the ground was gravelly and hard, and my boots had left no distinct traces on it.

At length, satisfied I suppose they had seen nothing, or what they had seen was only a couple of wolves or wild hogs, those that had dismounted to examine the road for "sign" sprang into their saddles, and they all rode on at a gallop. As soon as I saw they were fairly off, I drew a long breath, and I think Scout did so too, for I had choked him until his tongue lolled out.

When the lancers had got to a safe distance, I loosened my grasp from his neck and let him up. But he never forgot the lesson I gave him on that occasion, and whenever I wished him to lie down and keep quiet, I had only to place my hand on his neck, when he would crouch down and remain as still as a mouse until I told him to rise. Thankful for what under the circumstances seemed to me almost a miraculous escape, I took my way back to the timber* resolved that henceforth I would keep a better look out, and travel as little as possible in daylight, through the open prairies.

CHAPTER XIX

BEARS AROUND THE CAMP RETURN TO "MY" DOMICIL, MAKE MYSELF AT HOME AGAIN
LOBO WOLVES, ANOTHER START, JOINTED SNAKE,
A DROVE OF MUSTANGS
DISSAPPOINTED AND RETURN "HOME",
DIFFICULTY OF TRAVELING THROUGH THE PRAIRIES
MOCCASINS VERSUS SHOES, AGAIN IN SEARCH OF A ROAD
NIGHT ALARMS, IMMENSE TURKEY ROOST.

When ! reached the woods the sun was about setting and as it was too far to think of returning to my "domicil," selected a suitable locality and encamped for the night. During the night several large animals which I supposed to be bear came around camp, and the noise they made in the cane, kept Scout in such constant state of excitement, that I am sure he got but little sleep.

The next morning, I retraced my way down the brake, and about midday reached my quarters, where I found everything as I had left it the day before. After feasting again on fried chicken, sweet potatoes and hot coffee, I took a seat on the porch, with a volume of Don Quixote (which I read for the first time at this house), and cocking my feet up on the bannisters, I made myself comfortable for the rest of the evening.

Whilst I was thus taking "mine ease in mine inn," it occurred to me that if Mahomet couldn't get to the mountain, perhaps the mountain might come to Mahomet in other words, if I couldn't get to the Texan army, perhaps it would be just as well to remain where I was until the Texans whipped the Mexicans and re-occupied the country.

That they would do so eventually I had not the slightest doubt, although the Mexicans had told us when prisoners at Goliad (for the purpose of discouraging us and preventing us from making any atempt to escape), that Santa Anna had defeated Gen. Houston's army, and that the whole country was virtually in their possession. But in fact I did not seriously entertain for a moment the idea of remaining any longer where I was, comfortable as were my quarters, than I could possibly help; for I knew very well I would not be satisfied with such an inactive life, when my countrymen were all in the field battling against the merciless foe. So I retired to my sleeping apartment that night with the determination of renewing my search for a road the next morning, and to persevere in it until I succeeded.

During the night I heard the howl of several "lobo" wolves very near the house, but of course I did not fear them within the walls of my castle. The fact is, I did not fear anything except a visit from marauding parties of Mexicans or Indians, against whom neither the log walls of my castle nor my two formidable looking carving knives would have afforded me much protection. Audubon, who is a recognized authority upon the subject of birds, if not of beasts, told me that the lobo was the largest known species of wolf in the world, and certainly they are much larger than any on the American continent.

They resemble the hyena in form as much or more than they do that of the common wolf. Their howl is also very different, and when camping out alone on the prairies, it always seemed to me to be the most mournful, doleful and "lonesome" sound I ever heard. Several instances have been known since the settlement of Texas of their attacking travelers when benighted on the prairies, and I was once myself with a party of rangers who rescued a way-farer from their clutches, and who, but for our timely arrival, would undoubtedly have been torn to pieces by them.

Nothing else occurred to disturb me during the night, and the next morning I rose betimes, and as soon as breakfast was over I shouldered my knapsack and set out, intending to make a thorough search for a road along the edge of the brake below. In the bottom to-day I noticed that many of the trees were putting forth their leaves, an indication that spring had fairly set in, and a variety of wild flowers were also beginning to make their appearance on the prairie.

To-day I came across a specimen of the jointed snake, the first I had ever seen. It was a small snake, not more than fifteen or twenty inches in length, and its skin had a vitrified or glassy appearance. It seemed to be rather sluggish and unwieldly, and when I struck it a slight tap with a small stick, to my great astonishment, it broke into half a dozen pieces, each piece hopping off in a very lively way "on its own hook." I have since heard it asserted, that after a time the broken parts of the snake will come together and reunite and then crawl off as if nothing had happened to it; but I shall always be doubtful of the story until satisfactory vouch ers of its truth, duly authenticated and sworn to, are produced.

About midday I noticed a cloud of dust rising in the prairie way off to my right, caused, as I at first supposed by a large body of troops in motion. I was traveling near the edge of the cane brake, both for greater security and for fear I might pass by without observing it, some road leading across. I therefore quickly concealed myself behind a small thicket, from whence I could see all that was passing on the prairie.

Presently I saw issue from the cloud of dust a dense body of horses, which, on their nearer approach, I perceived were "uncurbed by bit and riderless." I supposed there were at least six or seven hundred in the drove. I saw they would pass within a short distance of the thicket where I was concealed, and when nearly opposite, I suddenly sprang out in full view of them and gave a loud whoop. They halted at once and with heads erect, stood for an instant looking at me in astonishment, then with the precision of a troop of cavalry, they wheeled about and went back in the direction they had come.

I continued on my way, and when I supposed I had traveled at least six or seven miles from where I had started, to my great joy I came to a plain road, running from the prairie into the brake. I felt confident it would take me through it, but when I followed it a hundred yards or so into the brake, it came to an abrupt termination at a place where a large tree had been cut down and split into boards! There was not a vestige of a road beyond that point nothing but almost solid walls of tall canes matted together with green briers and vines.

Sadly disappointed and dispirited, I retraced my steps to the prairie, and thence back towards what I began now to regard as my permanent home, where I arrived a little after sunset, so "beat out" with my day's tramp that I turned into my bed supper-less, and slept like a log until roused at daylight by the crowing of my chickens and the squealing of my pigs.

It may seem strange to some, that one accustomed to walking as I was, and after living upon the "fat of the land" as I had of late, should have been so much fatigued by a little tramp of twelve or fifteen miles but that was precisely "what was the matter with Hanna." After starving for so long a time, I had indulged too freely in "fried chicken;" and besides, walking through the woods and prairies is not like traveling on a well beaten road.

In the former your progress is often necessarily slow and laborious on account of having to force your way through rank grass and many creeping vines, that are constantly entangling one's legs, and occasionally tripping one up. Moreover the soles of your shoes soon become as slick as glass by rubbing on dry leaves and grass, so that you are frequently slipping backward instead of going forward.

Being determined to persevere in my attempt to find a road that would enable me to cross the brake, the next morning I shouldered my knapsack, and set out again in the direction I had taken two days previously when I made such a narrow escape from the lancers. Scout evidently seemed to think I was wandering about in a very aimless way, nevertheless he trotted along after me without asking any questions.

I traveled up the brake a mile or so beyond the point where I had turned back on the former occasion, examining closely every nook and bend for trails or roads. In this way I discovered one or two that had escaped my observation on my previous trip, but they "petered out" after going a short distance into the cane. Finding no road or trail to answer my purpose, and night coming on, I encamped in some timber near the edge of the cane.

A little after dark I heard a great many turkeys flying up to roost in the trees around my camp. The wolves howled incessantly, and once the sharp scream of a panther close by roused Scout from his slumber and he dashed off in the direction of the sound, but very soon came running back with his tail between his legs. It was evident he wanted my "moral support," but I declined hunting panthers in the night with a carving knife. I felt no fear of them, however, in camp, as I had a blazing fire, which I took especial care to keep well supplied with fuel.

I have been told that in India tigers have been known to come up to camp-fires and seize upon persons sleeping near them. This may be true, but there is no wild beast (with the exception perhaps of the grizzly bear) on the North American continent that will venture so near a blazing fire at least I have never heard of an instance of the kind during the many years I have lived on the frontiers.

At daylight I was aroused from my slumbers by the clucking and gobbling of turkeys. There must have been several hundred of them upon the trees within fifty yards of where I was lying. One fat old fellow was sitting upon a limb not more than thirty feet from me, strutting and gobbling in the most impudent way. It seemed to me he knew I was particularly fond of roast turkey, and that he was "cutting up his didoes" for no other purpose than to tantalize me with the display of his goodly proportions. Even when

I got up and walked towards him, he took no notice of me, until I threw a stick at him, when he uttered an exclamation something like "what!" and soared away to his feeding grounds.

After breakfast I continued my route along the edge of the brake. When I had gone about two miles, I noticed a house on the prairie near a small grove of timber, half a mile or so to my left, and I concluded to go out and examine the premises. The house was a small log cabin, surrounded by an enclosure containing perhaps a dozen or fifteen acres. It was poorly furnished and I saw nothing about the premises except some ducks and chickens.

CHAPTER XX.

I LAY IN A SUPPLY OF POULTRY,
I ASTONISH A COUPLE OF MEXICANS,
THEIR SINGULAR MOVEMENT
"COUNCIL OF WAR"-, RETURN AGAIN TO MY OLD QUARTERS ,
FIND A ROAD ACROSS THE BRAKE AT LAST ENCOUNTER WITH TWO BEARS.

As I did not know how long it might be before I should have a chance at "fried chicken" again, I determined to take toll out of the poultry about this house. With the assistance of Scout I soon caught and killed two fat pullets and a duck, which I tied on the outside of my knapsack. I then took a plain road running near the house and nearly parallel with the brake, and when I had gone about a mile I met with an adventure that terminated in the most singular and unaccountable manner.

The road at that point was about a quarter of a mile from the brake. How it happened I did not see them sooner, I cannot imagine, unless I had fallen into what the negroes call a "fit of the mazes," but at any rate I suddenly found myself nearly opposite to two Mexican soldiers who were seated on the grass about forty paces to the left of the road.

One of them was armed with a musket and the other with a lance, similar to those I had seen used by Mexican cavalry. Near them a horse, saddled, was grazing, and one of the soldiers held the end of his lariat in his hand. I have since thought the horse must have been lying down until I came near them, as otherwise I think I would have seen him sooner. As I have stated, it was a quarter of a mile at least to the nearest part of the brake, and the idea flashed across my mind that after all my narrow escapes I was certainly caught at last.

Retreat to the brake I knew was impossible, as they could easily overtake me on the horse, and for a moment I stood irresolute not knowing what course to pursue. But the very hopelessness of the case produced a feeling of recklessness as to consequences, and I leisurely continued my way along the road; at the same time trying to look as unconcerned as possible and as if I didn't know (and didn't care) that a Mexican soldier was within five miles of me. All the while however, I was watching them closely.

As I passed them, they made no movement except to turn eir heads and gaze at me apparently in the utmost astonishment, which considering the figure I cut, just at that time, is not to be wondered at. There is not the slightest doubt that I presented a very singular and anomalous appearance. I was tanned by long exposure to sun and weather until I was nearly as dark as an Indian; my cap resembled a Turkish turban, the leather front having been long since carried away in some of its frequent encounters with green briers and other thorny shrubs; my hunting shirt was ragged and blackened with smoke, and my pantaloons, or what remained of them, were buttonless, and held up by a broad leather belt, from which a tin cup hung dangling on one side and two long carving knives on the other, and to complete this unique costume, my shoulders were surmounted by a portly knapsack, to which were tied the two pullets and the duck I had just killed.

This "tout ensemble" of course accounts reasonably enough for the astonishment with which the soldiers gazed upon me as I passed, but still it does not satisfactorily explain their subsequent movements, especially as they could plainly see that with the exception of my two carving knives, I had no arms.

However, they did not move until I had gone forty or fifty yards beyond them, when both suddenly rose to their feet and hastily mounted their horse, one behind the other. I of course supposed they intended to pursue me, but to my great wonder and astonishment as well as relief, they went off in the opposite direction, across the prairie, as fast as they could urge their horse on with whip and spur.

The one mounted behind had a short heavy whip called a "quirt," and as far as I could see them distinctly, his quirt was incessantly and vigorously applied to the flanks of their steed, and every now and then I could see them looking back as if they expected me to pursue them. What they took me for I am at loss to imagine, but if they had taken me for Old Nick himself I would not have quarreled with them on that score, in consideration of the expeditious manner in which they had left the field not staying even to say "adios." For fear, however, I might not prove to be such a terrible object to other straggling parties of Mexicans whom I might possibly meet with on this road, I left it, and did not halt until I came to the brake.

There I stopped to rest a while, and hold a "council of war" with Scout, as to what was to be done next. Scout, although he expressed no opinion on the subject, I know was strongly in favor of going back to the "flesh pots of Egypt," and finally we agreed to return to our old quarters. I had noticed an old axe there in one of the outhouses on the place, and I determined to set to work regularly and cut my way with it through the brake, if it took me a month to do it.

It seemed very strange to me at the time, that the settlers on old Caney did not cut roads through it when they retreated before the Mexican army. But subsequently, when I mentioned the matter to one who lived on Caney when the settlers abandoned their homes there, he told me that all living on the south side had cut roads from their houses across the brake, but that in every instance they had some circuitous way to reach them, and that no sign of a road was visible on the edge ot the brake. This statement was confirmed to some extent by the fact that no one, unless closely searching for it as I was, would have suspected the existence of a road where I found one.

In pursuance of the course I had determined to follow, after resting a while, Scout and I started back to our old quarters, and about an hour before sunset I crossed the strip of ca.ne and the bridge of logs over the bayou and entered the little prairie in which my domicil was situated. As I was proceeding leisurely towards the house, it occurred to me that it might be well to examine again the north side of the prairie bordering the main brake which heretofore I had only partially done. With this intention I left the path I was following, and when I had gone a few hundred yards I came to a trail leading towards the brake along which the marks of wagon wheels were dimly visible. This I followed until it led me into an indentation in the brake, which was so narrow and so well concealed by bushes and eane as to be barely perceptible at the distance of a few paces.

Still following the traces of wagon wheels, I came on the farther side of this nook to a newly cut road wide enough for the passage of a wagon and team. I was satisfied that at last I had found what I had been so long in search of, but in order to assure myself of the fact, I followed the road for nearly half a mile into the brake, and as it still ran on in the same direction I was convinced it would take me through. By this time the sun had set, and I concluded to return once more to my old quarters, and make an early start in the morning.

As I walked along my attention was suddenly drawn to two large black objects in the road a short distance ahead of me. I stopped a moment to ascertain what they were, and as I did so, Scout gave a low growl and retreated behind me. By the dim light that struggled through the overlapping canes I at length discovered that these black objects were two large bears, standing perfectly still in the road, and apparently waiting for us to come up. For an instant I thought of retreating, but on reflection, as I had never heard of any one being attacked by black bears unless wounded, I screwed up my courage (nearly breaking the screw-driver in the attempt) and resolved to pass them if I could.

There was no chance to go around them, for the cane was so thick on both sides of the road, I might almost as well have tried to penetrate a solid wall. So I drew my longest carving knife, and boldly (apparently) advanced towards them.

They stood perfectly still until I was within eight or ten feet of them, when they commenced growling, and looked so large and ferocious, and so bent on disputing my right of way, I felt more than half inclined (as Scout had done already) "to tuck my tail" and beat a hasty retreat. But I knew it was too late to turn back, and that any show of timidity would embolden them to attack* if they had not intended doing so.

I therefore continued to advance, and my apparent boldness seemed to daunt them a little (if they had only known how badly I was scared I am sure they would have seized me) and when almost near enough to have touched them, one of them sullenly drew off to one side of the road and one to the other, and Scout and I passed between them. As we went between them, they showed their white teeth and growled so fiercely that every instant I expected they would rush upon us, but they did not, nor did they attempt to follow us. All the while Scout kept close at my heels with his tail between his legs the first and last time I ever saw him completely cowed.

It is asserted that the black bear never attacks a man, unless wounded or brought to bay, and I do not say positively that these two had any intention of making their supper on us, but to say the least of it, their bearing towards us was exceedingly suspicious; and besides, I thought they might just as well kill a fellow at once as to scare him to death. At any rate Scout and I congratulated ourselves (at least I know I did), when we were once more safe within the four walls of our house.

CHAPTER XXI

PREPARATIONS FOR THE ROAD SETTLE MY BOARD BILL WITH A PROMISORY NOTE
BID FAREWELL TO MY COMFORTABLE QUARTERS AND LEAVE IMMENSE CANE BRAKE/ DODGE A MEXICAN SOLDIER AND A PARTY OF INDIANS BEARS AROUND CAMP.

I rose early next morning to prepare as much provision for the road as I could conveniently carry. I cooked the duck and one of the pullets I had killed the day before (Scout and I had demolished the other for breakfast), and ground a gallon or so of meal on on a steel mill. Besides these, my supplies consisted of five or six pounds of bacon, several pounds of sugar, two pounds of coffee parched and ground, some salt and pepper, and two bottles of honey.

This, I thought, with care would last us eight or ten days, even if we found nothing on the road. I also had a tin cup for making coffee, and of course my two carving knives which I had sharpened on a whetstone were as keen as razors. For these I had made scabbards out of a piece of leather and sewed them to my belt. When ready to start I scribbled with a bit of charcoa the following "due bill" upon the wall of my sleeping apartment: ' an American captured by the Mexicans but escaped from them at Goliad, is indebted to the proprietor of this house for one week's board and lodging and some extras, and will pay the same on demand." The extras referred to consisted of articles of clothing, pipe and tobacco, etc.

This note has never been presented for payment, and I suppose it is barred now by the statute of limitation nevertheless, I would cheerfully pay the principal now but not the interest, for that would put the amount far above my present assets, and I should be compelled to take the benefit of the Bankrupt Act. Having thus settled my board bill on such easy terms, I shouldered my knapsack, stuck my carving knives into my belt, and followed by Scout, I took my way towards the road I had found the evening before.

Just as I was entering the brake, I turned to take a last look at the house that had been a haven of rest to me after my wanderings in the wilderness, and I experienced a feeling of regret when I thought that in all probability I should never see it again. There I had truly been "the monarch of all I surveyed." I could loll upon the sofas tumble up the beds wipe the mud from my boots on the rugs and carpets smoke tobacco (by no means of the best quality) in the drawing room select my own "menu" from the well stored pantry and the poultry in the yard and there was none to say me nay. Even now I look back with pleasant recollections to my sojourn in those comfortable quarters, for it was the only time I ever had complete and undisputed control of such an establishment. "Peace to its ashes" if, as is highly probable* it was subsequently burned by the Mexicans.

As I passed the place where I had encountered my doubtful friends the two bears the evening before, I noticed many of their tracks in the mud on the side of the road. They were very much like the tracks made by a bare-footed man (no pun intended for I detest puns), except that the heel part was as long as the toe. After traveling I suppose between three and four miles, crossing on the way a sluggish bayou, over which I "cooned it" on a fallen tree, to my great satisfaction I saw light ahead, and in a short time came to the open prairie.

At that day as I have before stated, nearly the whole of the bottoms on old Caney was covered by an unbroken cane- brake sixty or seventy miles long and from three to five in width. This I had from others who were settlers in that portion of the country at an early day, and the statement is probably correct. The soil of this brake is exceedingly fertile, and the time will come no doubt, when it will be converted into one continuous sugar and cotton plantation. At the points where I saw it, it was a dense mass of cane, briers and vines, with here and there a scattering tree growing in their midst. Bears, panthers, wild hogs and other "varmints" were very numerous in it and along its borders.

About half a mile below the place where I came out into the open prairie, I saw a house near the bottom, and as I had made it a rule to search every one I passed for guns and ammunition, I started with that intention towards the one in question. I kept well under the shelter of some timber bordering the brake, to screen myself from the view of any one who might be about the premises. In this timber I struck a plain trail leading towards the house, which I took. I followed it perhaps a hundred yards or so, when as I turned a short bend in the path, I caught sight of a Mexican soldier, with his gun on his shoulder, walking rapidly towards me. Luckily a dense growth of bushes bordered the path at the point where I then was, and although I had but little hope the Mexican had not seen me.

I instantly sprang into the bushes and laid down among them. Scout, who evidently had not forgotten the choking I gave him on a previous occasion, quickly followed me, and took his station by my side. It seems, however, the Mexican did notice us, for he came on, and passed within six feet of us without halting. I could almost have touched him with my longest carving knife, and if he had been a little weakly chap I think I would have been tempted to spring suddenly upon him as he passed and give him a tussle for his gun, but he was a big strapping fellow, and I knew I would have no chance of coming off winner in a hand to hand encounter with him, even if I had not been hampered with a heavy knapsack, and other "impedimenta." I coneluded therefore that "discretion was the better part of valor," and did not move until he was hidden from my view by a turn of the path.

As it was evident he came from the house I had seen, and as I thought it highly probable there were "more ot the same sort" there, I gave up the idea of searching it for guns, for fear I might find more of them there than was desirable; so I gave it a wide berth, and striking off through the woods to the right I came out again to the prairie two or three miles below. The day was cloudy and dark, and I couldn't see the timber on the opposite side; consequently. I could form no idea of its extent. Besides (having made a late start on account of being delayed in preparing provisions for the road), the sun was by this time getting pretty low, and I thought it best to encamp for the night and start anew in the morning.

In a little open space just within the brake, separated from the prairie by a very narrow strip of cane I pitched my camp; in other words, I pulled off my knapsack, and stretched myself upon a bed of dry grass which I had cut with a knife. It was too early to cook supper, and as I had no dread of wild beasts till dark, I did not start a fire, and very fortunate it was for me I had not done so. I was just falling into a doze, when Scout gave a low growl and at the same moment I heard the tramping of horses' hoofs.

I looked through an opening in the strip of cane between me and the prairie and saw five or six Indians who were driving a number of horses, coming along the edge of the brake. Just as they were opposite to the spot where Scout and I were lying, two of the horses broke away from the "caballada," ran through the strip of cane and nearly over us. One of the Indians started after them, and was crossing the strip of cane, when the two runaways seeing Scout and I lying upon the ground, suddenly wheeled and ran back to the prairie, and the Indian turned also and followed them. If he had come six feet further he must inevitably have seen us. As it was he did not discover us, and the Indians and their drove of horses soon passed out of sight.

These two "close calls" both occurring the same day, convinced me that I had but little chance to make my way safely through a country swarming with roving bands of Mexicans and Indians; and yet, although I passed their recent encampments at several places, I never saw an Indian afterwards, nor a Mexican, except some squads of cavalry a long way off on the prairie.

During the night I heard bears crashing through the cane, and splashing in the water of the pool near which I was encamped. The number of bears at that day on old Caney was so great I cannot imagine how the settlers there managed to raise hogs unless they kept them constantly penned up. The next morning I saw many of their tracks on the edge of the pool, where they had been digging up some kind of plant with a bulbous root.

CHAPTER XXII

PRAIRIE ON FIRE I START A COUNTER FIRE, NARROW ESCAPE FROM DROWNING
LOSE MY KNAPSACK AND ALL MY "GRUB", BUT SAVE MY TINDER
A MIRAGE MEXICAN CAVALRY, MEET WITH TWO SPIES FROM THE TEXAN ARMY AND RETURN WITH THEM.

The next morning as soon as I had eaten breakfast and Scout had "cleaned up" the fragments. I set off towards a long line of timber that was just barely visible on the farther side of the prairie. Not a great while after I had started I noticed a long way off to the west, a column of smoke rising up, which I supposed indicated an encampment of Mexicans or Indians in that quarter. When I had traveled perhaps three or four miles, I observed that this smoke was increasing rapidly in volume and extent, and that it appeared to be approaching the direction I was going.

Then, for the first time it occurred to me that the prairie was on fire, and I began to be seriously apprehensive that the fire might overtake me before I could reach the timber. The grass of the last season's growth was from fifteen to eighteen inches in height, and as dry as tinder, and it seemed very probable, with the stiff breeze blowing at the time, that the fire would overtake me before I could gain the opposite side of the prairie, still five or six miles distant.

I hurried on as fast as I could, but before I had gone two miles further, I was convinced that escape by flight was impossible. I had heard old frontiersmen say, that the only thing to be done in a case of this kind, was to "fight fire with fire." I took my flint and steel from my pocket, ignited some tinder which I wrapped in a wisp of dry grass, and swinging it quickly backwards and forwards in my hand, it was soon in a blaze. With this I set fire to the grass ahead of me, and in a few moments I had the satisfaction of seeing my counter fire sweeping the grass that grew in the direction I was going.

By this time the wall of fire extending in a long line across the prairie behind me, was swiftly moving towards me. Already I could see bright tongues of flame flashing out at intervals through the dense column of smoke, and a dull continuous roar, like the distant beating of surf on a rock bound shore, was distinctly audible. Hundreds of deer, antelope and other animals came scampering by me in the wildest terror, and numerous vultures and hawks were seen hovering over the smoke, and occasionally pouncing down upon rabbits and other small animals, roused from their lair by the advancing flames.

The nearer it came the faster it seemed to come, and I could see blazing tufts of grass borne along by the wind setting fire to the prairie sometimes fifty or a hundred yards ahead of the main fire. But by the time it had reached the place where I had set my counter fire going, the grass for several hundred yards was burnt off, and of course the fire was arrested there for want of fuel. I had nothing to do but follow the track of the fire I had started, which cleared the way before me as I went, and rendered walking much less fatiguing than it otherwise would have been verifying the truth of the old saying "that it is an ill wind that blows no good."

In about two hours after I had set my counter fire going, I came to the outskirts of the timber for which I had been steering, and through which I continued my course until I was stopped by a deep bayou. On the bank of this bayou in a little open space not twenty feet square, I pitched my camp, and from the fallen trees around I collected fuel enough to keep my fire going all night. There I soon prepared a meal from the provisions I had in my knapsack, to which 1 and Scout did ample justice as we had not tasted food since early in the morning. As it was still several hours till night ?

I employed myselt in repairing my dilapidated wardrobe with a needle and some thread I had found in my house on Old Caney.

Nothing occurred to disturb my slumbers during the night. The next morning after breakfast I shouldered my knapsack and started again. The bayou on which I had camped, though the current was very strong, looked so narrow I thought I could easily swim it without taking off my knapsack; so I plunged in at once, but unfortunately when I had about reached the middle of the stream, one of the straps that held it in position gave way, and in an instant the rapid current twisted it around my neck, and I went down with it like a stone to the bottom. I exerted myself to the utmost to free myself from it but without success, until I thought of my carving knives.

With great difficulty I drew one of them from the scabbard (it seemed to me that everything about me was tangled up) and cut the strap that fastened the knapsack around my neck. The moment I was freed from it I rose to the surface, puffing and blowing; like a porpoise, and half strangled with the water I had swallowed much against my will, for I was not in the least thirsty. Scout having no knapsack to encumber him, had already reached the opposite shore, and was running up and down the bank, whining most dolorously, and showing plainly his anxious concern for my safety. I quickly gained the shore myself after coming to the surface, but alas! I was compelled to leave my precious knapsack which contained our whole supply of provisions, at the bottom of the bayou. However, I was very glad to get out of the scrape as well as I had done.

The first thing I did after reaching the shore, was to examine the condition of my tinder, and I was glad to find that but little water had penetrated the greased cloth in which it was wrapped. I took it out and spread it in the sunshine, so that what little moisture it had imbibed might evaporate. If I had lost my tinder as well as my provisions, I would have been in a truly pitiable condition.

When I had partially dried my clothes, I set out again in my usual direction, which led me for some distance through a thick growth of underbrush, from which I finally emerged into open post oak woods. I went on through these until nearly sunset, when the howling of wolves warned me that it was time to select a suitable place to encamp.

I chose a spot in a thick grove on the margin of a pond. There I started a fire, and as I had to go to bed supper-less, I determined that at any rate my bed should be a good one. With one of my carving knives I cut a quantity of long dry grass, which I spread before the fire, on which I and Scout after the mishaps and fatigues of the day slept soundly till morning.

As soon as it was daylight, as I had no breakfast to cook and eat, I was on my way again, and in a little while I came to a prairie, on the farther side of which I saw a forest and a large lake near it. Towards this lake and forest I steered my course, but after traveling some distance, I was astonished to find that apparently they were as far off as when I first saw them.

Whilst I was wondering at this, I noticed that the lake and forest were each moment growing more indistinct, and at length they vanished altogether, and in their place nothing was visible but the level expanse of the open prairie. I knew then that the appearance of this lake and woods was an optical illusion termed a "mirage." produced by some peculiar state of the atmosphere. I have frequently seen them since, on the plains in the west, and on several occasions have been cruelly tantalized when suffering from thirst, by the sight of lakes that disappeared before I could reach them.

After traveling a while longer, I saw some distance ahead of me a grove, and still further on a forest was dimly visible. At first I thought it probable that these also were only the illusion of a grove and forest, and that they too would disappear and give me the slip, but they proved to be the "genuine articles." To this forest I steered my course, guided by the intervening grove.

I saw several squads of Mexican cavalry on the way, but they did not come near me, and I avoided observation simply by lying down on the ground, until they had passed by. But what astonished me much was, that these squads were all traveling in a disorderly manner towards the west. It soon occurred to me, however, that the Mexican army must have met somewhere with a signal defeat, and that those I saw were straggling detachments from their routed forces. I have no doubt this supposition was correct, for the battle of San Jacinto, in which Santa Anna was taken prisoner, was fought and won by the Texans under Gen. Houston, a few days previously.

About noon, I came to the grove that had served me as a landmark to guide me on my course, and feeling somewhat fatigued, I laid down just outside of it to rest a while. I had been there but a few moments when I had practical evidence that the vast distance at which the buzzard is said to see a carcass on the ground, had not been exaggerated. When I laid down not a buzzard was in sight, although I had an unbroken view for miles in every direction, but in less than five minutes, half a dozen of them were wheeling and circling above my head, and coming lower and lower, evidently for the purpose of ascertaining if Scout and I had been killed long enough to suit their fastidious taste. "My friends," said I, "on this occasion you are a little too 'previous' you have come very near several times having the satisfaction of picking my bones, but to prove to you that I am not as yet a fit subject for*a 'post mortem' feast, I'll move on." The first movement I made satisfied them on that point, and they departed as quickly as they had -come.

Continuing my course, about sun set I came to a deep and rapid stream, which I know now was the San Bernard, and I encamped for the night on the bank. By this time I was suf fering much from hunger, but there was nothing in camp to eat, and I and Scout were compelled to satisfy the cravings of our appetites as well as we could, by going to sleep.

The poet calls sleep "tired nature's sweet restorer," and under ordinary circumstances, no doubt there is some truth as well as poetry in the saying, but when a fellow has had nothing to eat for several days, and his bed is the naked ground, sleep as a restorer isn't a marked success at least I was just as tired and hungry when I woke up the next morning as I was when I laid down. However, I was in hopes that I might find a settlement on the other side of the river where something to eat could be had, and without any preparation except simply tying my cap on my head securely to keep my precious tinder from getting wet, I plunged into the turbid stream closely followed by Scout. The water was very cold, but I soon crossed over and ascended the bank that rose up almost perpendicularly thirty or forty feet on that side of the stream.

When I got to the top of the bluff, I discovered a house a few hundred yards above me, to which I turned my course. As it was all open prairie on that side of the river except a few scattering groves, I had a good chance to reconoitre the premises before approaching them, and seeing nothing to indicate that the house was occupied, I went up. It proved to be a single log cabin, in rather a dilapidated condition, and had been ransacked by some plundering party of Mexicans who had taken or destroyed any provisions that might have been there, except a handful of corn I found in a barrel. As I was thoroughly chilled after swimming the 'river, I concluded I would build a fire in the chimnev for the double purpose of drying my clothes and parching the corn I had found.

There was but one door and one window to the cabin, both on the same side, and while 1 was busily engaged in parching corn, my attention was drawn to a grating sound in the direction of the window, and turning to look, I saw the muzzle of a gun protruding through it. But Scout had noticed it, too, and giving a savage growl, he sprang at one bound through the window, and at the same instant almost I heard some one rip out an oath in good, King's English, and exclaiming "come take your dog off," in such choking accents as convinced me there was urgent need of haste. I ran out immediately, and with some difficulty forced Scout to let go the grip he had taken upon a thick woolen comforter, which fortunately for him, my visitor had wrapped around his neck.

After he had somewhat recovered from the surprise and alarm into which the unexpected onset of Scout had thrown him, he asked me where I was from, and how I came to be out there all alone among the Mexicans and Indians. When I had satisfied him on this point, he told me that he and a Capt. D were out on a spying expedition, and seeing a smoke coming out of the cabin chimney where I was carry ing on my culinary operations, they had come to the conclusion that a party of Mexicans had halted there.

After a consultation as to the best mode of proceeding, it was determined that Capt. D should remain with the horses under cover of a grove a few hundred yards from the cabin, whilst his companion, Mr. H , should cautiously approach it on foot, and ascertain the strength of the party within. If too strong for them to contend with, he was to fire upon them through the door or window and then make his retreat as fast as possible to the grove where he had left Capt. D and the horses. But in arranging this programme, they did not consult Scout, who revenged himself in the manner I have stated. Afrer giving me this information and telling me that the Texans had whipped the Mexicans at San Jacinto,

Mr. H gave a whoop (the preconcerted signal for Capt. D to come on), and in a few moments he rode up, leading H 's horse and another one, which to my great satisfaction I found was well packed with provisions. As I have stated, I already had a fire under way, and in a little while a pot of coffee was simmering on it, and a haversack of eatables, biscuits, potatoes, cold ham, etc., was spread upon the floor. Those biscuits! I shall never forget them! None of your little thin flimsy affairs, such as are usually seen on fashionable tables, but good solid fat fellows, each as big as a saucer, and with dark colored spots in the center, where the "shortening" had settled in the process of baking.

When the coffee was ready I was invited to "pitch in," which I did promptly and without any pressing, after casting a contemptuous look towards the little pile of parched corn on the hearth, which I had previously prepared for my breakfast.

As well as I remember, I think I was dealing with my fifth biscuit, and was looking longingly toward the sixth, when Capt. D mildly suggested that in his opinion I had better "knock off" for a while for fear of consequences. To this I made no reply except to seize the sixth biscuit, and while 1 was disposing of that, Capt. D expeditiously cleared the board, and deposited the remainder of the provisions in the haversack. We then mounted the horses (the pack animal having been turned over to me) and in a day or so we reached the Brazos, where a portion of the Texan army was encamped.

I have nothing further to add, except that when I left for the "States" a month or so subsequently, finding it impossible, owing to the crowded condition of the schooner in which I sailed to take Scout with me, I gave him to my friend H , who promised me he should be well taken care of. Many years afterwards I met with H at Austin, and he told me that Scout lived to a good old age, and died the respected progenitor of a breed of dogs that were highly prized for their valuable qualities.

THE END.

EDWARD BURLESON

Edward Burleson - *Vice President of Texas*

Edward Burleson (December 15, 1798 – December 26, 1851) was a soldier, general, and statesman in the state of Missouri, the Republic of Texas, and later the U.S. state of Texas.

Burleson, soldier and statesman, son of Capt. James and Elizabeth (Shipman) Burleson, was born at Buncombe County, North Carolina, on December 15, 1798. He served as a private in the War of 1812 in his father's company, part of Perkin's Regiment, Alabama. He married Sarah Griffin Owen on April 25, 1816, in Madison County, Missouri Territory; they had nine children. On October 20, 1817, Burleson was appointed a captain of militia in Howard County, Missouri; he was commissioned colonel on June 13, 1821, in Saline County, and was colonel of militia from 1823 to 1830 in Hardeman County, Tennessee.

He arrived in Texas on May 1, 1830, and applied for land in March 1831; title was issued on April 4, 1831. On August 11, 1832, at San Felipe de Austin, he was a member of the *ayuntamiento* governing the counties of Austin, Bexar, Goliad, and Guadalupe. On December 7, 1832, he was elected lieutenant colonel of the militia of Austin Municipality. In 1833 he was elected a delegate to the Second Convention in Mina. From 1830 to 1842 he defended settlers in numerous engagements with hostile Indians.

On May 17, 1835, in Bastrop he was elected to the committee of safety and was therefore unable to attend the Consultation of In October 1835 he was appointed a lieutenant colonel in the Texas army and served under Stephen F. Austin in the opening stages of the Texas Revolution. During the Siege of Béxar, Burleson served as the second-in-command to Gen. Austin and in November, 1835 he was elected Major General of Texas Volunteers and took command of the Volunteer army besieging San Antonio de Béxar and received the surrender of Mexican general Martín Perfecto de Cos. In March, he was appointed a Colonel of Texas Regulars and led the First Volunteer Infantry Regiment during the Runaway Scrape and at the Battle of San Jacinto Stephen F. Austin's Trip to Mexico City

For various causes, Austin was the only one of the commissioners that went to Mexico. He set out shortly after the adjournment of the convention, and reached the capitol in time to see it the scene of confusion and intrigue. As his stay in Mexico was lengthy, and greatly prolonged by political events, it will be proper to refer to them in this place.

On the first of June next following the installation of Santa Anna, General Duran pronounced in favor of the church and the army,that is, a strong government—at the same time nominating Santa Anna as dictator. It has been suggested that Santa Anna was at the bottom of this movement, though without any other evidence than that of his subsequent conduct. He sent out a strong force, under the command of Arista, for its suppression, accompanying the expedition in person. On the march, Arista himself declared for the plan of Duran, and secured the person of the president, at the same time proclaiming him dictator.

This declaration was echoed back from the army in the city. But Gomez Farias, a civilian, and an honest supporter of the constitution of 1824 —the vice-president, and acting as president in the absence of Santa Anna, suspecting that the latter had some hand in this matter, proceeded, with the aid of Lorenzo de Zavala, then governor of Mexico, to raise a force of republicans, and in a short time put down this attempt upon the constitution. Santa Anna appears to have remained a willing captive in the hands of Arista. It was only when he found that the movement was abortive, that he pretended to escape from his captors, and returned to the capitol. Arista was pardoned, and Duran banished. This little farce is an epitome of the life of Santa Anna, and coordinate with the Mexican mind.

Upon these new laurels Santa Anna retired to his estates, leaving the government in the hands of Farias and a republican Congress. The country was deeply in debt, the revenues were exhausted; the active means and resources of the nation had fallen into the possession of the clergy. To lighten the public burden, the army was reduced; and, to raise further means to meet the public wants, a part of the revenues of the church were appropriated. These admirable decrees of the Congress were duly approved by Farias.

The church was thereupon aroused, and, uniting with those opposed to the federal form of government, poured in their petitions for the repeal of these laws. Santa Anna, while in his retirement, meditating on his ambitious projects, had determined in his mind to abandon the republican party, overturn the constitution, and establish an absolute government. His instruments to be used for the accomplishment of these ends were the church and the army,acting at once on the superstitions and fears of the people. Hence he gave countenance and encouragement to these petitions, and openly expressed his dislike for Farias and his administration.

Stephen F. Austin Arrested in Mexico

While Santa Anna was thus fanning the flame of a civil war, in which he expected to reap the principal harvest, Austin, the faithful representative of Texas, was endeavoring in vain to obtain the action of the government upon the matters by him laid before it. His petitions were referred to a committee of Congress, where they slept, while a revolutionary contest was raging in many parts of the republic, and especially about the capitol. To add to the confusion, the cholera broke out with great virulence, and in a few weeks carried off ten thousand of the inhabitants in the metropolis alone.

The epidemic had deranged the meetings of Congress; and so desponding were the hopes of Austin, that, in his letter of the 2d of October, 1833, to the municipality of Bexar, he recommended that all the municipalities of Texas should unite in organizing a state, under the provisions of the Acta Constitutiva of May 7, 1824, and, by union and harmony, prepare for a refusal of their application by the supreme government. He further advised them that, if they did not take matters into their own hands, Texas was ruined forever.

While this letter was on its way, Austin succeeded in procuring the repeal of the law of April 6, 1830, prohibiting natives of the United States from immigrating hither as colonists, and set out for home on the 10th of December, 1833. But his letter of the 2d of October was transmitted by the municipality of Bexar to Vice-President Farias, who, finding in it what he believed to be treasonable matter, dispatched an express for Austin, had him arrested at Saltillo, and taken back to Mexico and imprisoned.Farias, though in principle a republican, was not accustomed to the freedom of speech natural to the Texans.

In the October previous, Austin had told him very plainly that the Texans had determined, if the federal government did not remedy the evils which threatened them, "to remedy them themselves, without waiting any longer on the ground that self-preservation rendered such a step necessary, and would justify it." Farias construed this into a threat and personal insult; and, though he had become partially reconciled to Austin before he departed on the 10th of December, the letter to the corporation of Bexar renewed and increased his exasperation.

Austin was shut up in prison on the 13th of February, 1834, where he remained in close confinement for three months, excluded from the use of books or writing-materials, or even the light of day. We will now turn our attention to the state legislature and its proceedings. This body, having met on the first of January, 1833, reaffirmed its recognition of Pedraza as president of the republic; at the same time it declared that the state would not support any agreement (convenio) tending directly or indirectly to attack the federal form of government and the state sovereignty.

It further declared that it recognized as the will of the nation only what was approved by a majority of the legislatures. It shortly after proceeded to attack the right of petition, and declared that any person or corporation, who assumed the voice of the people by making any petition, usurped the rights of society, and excited disorder. More than three persons were forbidden to join in a petition! The entire law is an attack upon the liberties of freemen; but we must see and know the people of Coahuila before we pronounce too strongly against the legislature.

On the 9th of March following, the legislature carried into effect its previously-expressed determination to remove the seat of government to Monclova, and ordered the state officers to appear at the latter place by the first of April. This exasperated the people of Saltillo, and they were found ready to join any party that might rise in opposition to the legislature. This feeling was increased, on the meeting of that body, by the enactment of a decree disbanding the civic force of thirty men at Saltillo which had been supported by the state.

Among other things, the legislative body repealed the law of the 9th of April, 1832, prohibiting persons not born in the republic of Mexico from retailing goods in the state. This law had given just cause of offence to the people of Texas, and its repeal was due to their growing influence. To this influence may likewise be assigned the law granting to Madero (a favorite with the Texans) the exclusive right to navigate the Trinity. Settlement of the Beales and Grant Colony in Texas

During the latter part of the year 1833 began the settlement of the colony of Beales and Grant. They had obtained a con-cession for eight hundred families, to be located between the Rio Grande and the Neuces. In the last days of December, about sixty colonists, under Mr. Beales, reached the new settlement, and laid off the town of Dolores, on Las Moras, a small stream about ten feet wide and two feet deep. They remained there about a year, when they dispersed.

They were Europeans, and but poorly qualified for such an enterprise. Kennedy, himself an Englishman speaking of the failure of this colony, says it supplied "further evidence of the superiority of the Anglo-Americans in forming colonies. The North Americans are the only people who, in defiance of all obstacles, have struck the roots of civilization deep in the soil of Texas. Even as I trace these lines, I reflect upon their progress with renewed wonder and admiration. They are indeed the organized conquerors of the wild, uniting in themselves the threefold attributes of husbandmen, lawgivers, and soldiers.

Santa Anna Becomes Dictator of Mexico: 1834

The year 1834 was occupied in Mexico in changing the form of government from that of a republican confederation of states to a purely national government, controlled by a single man, without any other restraint than he might chose to place upon his own actions. Farias met the complaints and petitions of the clergy and the monarchists by banishment and the prison. Santa Anna, who had been watching the progress of things, now discovered that the combined influence of the clergy, the army, and the monarchists, would be sufficient to answer his purpose.

He accordingly came out from his retreat, and resumed his seat as president of the nation. It was very soon ascertained that the Congress would not repeal the obnoxious laws lately enacted; and that body, in consequence, became very odious to the centralists. By the constitution of the republic, Congress was required to close its sessions annually on the 15th day of April, with liberty to continue its sittings thirty days longer (Sundays and solemn festivals excepted), should they deem it necessary, or if the president should require it.

At the close of its regular session in 1834, Congress declared it necessary to sit thirty days longer. This period would expire on the 14th of May following. It appears that, on the day preceding, Santa Anna notified the members that, if they did not disperse, he would use a military force to turn them out of their hall. The Congress adjourned on the 14th, declaring that its rights had been invaded by Santa Anna. The latter, however, immediately appealed, by a proclamation, to the people, setting forth the alleged tyranny of the vice-president Farias and the majority of the Congress.

The popularity of Santa Anna gave weight to his address; and the consequence was a pronunciamento and plan, drawn up at Cuernavaca, on the 25th of May. It proposed, first, that the late laws against the church, and those for the banishment of the monarchists, who had taken an active part against the federalists, should be declared void; second, that the Congress should be dismissed, and another convened with power to form a new constitution; and, third, that Santa Anna should be sustained in carrying into execution the views he had published. This plan was almost universally adopted; but, as it required some time to go through a decent formality, and to elect a new Congress, we will return to other events.
Burleson continued to serve in the army after the war and was eventually promoted Brigadier General of Texas Regulars.

Political service

He served as vice president of the Republic of Texas in President Sam Houston's second term from 1841-1844. He was a Presidential candidate in the Texas Presidential Election of 1844, but was defeated by Anson Jones. Although he served under Sam Houston, the two despised each other. Burleson was also involved in the Mexican-American War after Texas was annexed by the United States in 1846. The location of his grave in Austin is in what later became the Texas State Cemetery.

Bibliography

Eugene C. Barker, ed., *The Austin Papers* (3 vols., Washington: GPO, 1924 28). Mary Starr Barkley, *History of Travis County and Austin, 1839–1899* (Waco: Texian Press, 1963).

Edward Burleson,Jr., Papers, Dolph Briscoe Center for American History, University of Texas at Austin.

John H. Jenkins and Kenneth Kesselus, *Edward Burleson: Texas Frontier Leader* (Austin: Jenkins, 1990).

Frank W. Johnson, *A History of Texas and Texans* (5 vols., ed. E. C. Barker and E. W. Winkler [Chicago and New York: American Historical Society, 1914; rpt. 1916]).

Kenneth Kesselus, *History of Bastrop County, Texas, Before Statehood* (Austin: Jenkins, 1986).

Frances Stovall et al., *Clear Springs and Limestone Ledges: A History of San Marcos and Hays County* (San Marcos: Hays County Historical Commission, 1986).

Dudley Goodall Wooten, ed., *A Comprehensive History of Texas* (2 vols., Dallas: Scarff, 1898; rpt., Austin: Texas State Historical Association, 1986).

Source;; Helen Burleson Kelso - *Handbook of Texas Online*, s.v. http://www.tshaonline.org/handbook/online/articles/BB/fbu40.html

STEPHEN FULLER AUSTIN

Stephen F. Austin, (November 3, 1793 – December 27, 1836), known as the Father of Texas, was born in the mining regions of southwestern Virginia (Wythe County), in what is now known as Austinville, some 195 miles west of Richmond, Virginia. He was the second child of Moses Austin and Mary Brown Austin, the first, Eliza Austin, having lived only one month. On June 8, 1798, when he was four years old, his family moved forty miles west of the Mississippi River to the lead mining region in present-day Missouri. His father Moses Austin received a *Sitio* from the Spanish government for the mining site of Mine á Breton.

When Austin was eleven years old, his family sent him to be educated at Bacon Academy in Colchester, Connecticut and then at Transylvania University in Lexington, Kentucky, from which he graduated in 1810. After graduating in Kentucky, Austin began studying to be a lawyer, at age twenty-one he served in the legislature of the Missouri Territory. As a member of the territorial legislature, he was "influential in obtaining a charter for the struggling Bank of St. Louis."

Austin was left penniless after the Panic of 1819, and decided to move south to the new Arkansas Territory. He acquired property on the south bank of the Arkansas River, in the area that would later become Little Rock. After purchasing the property he learned that the area was in consideration as the location for the new territorial capital, which could make his land worth a great deal more.

He made his home in Hempstead County, Arkansas, before moving to the Texas territories. Two weeks before the first territorial elections in 1820, Austin declared his candidacy for Congress.

He was later named a judge for the First Circuit Court. Over the next few months, Little Rock did become the territorial capital, but Austin's claim to land in the area was contested and the courts ruled against him. The Territorial Assembly also reorganized the government, abolishing Austin's judgeship. Austin then moved to Louisiana. He reached New Orleans in November 1820, where he met and stayed with New Orleans lawyer and former Kentucky congressman Joseph H. Hawkins and made arrangement to study law.

During Austin's time in Arkansas, his father had traveled to Spanish Texas and received an empresarial grant that would allow him to bring 300 Americans to Texas (they would later become known as *The Old 300*). Moses Austin was attacked on his way back to Missouri. Upon returning home Moses became ill with what was believed to be pneumonia and died on June 10, 1821. He left his empresario grant to his son, Stephen. Though Austin was reluctant to carry on his father's Texas venture, Austin was ultimately persuaded to pursue the colonization of Texas by a letter from his own mother, Mary Brown Austin, written just two days before Moses Austin died.

Austin had boarded the steamer *Beaver* and departed to New Orleans to meet Spanish officials led by Erasmo Seguín. He was at Natchitoches, Louisiana, on June 31, 1821, when he learned of his father's death. "This news has effected me very much, he was one of the most feeling and affectionate Fathers that ever lived. His faults I now say, and always have, were not of the heart."

His party traveled the 300 miles (480 km) in three weeks to San Antonio with the intent of reauthorizing his father's grant, arriving on August 12. While in transit, they learned that Mexico had declared its independence from Spain, and Texas had become a Mexican province rather than a Spanish territory. In San Antonio, the grant was reauthorized by Governor Antonio María Martínez, who allowed Austin to explore the Gulf Coast between San Antonio and the Brazos River in order to find a suitable location for a colony. As guides for the party, Manuel Becerra, along with three Aranama Indians, went with the expedition.

Austin advertised the opportunity in New Orleans, stating that the land was available along the Brazos and Colorado rivers. A family of a husband, wife and two children would receive 1,280 acres (520 ha) at twelve and a half cents per acre. Farmers could get 177 acres (0.72 km^2) and ranchers 4,428. In December 1821, the first U.S. colonists crossed into the granted territory by land and sea, on the Brazos River in present day Brazoria County, Texas.

Immigration Agent

Austin's plan for a colony was thrown into turmoil by the independence of Mexico from Spain in 1821. Governor Martínez informed Austin that the *junta instituyente*, the new rump congress of the government of Agustín I of Mexico, refused to recognize the land grant authorized by Spain, based on a new policy of using a general immigration law to regulate new settlement in Mexico. Austin traveled to Mexico City and managed to persuade the *junta instituyente* to authorize the grant to his father, as well as the Law signed by the Spanish Emperor on January 3, 1823.

The old Imperial Law offered heads of families a league and a labor of land, 4,605 acres (19 km^2), and other inducements. It also provided for the employment of agents, called empresarios, to promote immigration. As empresario, Austin himself was to receive 67,000 acres (270 km^2) of land for each two hundred families he introduced. According to the law, immigrants were not required to pay fees to the government. This fact soon led some of the immigrants to deny Austin's right to charge them for services at the rate of 12½ cents an acre.

When the Emperor of Mexico, Agustín de Iturbide, abdicated in March 1823, the law was annulled once again. In April 1823, Austin induced the congress to grant him a contract to bring 300 families into Texas. He wanted honest, hard-working, people who would make the colony a huge success. In 1824 the congress passed a new immigration law that allowed the individual states of Mexico to administer public lands and open them to settlement under certain conditions. In March 1825 the legislature of the Mexican state of Coahuila y Tejas passed a law that was similar to the one authorized by Iturbide. The law continued the system of empresarios, as well as granting each married man a league of land, 4,428 acres (18 km^2), with the stipulation that he must pay the state thirty dollars within six years.

By late 1825, Austin had brought the first 300 families to his settlement (The Austin Colony), now known in Texas history as the Old Three Hundred, to the grant. Austin had obtained further contracts to settle an additional 900 families between 1825 and 1829. He had effective civil and military authority over the settlers, but he was quick to introduce a semblance of American law - the Constitution of Coahuila y Tejas was agreed on in November 1827. Also, Austin organized small, informal armed groups to protect the colonists, which evolved into the Texas

Rangers. Despite his hopes Austin was making little money from his endeavors; the colonists were unwilling to pay for his services as *empresario* and most of the money gained was spent on the processes of government and other public services.

It was during these years that Austin, a member of Louisiana Lodge No. 111 at Ste. Genevieve, Missouri, sought to establish Freemasonry in Texas. Freemasonry was well established among the educated classes of Mexican society. It had been introduced among the aristocracy loyal to the House of Bourbon, and the conservatives had total control over the Order. By 1827 Americans living in Mexico City had introduced the American York Rite of Freemasonry as a liberal alternative to the established European-style Scottish Rite.

On February 11, 1828, Austin called a meeting of Freemasons at San Felipe for the purpose of electing officers and petitioning the Masonic Grand Lodge in Mexico City for a charter to form a lodge. Austin was elected Worshipful Master of the new lodge. Although the petition reached Matamoros, and was to be forwarded to Mexico City, nothing more was heard of it. By 1828, the ruling faction in Mexico was afraid that the liberal elements in Texas might try to gain their independence. Fully aware of the political philosophies of American Freemasons, the Mexican government outlawed Freemasonry on October 25, 1828. In 1829, Austin called another meeting where it was decided that it was "impolitic and imprudent, at this time, to form Masonic lodges in Texas."

He was active to promote trade and to secure the good favor of the Mexican authorities, aiding them in the suppression of the Fredonian Rebellion of Haden Edwards. However, with the colonists numbering over 11,000 by 1832 they were becoming less conducive to Austin's cautious leadership, and the Mexican government was also becoming less cooperative - concerned with the growth of the colony and the efforts of the U.S. government to buy the state from them. The Mexican government had attempted to stop further U.S. immigration as early as April 1830, but again the skills of Austin had gained an exemption for his colonies. He gave 640 acres (2.6 km^2) to the husband, 320 to the wife, 160 for every child, and 80 for every slave.

Relations with Mexico

The application of the immigration control and the introduction of tariff laws had done much to dissatisfy the colonists, peaking in the Anahuac Disturbances. Austin then felt compelled to involve himself in Mexican politics, supporting the upstart Antonio López de Santa Anna. Following the success of Santa Anna, the colonists sought a compensatory reward, proclaimed at the Convention of 1832,resumption of immigration, tariff exemption, separation from Coahuila, and a new state government for Texas. Austin was not in favor of these demands, he considered them ill-timed and tried his hardest to moderate them.

When they were repeated and extended at the Convention of 1833, Austin traveled to Mexico City on July 18, 1833, and met with Vice President Valentín Gomez Farías. Austin did gain certain important reforms; the immigration ban was lifted, but not a separate state government. Separate statehood required a population of 80,000 before it could be granted, and Texas had only 30,000.

Texas Revolution

In his absence, a number of events propelled the colonists toward confrontation with Santa Anna's centralist government. Austin took temporary command of the Texan forces during the Siege of Bexar from October 12 to December 11, 1835. After learning of the Disturbances at Anahuac and Velasco in the summer of 1835, an enraged Santa Anna made rapid preparations for the Mexican army to sweep Anglo settlers from Texas. War began in earnest in October 1835 at Gonzales. The Republic of Texas, created by a new constitution on March 2, 1836, won independence following a string of defeats with the dramatic turnabout victory at the Battle of San Jacinto on April 21, 1836, and the capture of Santa Anna the following morning.

Austin in the Republic of Texas

In December 1835 Austin, Branch Archer and William H. Wharton were appointed commissioners to the U.S. by the provisional government of the republic. On June 10, 1836, Austin was in New Orleans when he received word of Santa Anna's defeat by Sam Houston at San Jacinto. Austin returned to Texas to rest at Peach Point in August. On August 4, he announced his candidacy for president of Texas. Austin felt confident he

could win the election until with two weeks before the election, on August 20, Houston entered the race. Austin wrote, "Many of the old settlers who are too blind to see or understand their interest will vote for him." Houston carried East Texas, the Red River and most of the soldier vote. Austin polled 587 votes to Sam Houston's 5,119 and Henry Smith's 743 votes.
Houston would appoint Austin as the first Secretary of State of the new Republic; however, Austin had served only around 2 months before his death.

Death and Estate

In December 1836, Austin was in the new capital of Columbia (now known as West Columbia) where he caught a severe cold; his condition worsened. Doctors were called in, but could not help him. Austin died of pneumonia at noon on December 27, 1836, at the home of George B. McKinstry right outside of what is now West Columbia, Texas. Austin's last words were "The independence of Texas is recognized! Don't you see it in the papers?..." Upon hearing of Austin's death, Houston ordered an official statement proclaiming: "The Father of Texas is no more; the first pioneer of the wilderness has departed." Austin was originally buried at Gulf Prairie Cemetery in Brazoria County, Texas. Austin's body was moved, however, in 1910 to the Texas State Cemetery in Austin, Texas.
Austin died without having produced offspring, and bequeathed all his land, titles, and possessions, to his sister, Emily Austin

Mirabeau Buonaparte Lamar

Mirabeau Buonaparte Lamar (1798–1859) was a Texas politician, diplomat and soldier who was a leading Texas political figure during the Texas Republic era. He was the second President of the Republic of Texas, after David G. Burnet (1836 as ad-interim president) and Sam Houston. He grew up at Fairfield, his father's plantation near Milledgeville, Georgia. As a child, he loved to read and educated himself through books. Although he was accepted to Princeton University, Lamar chose not to attend. Instead, he worked as a merchant and then ran a newspaper, but both of those enterprises failed.

In 1823, Lamar's family connections helped him to gain a position as the private secretary to Georgia Governor George M. Troup. In this position, Lamar issued press releases and toured the country giving speeches on behalf of the governor. On one of his trips, he met Tabatha Jordan, whom he married in 1826.

When Troup lost his reelection bid in 1828 Lamar established a newspaper in Columbus, Georgia, the Columbus Enquirer. This venture was much more successful than his previous business attempts. The year after its establishment Lamar's wife died, and he withdrew his name from consideration for election to the Georgia Senate and began to study law. He was admitted to the bar in 1833 and ran an unsuccessful campaign for a seat in the U.S. Congress. He was also a Master Mason, and a member of Harmony Lodge No. 6 in Galveston, Texas.

Texas Revolution

Lamar's brother Lucius committed suicide in 1834, and a grief-stricken Lamar began traveling to escape his memories. In the summer of 1835 he ended up in Texas and decided to stay. He joined Sam Houston's army in spring 1836 and distinguished himself with bravery at the Battle of San Jacinto. On the eve of the battle, Lamar made a courageous rescue of two surrounded Texans in a move that drew a salute from the Mexican lines. He was promoted to colonel and was the commander of the cavalry during the battle the following day. His actions led to an appointment as the

Secretary of War in the interim Texian government. In 1836, he was elected vice-president of the Republic of Texas under Houston.

Texas President

Lamar was the unanimous choice to replace Houston as president in 1838, and was inaugurated on December 1, 1838. It was said that when Houston was giving his farewell speech, Houston was dressed like George Washington. When he went on the stage he noticed that Lamar was eager to give his speech. Houston lengthened his farewell address to 3 hours, "which so unnerved Lamar that he was unable to read his inaugural speech." Instead, it was given by his aide, Algernon P. Thompson.

In his first formal address to the Texas Congress several weeks later, *Lamar urged that the Cherokee and Comanche tribes be driven from their lands in Texas,* even if the tribes must be destroyed. He also proposed to create a national bank and to secure a loan from either the United States or Europe. Finally, *he opposed annexation to the United States* and wished to entice European countries to recognize the Republic of Texas. His first actions were against the Indian tribes. In 1839 his troops drove the Cherokee tribes from Texas. The battle resulted in the death of Houston's friend Chief Bowles, leaving Houston angry at Lamar. A similar campaign was fought against the Comanche. Although many lives were lost, the Comanches could not be forced from the area. Lamar believed that *it was necessary to bring about the "total extinction" of the Indian tribes.*

Lamar appointed a commission to select a permanent site for the capital of the Republic. After two months of debate, they recommended the small town of Waterloo along the Colorado River. The town was renamed Austin, and by October 1839 all of the records and employees were transferred from Houston to Austin. That same year, he founded the Texas State Library (presently known as the Texas State Library and Archives Commission).

During his administration, Lamar sent three separate agents to Mexico to enact a peace settlement. All failed. Nonetheless, Lamar succeeded in gaining official recognition for his country from the United Kingdom, France, and Belgium. Despite several attempts to secure a loan from these and other countries, Lamar was unsuccessful. To fill the treasury, he authorized issuance of a large amount of Republic of Texas paper money, known as Redbacks. The paper money was virtually worthless. Spending doubled during Lamar's term, and combined with the worthless currency, caused financial difficulties for the government.

Lamar believed that the Rio Grande River was the western boundary of Texas, and wanted to send an expedition to New Mexico to establish trade, and convince the residents to join the Republic. The Texas Congress refused to fund the expedition in 1839 and 1840. In June 1841, Lamar took $89,000 from the treasury and sent an expedition on his own. The members were arrested when they reached Santa Fe, and told they would soon be released. Instead, they were marched to Mexico City, and many died during their enforced march.

He has been called "the Father of Texas Education." During his administration, he convinced the legislature to set aside 3 leagues of land for each county to equip schools. He also allotted 50 leagues of land for the support of two universities, later Texas A&M University and the University of Texas. Although nothing was implemented during his term, these actions led to the foundation for a statewide public school system.

One of his quotes sums up his views *"Cultivated mind is the guardian genius of democracy and, while guided and controlled by virtue, the noblest attribute of man. It is the only dictator that freemen acknowledge and the only security that freemen desire."*

Later years

Houston became president again when Lamar's term completed in late 1841. Lamar returned to service in the army, and distinguished himself at the Battle of Monterrey during the Mexican-American War. In late 1847, he became a post commander at Laredo, but disliked the job as he wished to see more action. While in Laredo, he married again. He represented Eagle Pass in the Texas State Legislature for several years after Texas was annexed to the United States. In 1857, President James Buchanan appointed Lamar to be the Minister to Nicaragua. Lamar served in Managua for twenty months before returning to Texas in October 1859 because of poor health. He died of a heart attack at his Richmond plantation on December 19, 1859.

DAVID (DAVY) CROCKETT

His personal motto - *"Be always sure you're right-then go a-head!"*

David "Davy" Crockett, was born in Greene County, Tennessee, August 17.1786. He died at the battle of the Alamo in 1836. In his youth he worked as a wagoner also as a labourer. In 1806 he married Mary 'Polly' Finlay of Jefferson County. From 1811 after having two sons they moved to live in the mountains of East Tennessee. Later he enlisted in the militia as a scout in the Creek Indian War campaign in November 1813

He was under the command of the celebrated future President, Andrew Jackson. After his enlistment in the Creek Indian War expired he returned home. When the 'Battle of New Orleans occurred he re-enlisted September 28.1814 as a sergeant. He arrived on November 7th the day after Jackson took Pensacola, and spent his time trying to ferret out the British-trained Indians from the Florida swamps. After his discharge in 1815 he returned home.His wife Polly died the following year.

On May 21, 1815, Crockett was elected a lieutenant in the Thirty second Militia regiment of Franklin County. He married Elizabeth Patton, a widow with two children. The following year the family moved to Lawrence County, Tennessee and settled at the head of Shoal Creek. He then continued his political and military career .After a brief stint at being a Justice of the Peace, he then became the town commissioner of Lawrenceburg, and was elected colonel of the Fifty-seventh Militia regiment in the county that same year. (1818)

New Year's Day 1821 marked a turning point in Crockett's career. He ran as commissioner for a seat in the Tennessee legislature as the representative of Lawrence and Hickman counties. He won the August election and, from the beginning, took an active interest in public land policy regarding the West. After the session concluded he moved his family to Gibson County in West Tennessee and was reelected in 1823, He was defeated in August 1825 in his first bid for a seat in Congress. In 1826 after returning to private business, Crockett nearly died when his boats carrying barrel staves were wrecked in the Mississippi River.

He was brought to Memphis and encouraged to run again for Congress by Maj. M. B. Winchester and won election over Gen. William Arnold and Col. Adam Alexander to the United States House of Representatives in 1827. He was re-elected to a second term in 1829 and differed with President Andrew Jackson and the Tennessee delegation on several issues, including land reform and the Indian removal bill. In his 1831 campaign for a third term, Crockett openly and vehemently attacked Jackson's policies and was defeated in a close election by William Fitzgerald.

Crockett's reputation as a sharpshooter, hunter, and yarn-spinner had attracted national attention. *Life and Adventures of Colonel David Crockett of West Tennessee* was published in 1833 and reprinted the same year under the more accurate title of *Sketches and Eccentricities of Colonel David Crockett of West Tennessee*. His extraneous life was described in popular books during the period 1835 to 1856 often published under his own name. They were responsible for creating the legend of 'Davy Crockett'. He was discontent due to *the inaccuracy of events attributed to him*, he made reference to these false character portrayals in his own autobiography.

In 1833 he returned to Congress. In 1834, he published his autobiography, (assisted by Thomas Chilton) *A Narrative of the Life of David Crockett of the State of Tennessee, the only book that Davidever wrote. In 1835 he lost his* congressional campaign to a lawyer supported by Jackson and by Governor Carroll of Tennessee. By a small margin of votes he became discontent with political life and decided to move west with his family to Texas on November 1, 1835, departed and noted that he was planning, "to explore the Texas well before I return."

He had no plans in mind to join the popular fight to establish Texas independence (at that time, it was Spanish territory). At a farewell gathering in Memphis with friends he made his famous comment, "Since you have chosen to elect a man with a timber toe to succeed me, you may all go to hell and I will go to Texas." With several others he left the next day on a route down the Mississippi River to the Arkansas, from there up-river to Little Rock, then overland to Fulton, Arkansas, up the Red River along the northern boundary of Texas; across the Red River, through Clarksville, to Nacogdoches and San Augustine; and on to San Antonio.

At San Augustine the party evidently divided. *Crockett and Patton signed the oath of allegiance*, but only after Crockett insisted upon the insertion f the word "republican" in the document. They then swore their allegiance to the "Provisional Government of Texas or any future *republican* Government that would be declared." His enthusiasm over Texas is evident in a letter he wrote, January 9, 1836,

"I must say as to what I have seen of Texas it is the garden spot of the world. The best land and the best prospects for health I ever saw, and I do believe it is a fortune to any man to come here. There is a world of country here to settle." I have taken the oath of government and have enrolled my name as a volunteer and will set out for the Rio Grand in a few days with the volunteers from the United States. But all volunteers is entitled to vote for a member of the convention or to be voted for, and I have but little doubt of being elected a member to form a constitution for this province. I am rejoiced at my fate. I had rather be in my present situation than to be elected to a seat in Congress for life. I am in hopes of making a fortune yet for myself and family, bad as my prospect has been." He intended to become land agent for the new territory.

The Alamo - Crockett's Diary

Crocket kept a diary[1] and his notes whilst he was in San Antonio showed him in high spirits, although he indicated that he had a shortage of provisions. On February 23rd the enemy were in sight and the little garrison resolved to defend the Alamo. They made a large flag with thirteen stripes, red and white alternately on a blue background, with a large white star in the center, and between the points, the word "Texas. Over the next nine days he wrote an account of their sufferings and their brave defense. On March 3rd, he indicated that they had given up all hope of receiving any assistance from outside so they promised to fight on to the end and in dying to kill as many of the enemy combatants.

His diary entry for March 4th, stated; "Shells have been falling into the fort like hail during the day, but without effect. About dusk this evening we saw a man running toward the fort pursued by about a dozen Mexican

[1] Refer to the diary found in 1975 which is an accurate account of the death of Crockett and his fellow defenders, as recorded by a contemporary Mexican combatant.

cavalry "The bee hunter[2] immediately recognized him as the old pirate who had gone to Goliad for assistance, and calling to two others, the bee hunter sallied out of the fort to the relief of the old man, I followed close after. Before we reached him the Mexicans were close upon his heels. He stopped suddenly, turned short upon his pursiuers, discharge his rifle, and saw one of his enemies fall from his horse. After running a short distance again, the old pirate, finding that he would be taken and cut to pieces, turned fiercely, and to the amazement of the enemy clubbed his gun and dashed among them like a wounded tiger.

By the time we reached him, his pursuers had fled like sparrows and in the ardour of the moment we followed them some distance, not seeing that our retreat was cut off by another detachment of cavalry. Nothing was to be done but to fight our way through. They were about twenty in number and stood their ground while we dashed among them, and for about five minutes a bloody conflict ensued. Then a detachment was seen coming from the fort to our relief, and the Mexicans scampered away, leaving eight of their men dead upon the field. We did not escape unscathed, for both the pirate and the bee hunter were mortally wounded, and I received a saber cut across the forehead."

"The old man died without speaking as soon as we entered the fort. We bore my young friend to his bed, dressed his wounds, and I watched beside him. He lay without complaint or manifesting pain, until about midnight, when he spoke. I asked him what he wanted. "Nothing," he replied with a sigh that seemed to rend his heart, and his eyes filled with tears as he continued his 'Poor Kate of Nacogdoches; her words were prophetic, Colonel.' Then he sang in a low voice,- 'But toom' cam' the saddle, all bluidy to see, and hane cam' the steed, but hame never cam' he, "He spoke no more, an a few minutes afterward died. Poor Kate, who will tell this to thee?"

The last entry in Crockett's diary bears the date March 5th. It is as follows: "Pop, pop, pop! Bom, bom, bom! throughout the day.- No time for memorandums now,- Go ahead!-Liberty and independence forever!"

The true account of his death - From a contemporary enemy combatant's diary, discovered in 1975

---

[2] Earlier they had mentioned another in their group of defenders was a bee hunter from Nacogdoches, Texas also another person, a fierce old man, had been a pirate but had abandoned the sea.

In early February Crockett arrived at San Antonio de Béxar; Antonio López de Santa Anna arrived on February 20. He joined up with Col. William B. Travis, who had disregarded Sam Houston's orders to withdraw from the Alamo, rather than support Houston, a Jackson sympathizer. He saw his future path in the struggle to establish an independent Texas.

Crockett died in battle of the Alamo on March 6, 1836. The manner of his death was uncertain, however, until the publication in 1975 of *the diary of Lt. José Enrique de la Peña. Susanna Dickinson, wife of Almaron Dickinson,* an officer at the Alamo, said Crockett died on the outside, one of the earliest to fall.

Joe, *Travis's slave and the only male Texan to survive the battle, reported seeing Crockett lying dead with slain Mexicans around him a*nd stated that only one man, named Warner, surrendered to the Mexicans (Warner was taken to Santa Anna and promptly shot).

When Peña's eyewitness account was placed together with other corroborating documents, Crockett's central part in the defense became clear. Travis had previously written that during the first bombardment Crockett was everywhere in the Alamo "animating the men to do their duty."Other reports told of the deadly fire of his rifle that killed five Mexican gunners in succession, as they each attempted to fire a cannon bearing on the fort, and that he may have just missed Santa Anna, who thought himself out of range of all the defenders' rifles. Crockett and five or six others were captured when the Mexican troops took the Alamo at about six o'clock that morning, even though Santa Anna had ordered that no prisoners be taken.

The general, infuriated when some of his officers brought the Americans before him to try to intercede for their lives, ordered them executed immediately. They were bayoneted and then shot. Crockett's reputation and that of the other survivors was not, as some have suggested, sullied by their capture. Their dignity and bravery was, in fact, further underscored by Peña's recounting that "these unfortunates died without complaining and without humiliating themselves before their torturers."

SOURCE;
Michael A. Lofaro, "Crockett, David," *Handbook of Texas Online*, s.v."
http://www.tshaonline.org/handbook/online/articles/CC/fcr24.html
by kind permission of “Handbook of Texas Online.”

JAMES (JIM) BOWIE

April 10, 1796 – March 6, 1836. Kentucky born; Jim Bowie a 19th-century American pioneer and soldier spent most of his life in Louisiana, where he was raised and later worked as a land speculator. His rise to fame began in 1827 on reports of the Sandbar Fight. What began as a duel between two other men deteriorated into a melee in which Bowie, having been shot and stabbed, killed the sheriff of Rapides Parish with a large knife. This, and other stories of Bowie's prowess with the knife, led to the widespread popularity of the Bowie knife. He played a prominent role in the Texas Revolution, culminating in his death at the Battle of the Alamo.

Countless stories of him as a fighter and frontiersman, both real and fictitious, have made him a legendary figure in Texas history and a folk hero of American culture.Bowie's reputation was cemented by his role in the Texas Revolution. After moving to Texas in 1830, Bowie became a Mexican citizen and married the daughter of the vice governor of the province. His fame in Texas grew following his failed expedition to find the lost San Saba mine, during which his small party repelled an attack by a large Indian raiding party.

At the outbreak of the Texas Revolution, Bowie joined the Texas militia, leading forces at the Battle of Concepción and the Grass Fight. In January 1836, he arrived at the Alamo, where he commanded the volunteer forces until an illness left him bedridden. Bowie died with the other Alamo defenders on March 6. Despite conflicting accounts of the manner of his death, the "most popular, and probably the most accurate" accounts maintain that he died in his bed after emptying his pistols into several Mexican soldiers.

According to his older brother, John, James Bowie was born in Logan County, Kentucky, in the spring of 1796 Historian Raymond Thorp gave his birth date as April 10, but Thorp did not provide any documentation for that date Bowie was the ninth of ten children born to Elve Ap-Catesby Jones and Rezin Bowie.

His father had been injured while fighting in the American Revolution, and in 1782 married the young woman who had nursed him back to health. The Bowies moved frequently, first settling in Georgia, before moving to Kentucky. At the time of Bowie's birth, his father owned eight slaves, eleven head of cattle, seven horses, and one stud horse. The following year the family acquired 200 acres (80 ha) along the Red River. They sold that property in 1800 and relocated to Missouri before moving to Spanish Louisiana in 1802, where they settled on Bushley Bayou in Rapides Parish.

The Bowie family moved again in 1809, settling on Bayou Teche in Louisiana before finding a permanent home in Opelousas in 1812. The Bowie children were raised on the frontier and even as small children were expected to help clear the land and plant crops. All of the Bowie children learned to read and write in English, but Jim and his elder brother Rezin could also read, write, and speak Spanish and French fluently. The children learned to survive on the frontier, how to fish and run a farm and plantation. Jim Bowie became proficient with pistol, rifle, and knife, and had a reputation for fearlessness. As a boy one of his Indian friends even taught him to rope alligators.

In response to Andrew Jackson's plea for volunteers to fight the British in the War of 1812, Bowie and his brother Rezin enlisted in the Louisiana militia in late 1814. The war ended on December 24 of that year with the signing of the Treaty of Ghent, and the Bowie brothers arrived in New Orleans too late to participate in the fighting. After mustering out of the militia, Bowie settled in Rapides Parish, where he supported himself by sawing planks and lumber and floating it down the bayou for sale. In June 1819, he joined the Long expedition, an effort to liberate Texas from Spanish rule. The group encountered little resistance and, after capturing Nacogdoches, declared Texas an independent republic. The extent of Bowie's participation is unclear, but he returned to Louisiana before the invasion was repelled by Spanish troops.

Shortly before their father died in 1818 or 1819, he gave ten slaves, and horses and cattle to both Jim and his brother Rezin. For the next seven years the brothers worked together to develop several large estates in Lafourche Parish and Opelousas Louisiana's population was growing rapidly, and the brothers hoped to take advantage of its rising land prices through speculation. Without the capital required to buy large tracts, they entered into a partnership with a pirate, Jean Lafitte, in 1818 to raise money. By then, the United States had outlawed the importation of slaves, and most southern states allowed anyone who informed on a slave trader to receive half of what the imported slaves would earn at auction as a reward.

Bowie made three trips to Lafitte's compound on Galveston Island. On each occasion he bought smuggled slaves and then took them directly to a customhouse to inform on his own actions. When the customs officers offered the slaves for auction, Bowie purchased them and received back half the price he had paid, as allowed by the state laws. He then could legally transport the slaves and resell them at a greater market value in New Orleans or areas farther up the Mississippi River. Using this scheme the brothers collected $65,000 to be used for their land speculation.

In 1825, the two brothers joined with their younger brother Stephen to buy Acadia, a plantation near Alexandria. Within two years they had established the first steam mill in Louisiana to be used for grinding sugar cane. The plantation became known as a model estate, but, on February 12, 1831, they sold it and 65 slaves for $90,000. With their profits, Bowie and Rezin bought a plantation in Arkansas. Bowie and his brother John were involved in a major court case in the late 1820s over their land speculation.

When the United States purchased the Louisiana Territory in 1803, it promised to honor all former land grant claims, and for the next 20 years efforts were made to establish who owned what land. In May 1824, Congress authorized the superior courts of each territory to hear suits from people who claimed they had been overlooked. The Arkansas Superior Court received 126 claims in late 1827 from residents who claimed to have purchased land in former Spanish grants from the Bowie brothers.

Although the Superior Court originally confirmed most of those claims, the decisions were reversed in February 1831, after further research showed that the land had never belonged to the Bowies, and that the original land grant documentation had been forged. The U.S. Supreme Court upheld the reversal in 1833. When the disgruntled purchasers considered suing the Bowies, they discovered that the documents in the case had been removed from the court; left without evidence, they declined to pursue a case.

Bowie became internationally famous as a result of a feud with Norris Wright, the sheriff of Rapides Parish. Bowie had supported Wright's opponent in the race for sheriff, and Wright, a bank director, had been instrumental in turning down Bowie's loan application After a confrontation in Alexandria one afternoon, Wright fired a shot at Bowie. The uninjured Bowie was enraged and tried to kill Wright with his bare hands. Wright's friends intervened and stopped the attack, after which Bowie resolved to carry his hunting knife at all times. The knife he carried had a huge blade that was 9.25 inches (23.5 cm) long and 1.5 inches (3.8 cm) wide.

The following year, on September 19, 1827, Bowie and Wright attended a duel on a sandbar outside of Natchez, Mississippi. Bowie supported duelist Samuel Levi Wells III, while Wright supported Wells's opponent, Dr. Thomas Harris Maddox. The duelists each fired two shots, and, as neither man had been injured, resolved their duel with a handshake

Other members of the groups, who had various reasons for disliking each other, began fighting. Bowie was shot in the hip; after regaining his feet he drew a knife, described as a butcher knife, and charged his attacker. The attacker hit Bowie over the head with his empty pistol, breaking the pistol and knocking Bowie to the ground. Wright shot at and missed the prone Bowie, who returned fire and possibly, hit Wright. Wright then drew his sword cane and impaled Bowie.

When Wright attempted to retrieve his blade by placing his foot on Bowie's chest and tugging, Bowie pulled him down and disemboweled Wright with his large knife. Wright died instantly, and Bowie, with Wright's sword still protruding from his chest, was shot again and stabbed by another member of the group. The doctors who had been present for the duel retrieved the bullets and patched Bowie's other wounds.

Newspapers picked up the story, which became known as the Sandbar Fight, and described in detail Bowie's fighting prowess and his unusual knife. Witness accounts agreed that Bowie did not attack first, and the others had focused their attack on Bowie because "they considered him the most dangerous man among their opposition."

The fight cemented Bowie's reputation across the South as a superb knife-fighter.There is disagreement among scholars as to whether the knife used in this fight was the same kind of knife now known as a Bowie knife. Multiple accounts exist of who designed and built the first Bowie knife.

Some claim that Bowie designed it, while others attribute the design to noted knife makers of the time.

However, in a letter to *The Planter's Advocate*, *Rezin Bowie claimed to have invented the knife,* and many Bowie family members as well as "most authorities on the Bowie knife tend to believe it was invented by" Rezin Rezin Bowie's grandchildren, however, claimed that Rezin merely supervised his blacksmith, who was the creator of the knife.

After the Sandbar Fight and subsequent battles in which Bowie successfully used his knife to defend himself, the Bowie knife became very popular. Many craftsmen and manufacturers made their own versions, and major cities of the Southwest had "Bowie knife schools", which taught "the art of cut, thrust, and parry." His fame, and that of his knife, spread to England, and by the early 1830s many British manufacturers were producing Bowie knives for shipment to the United States. The design of the knife continued to evolve, but today a Bowie knife generally is considered to have a blade 8.25 inches (21.0 cm) long and 1.25 inches (3.2 cm) wide, with a curved point, a "sharp false edge cut from both sides", and a cross-guard to protect the user's hands.

After recovering from the wounds he suffered in the Sandbar Fight, in 1828 Bowie decided to move to Texas, at that time a state in the Mexican federation The 1824 Constitution of Mexico banned religions other than Roman Catholicism and gave preference to Mexican citizens in receiving land Bowie was baptized into the Roman Catholic faith in San Antonio on April 28, 1828, sponsored by the alcalde (chief administrator) of the town, Juan Martin de Veramendi and the wife of the administrator, Josefa Navarro.

For the next eighteen months Bowie travelled through Louisiana and Mississippi. In 1829, he became engaged to Cecilia Wells, who died in Alexandria on September 29, two weeks before their wedding. On January 1, 1830, Bowie left Louisiana for permanent residency in Texas. He stopped at Nacogdoches, at Jared E. Groce's farm on the Brazos River, and in San Felipe, where Bowie presented a letter of introduction to Stephen F. Austin from Thomas F. McKinney, one of the Old Three Hundred colonists.

On February 20, Bowie took an oath of allegiance to Mexico and then proceeded to San Antonio de Bexar. At the time the city, known as Bexar, had a population of 2,500 people, mostly of Mexican descent, and Bowie's fluency in Spanish helped him establish himself in the area. Bowie was elected a commander of the Texas Rangers later that year.

Although the Rangers would not be organized officially until 1835, Stephen F. Austin had founded the group by employing thirty men to keep the peace and protect the colonists from attacks by hostile Indians. Other areas assembled similar volunteer militias, and Bowie commanded a group of the volunteers.

Bowie *became a Mexican citizen* on September 30, 1830, after promising to establish textile mills in the province of Coahuila y Tejas. To fulfill his promise, Bowie entered into partnership with Veramendi to build cotton and wool mills in Saltillo. With his citizenship assured, Bowie now had the right to buy up to 11 leagues of public land. He convinced fourteen or fifteen other citizens to apply for land in order to turn it over to him, giving him 700,000 acres (280,000 ha) for speculation. Bowie may have been the first to induce settlers to apply for empresario grants which could then be sold in bulk to speculators like Bowie. The Mexican government passed laws in 1834 and 1835 that stopped much of the land speculation.

On April 25, 1831 Bowie married 19-year-old Maria Ursula de Veramendi, the daughter of his business partner, who had become the vice-governor of the province. At the time Bowie claimed to have a net worth of $223,000, mostly in land of questionable title. Bowie also lied *about his age, claiming to be 30 rather than 35*. The couple built a house in San Antonio, on land Veramendi had given them near the San José Mission. They decided, however, to move into the Veramendi Palace, and lived with Ursula's parents, who supplied them with spending money. The couple had two children, Marie Elve, born March 20, 1832, and James Veramendi, born July 18, 1833.

Shortly after his marriage Bowie became fascinated with the story of the "lost" Los Almagres Mine[1], said to be west of San Antonio near the ruin of

---

[1] “Silver! Enough for every person to own a silver mine!” Wrote Bernardo de Miranda y Flores in his 1756 report to the Spanish provincial governor after visiting Cerro del Almagre and Los Almagres, a mine near the Llano River. Almagre is a Spanish word that means "ochre", and Cerro del Almagre means a Mountain of Ochre. . In 1753, a small group of Spaniards excavated a pit at the. Later, the governor asked Miranda to visit and examine the site. Miranda cleared the pit, extracted samples, and had them assayed. Although the results showed

Santa Cruz de San Sabá Mission. The mine had been operated by local Indians before being seized by the Spanish. After Mexico won independence from Spain, government interest in the mines waned. A number of hostile Indian tribes roamed the area, including Comanche, Lipan Apache, and Karankawa, and without government troops to keep the tribes at bay, mining ceased. It was believed that after the Mexican citizens left the area, the Lipan Apaches took over the mines.

Bowie obtained permission from the Mexican government and mounted an expedition into Indian territory to search for the legendary silver mine, Bowie, his brother Rezin, and nine others set out for San Saba on November 2, 1831. Six miles (10 km) from their goal, the group stopped to negotiate with a large Indian raiding party following them. The attempts at parley failed and Bowie and his group fought for their lives for the next 13 hours. When the Indians finally retreated, Bowie reportedly had lost only one man, while over 40 Indians had been killed and 30 more wounded. In the meantime, a party of friendly Comanche Indians rode into San Antonio, bringing word of the raiding party, which outnumbered the Bowie expedition by 15–to–1.

The citizens of San Antonio believed the Bowie expedition must have perished, and Ursula Bowie began wearing widow's weeds. However, the surviving members of the group returned to San Antonio on December 6. Bowie's report of the expedition, written in Spanish, printed in several newspapers, further establishing his reputation. He set out again with a larger force the following month, but returned home empty-handed after two and a half months of searching.Bowie never talked of his exploits despite his increasing fame. Captain William Y. Lacey, who spent eight months living in the wilderness with Bowie, described him as a humble man who never used profanity or vulgarities.

Between 1830 and 1832 the Mexican Congress passed a series of laws that seemed to discriminate against Anglo colonists in the province of Coahuila y Tejas, increasing tension between the Anglo citizenry and Mexican officials. In response to the rumblings, Mexican troops established military

---

little silver, Miranda was convinced that silver-bearing ore could be extracted in large quantities because of the extent of iron oxide rock he observed. Unfortunately for the Spaniards, this land was occupied by Apaches who killed prospectors in the area during the next few years. *Source; Caran, S. Christopher - 2000 The Lost Spanish Silver Mines of Llano County, Texas. Texas Heritage 18 (4)l:9- 35, Texas Historical Foundation, Austin.*

posts in several locations within the province, including San Antonio de Béxar Although much of the military supported the administration of President Anastasio Bustamante, Antonio López de Santa Anna led an insurrection against him in 1832. Anglo colonists in Texas supported Santa Anna and General José Antonio Mexía, who led soldiers into Texas to oust commanders loyal to Bustamante

After hearing that the Mexican army commander in Nacogdoches, José de las Piedras, had demanded that all residents in his area surrender their arms, Bowie cut short a visit to Natchez in July 1832 to return to Texas. On August 2, 1832, he joined a group of other Texians and marched into Nacogdoches to "present their demands" to Piedras. Before the group reached the building housing the town officials, they were attacked by a force of 100 Mexican cavalry.

The Texians returned fire and, after the cavalry retreated, initiated a siege of the garrison. After a second battle in which Piedras lost 33 men, the Mexican army evacuated during the night. Bowie and eighteen companions ambushed the fleeing army and, after Piedras fled, marched the soldiers back to Nacogdoches Bowie later served as a delegate to the Convention of 1833, which formally requested that Texas become its own state within the Mexican federation.

Several months later, a cholera epidemic struck Texas. Fearing the disease would reach San Antonio, Bowie sent his pregnant wife and their daughter to the family estate in Monclova in the company of her parents and brother. The cholera epidemic *instead struck Monclova*, and between September 6 and September 14, Ursula, their children, her brother, and her parents all died of the disease. Bowie, on business in Natchez, heard of his family's deaths in November. From then on, he drank heavily and became "careless in his dress".

The following year the Mexican government passed new laws allowing land sale in Texas, and Bowie returned to land speculation. He was appointed as a land commissioner, tasked with promoting settlement in the area purchased by John T. Mason. His appointment ended in May 1835, when President Antonio López de Santa Anna abolished the Coahuila y Tejas government and ordered the arrest of all Texians (including Bowie) doing business in Monclova. Bowie was forced to flee Monclova and return to the Anglo areas of Texas.

The Anglos in Texas began agitating for war against Santa Anna, and Bowie worked with William B. Travis, the leader of the War Party, to gain support. Bowie visited several Indian villages in East Texas in an attempt to convince the reluctant tribes to fight against the Mexican government. Santa Anna responded to the rumblings by ordering large numbers of Mexican troops to Texas.

The Texas Revolution began October 2, 1835 with the Battle of Gonzales. Stephen F. Austin formed an army of 500 men to march on the Mexican forces in San Antonio with the cannon that had precipitated the fight. On October 22, Austin asked Bowie, now a colonel in the volunteer militia, and James W. Fannin to scout the area around the missions of San Francisco de la Espada and San José y San Miguel de Aguayo to find supplies for the volunteer forces. The scouting party left with 92 men, many of them members of the New Orleans Grays who had just arrived in Texas. After discovering a good defensive position near Mission Concepción, the group requested that Austin's army join them.

On the foggy morning of October 28, Mexican General Domingo Ugartechea led a force of 300 infantry and cavalry soldiers and 2 small cannons against the Texian forces Although the Mexican army was able to get within 200 yards (183 m), the Texian defensive position protected them from fire. As the Mexicans stopped to reload their cannon, the Texians climbed a bluff and picked off some of the soldiers. The stalemate ended shortly after *Bowie led a charge to seize one of the Mexican cannons*, at that time only 80 yards (73 m) away.

Ugartechea retreated with his troops, ending the Battle of Concepción. One Texian and ten Mexican troops had been killed One of the men under Bowie's command during the battle later praised him "as a born leader, never needlessly spending a bullet or imperiling a life, who repeatedly admonished ," " *Keep under cover boys, and reserve your fire; we haven't a man to spare."*

An hour after the battle ended, Austin arrived with the rest of the Texian army to begin a siege of San Antonio de Béxar, where General Martín Perfecto de Cós, the overall commander of Mexican forces in Texas, and his troops were garrisoned. Two days later, Bowie resigned from Austin's army because he did not have an official commission in the army, and he disliked the "minor tasks of scouting and spying".

On November 3, 1835, Texas declared itself an independent state, and a provisional government was formed with Henry Smith of Brazoria elected provisional governor. Austin requested to be relieved of his command of the army, and Sam Houston was named army chief. Edward Burleson was chosen as temporary commander of the troops in San Antonio. Bowie appeared before the council at some point and spoke for an hour, asking for a commission. The council refused Bowie's request, likely because of lingering animosity over his land dealings.

Houston offered Bowie a commission as an officer on his staff, but Bowie rejected the opportunity, explaining that he wanted to be in the midst of the fighting. Instead, Bowie enlisted in the army as a private under Fannin. He distinguished himself again in the Grass Fight on November 26. Cós had sent 100 soldiers to cut grass for the horses.

As they returned to San Antonio, Bowie took 60 mounted men to intercept the party, which they believed carried valuable cargo The Mexican troops quickened their pace in the hopes of reaching the safety of the city, but Bowie and his cavalry chased them. At the end of the fight, the Texians had two wounded men, but had captured many horses and mules.

Shortly after Bowie left San Antonio, Ben Milam led an assault on the city. In the ensuing fighting, the Texians suffered only a few casualties, while the Mexican army lost many troops to death and desertion. Cós surrendered and returned to Mexico, taking with him the last Mexican troops in Texas. Believing the war was over, many of the Texian volunteers left the army and returned to their families. In early January 1836, Bowie went to San Felipe and asked the council to allow him to recruit a regiment. He again was turned down, as he "was not an officer of the government nor army."

After Houston received word that Santa Anna was leading a large force to San Antonio, Bowie offered to lead volunteers to defend the Alamo from the expected attack. He arrived with 30 men on January 19 where they found a force of 104 men with a few weapons and a few cannons, but little supplies and gunpowder. Houston knew that there were not enough men to hold the fort in an attack and had given Bowie orders to remove the artillery and blow up the fortification. Bowie and the Alamo captain, James C. Neill, decided they did not have enough oxen to move the artillery, and they did not want to destroy the fortress. On January 26, one of Bowie's men, James Bonham, organized a rally which passed a resolution in favor of holding the Alamo. Bonham signed the resolution first, with Bowie's signature second.

Through Bowie's connections because of his marriage and his fluency in Spanish, the predominantly Mexican population of San Antonio often furnished him with information about the movements of the Mexican army. After learning that Santa Anna had 4,500 troops and was heading for the city,

Bowie wrote several letters to the provisional government asking for help in defending the Alamo, especially "men, money, rifles, and cannon powder". In another letter, to Governor Smith, he reiterated his view that "the salvation of Texas depends in great measure on keeping Bexar out of the hands of the enemy. It serves as the frontier picquet guard, and if it were in the possession of Santa Anna, there is no stronghold from which to repel him in his march toward the Sabine." The letter to Smith ended, "Colonel Neill and myself have come to the solemn resolution that we will rather die in these ditches than give it up to the enemy."

On February 3, William Travis arrived with an additional 30 troops, and several days later Davy Crockett appeared with twelve Tennesseans. Neill went on furlough on February 17 to visit his sick family, leaving Travis, a member of the regular army, in command Bowie was older than Travis with a better reputation and considered himself a colonel, thus outranking Travis, a major He refused to answer to Travis, who called an election for the men to choose their own commander.

They chose Bowie, infuriating Travis Bowie celebrated his appointment by getting very drunk and causing havoc in San Antonio, releasing all prisoners in the local jails and harassing citizens. Travis was disgusted, but two days later the men agreed to a joint command; Bowie would command the volunteers, and Travis would command the regular army and the volunteer cavalry.

Fearing for the safety of his wife's relatives in San Antonio, Bowie invited her cousins Getrudis Navarro and Juana Navarro Alsbury, as well as Alsbury's 18-month-old son, Alijo, to stay inside the walls of the Alamo. Bowie had been ill, and two doctors, including the fort surgeon, were unable to diagnose his illness.

Travis became the sole commander of the forces when Bowie was confined to bed. Santa Anna and his army reached the outskirts of San Antonio de Bexar several days later and began a siege of the Alamo on February 24. The Mexican army raised a red flag to warn the defenders that no quarter would be given

Travis sent Juan Seguin to recruit reinforcements on February 25, and 32 additional men arrived. Thirty-five years after the Alamo fell, a reporter identified Louis "Moses" Rose as the only man to have "deserted" the Texian forces at the Alamo. According to the reporter's version of Rose's account, when Travis realized that the Mexican army would likely prevail, he drew a line in the sand and asked those willing to die for the cause to cross the line.

At Bowie's request Crockett and several others carried the cot over the line, leaving Rose alone on the other side After its publication, several other eyewitnesses confirmed the account, but as Rose was deceased the story can only be authenticated by the word of the reporter, who admitted to embellishing other articles, "and thus many historians refuse to believe it."

Bowie *perished with the rest of the Alamo defenders* on March 6, when the Mexicans attacked. Most of the noncombatants in the fort, including Bowie's relatives, survived. Santa Anna ordered the alcalde of San Antonio, Francisco Antonio Ruiz, to confirm the identities of Bowie, Travis, and Crockett. After first ordering that Bowie be buried, as he was too brave a man to be burned like a dog, Santa Anna later had Bowie's body placed with those of the other Texians on the funeral pyre.

When Bowie's mother was informed of his death, she calmly stated, "I'll wager no wounds were found in his back." Various eyewitnesses to the battle gave conflicting accounts of Bowie's death. A newspaper article claimed that a Mexican soldier saw Bowie carried from his room on his cot, alive, after the conclusion of the battle.

The soldier maintained that Bowie verbally castigated a Mexican officer in fluent Spanish, and the officer ordered Bowie's tongue cut out and his still-breathing body thrown onto the funeral pyre. This account has been disputed by numerous other witnesses, and it is thought to have been invented by the reporter.

Other witnesses maintained that they saw several Mexican soldiers enter Bowie's room, bayonet him, and carry him, alive, from the room. Various other stories circulated, with some witnesses claiming that Bowie shot himself and others saying he was killed by soldiers while too weak to lift his head. Alcalde Ruiz said that Bowie was found "dead in his bed."

According to Wallace O Chariton, The "most popular, and probably the most accurate" version is that Bowie died on his cot, "back braced against the wall, and using his pistols and his famous knife." One year after the battle, Juan Seguin returned to the Alamo and gathered the remaining ashes from the funeral pyre. He placed these in a coffin inscribed with the names of Bowie, Travis, and Crockett. The ashes were interred at the Cathedral of San Fernando.

Despite his continual pronouncements of wealth, Bowie's estate was found to be very small. His possessions were auctioned for only $99.50. His larger legacy is his position as "one of the legendary characters of the American frontier". Bowie left a "frustratingly sparse paper trail" of his life, and, for many, "where history failed, the legends prevailed." Although Bowie's name and his knife were well-known during his lifetime, his legend grew after October 1852, when *DeBow's Review* published an article written by his brother John Jones Bowie, called "Early Life in the Southwest—The Bowies." The article focused primarily on the exploits of Jim Bowie. Beginning with that article, "romanticized stories" about Bowie began appearing in national press. In many cases, "these stories were pure melodrama, with Bowie rescuing some naïve planter's son or damsel in distress:*Both Bowie, Texas and Bowie County, Texas are named in honor of Jim Bowie.*

'BIG FOOT' WALLACE

Big Foot was born William Alexander Anderson Wallace in Lexington, Rockbridge Co., VA., on the 3rd of April, 1817.

Overview

William Alexander Anderson WALLACE
(Also known as "Big-Foot" Wallace)

In 1870 John C.Duval who had earlier published a record of his own earlier frontier expoits wrote about 'Big-Foot Wallace' his friend both who had considerable experience during the Texas War of liberation. The Battle of the Alamo is well recorded but the massacre of their own division by the Mexican Army at Goliad is not too well known.

William Anderson furnished notes of his own early experiences to John C.Duval and assisted put together a biography and stated that it was " the plain unvarnished story of the *Adventures of Big–Foot Wallace, the Texas Ranger and Hunter*, written with notes furnished by him and told as well "as my memory serves me, *in his own language*."

Anderson journeyed to Texas in 1836 for the purpose of avenging his brother and cousin, who had both been killed in the Goliad Massacre. Duval had fiest hand experience of the same massacre his own story is harrowing and gives a contemporary view of the foraging indigenous tribes and the continuous presence of Spanish cavalry and his own epic struggle to stay alive in the wilderness after the slaughter of his fellow troops after they had surrendered.

There were marked incursion by Mexican forces and counter attacks by Texans in 1842. A Mexican army under the command of General Adrián Woll invaded capturing San Antonio. Wallace was present at the battle of Salado in the same year when General Woll withdrew his forces back into Mexico. "Big-Foot" Wallace volunteered for the Somervell Expedition. under the command of Alexander Somervell (a large expedition of approximately 700 Texans). Somervell's forces captured Laredo also Guerrero.

The Mexicans were preparing a large counter attack and Somerville excercising caution ordered his troops to return to Texas.. Most of his men refused to return to Texas vowing revenge for Mexican atrocities (at Goliad) and their 'no prisoner's' policy that emenated from as high up as Santa Ana himself. Under the command of William S.Fisher they continued on and this campaign was later known as 'The Mier Expedition".Ben McCulloch was operating on the west bank of the Rio Grande with a small group of Texas Rangers who were acting as scouts for the Mier Expedition. but withdrew as Mexican regulars assembled across the river planning their assault on the town of Mier. McCulloch's group withdrew across the Rio Grande.

The Mexican army under command of General Pedro de Ampudia arrived and fortified the town. The Texans needed supplies and decided to force a crossing. The Texans faced an army that outnumbered them, ten to one and there occurred an enormous amount of casualties. The Mexicans had over 600 killed and reports of 200 wounded. The Texans amazingly only lost thirty killed and wounded. The Mexican commander General

Ampudia raisedthe white flag and talked the Texans into surrending.(The Texans were worn out and lacked water and supplies).

William (Wallace) recalled that whenever the Mexicans "hoisted the white flag and succeeded in persuading the Americans into a 'parley,' they invariably *got the better of them in some way or other* [sic]." Wallace recalled "never shall I forget the humiliation of my feelings, when we were stripped of all our arms and equipments, and led off ignominiously". Thus began a long march to Perote Prison for the Texans. They, however, were not content to accept captivity. Throughout the march, guards paraded the Texans through towns they stopped in, drawing a "hooting mob" who cursed them and pelted them with rocks, dirt clods, and stale eggs. Wallace remembered it made him "quite proud" that they were the cause of all this noise and excitement.

Wallace, however, gratefully remembered that some Mexican inhabitants gave them drinks and tortillas. During the march, supplies ran short and sometimes not even tortillas could be had, and the Texans spent many evenings shivering around meager campfires exposed to cold winds and rain. Conditions such as these quickly convinced the Texans that they had to escape.

On 11 February 1843, thirty miles from Saltillo, their chance came. The Texans led by Captain Ewen Cameron overwhelmed their guards, captured weapons, horses, and supplies and quickly departed. Initially following the road during their escape, the Texans left it and turned into the mountains to evade any possible pursuit. Food and water could not be found in that barren country. He recalled "all night long I could hear the men moaning ... and crying out for water! water!" While most of the men followed Captain Cameron onward, some men wandered away, searching for water. The debilitated Texas were easily recaptured by Mexican cavalry.

The recaptured Texans were put under close guard and marched back to Saltillo, where the infamous black bean episode occurred. In response to the escape attempt, President Antonio López de Santa Anna *ordered every tenth man shot.* To determine who of the remaining 176 Texans would be shot, an earthen pot was filled with 159 white beans and seventeen black beans. Those unfortunates *who drew the black bean were executed.* Wallace and the surviving Texans were marched to Perote Prison, where many died due to starvation, disease, and wounds. A few were occasionally released upon requests from United States, British, or French officials.

The last of the Mier prisoners were released by Santa Anna on 16 September 1844, finally bringing this ignoble chapter of Texas history to a close. “Big-Foot” Wallace never forgot the treatment that he suffered as a prisoner and after his release was quick to join the Texas Rangers under John Coffee (Jack) Hays with whom he served throughout the United States war with Mexico. Duval’s Adventures of Big–Foot Wallace gives learners a ‘you are there’ understanding of the events surrounding the Meir Expedition. Larger concepts that can be drawn from this PSA are the ill will that the black bean episode generated among Texans toward Mexico. Many Texans like Wallace saw the Mexican-American war as the time to take their revenge.

The Adventures of Big Foot Wallace

THE TEXAS RANGER AND HUNTER
BY
JOHN C. DUVAL

*Author of "Jack Dobell, or, A Boy's Adventures in Texas'**
"The Young Explorers, or Boy-Life in Texas," Etc.

MACON, GEORGIA
THE J. W. BURKE COMPANY

Entered, according to Act of Congress, in the year 1870. by

J. W. BURKE A CO.
in the Clerk's Office of the District Court of the United States for the Southern District of Georgia

Preface

The writer of this little book is well aware that it will not stand the test of criticism as a literary production. A frontiersman himself, his opportunities for acquiring information, and for supplying the deficiencies of a rather limited education, have of course been "few and far between;" and therefore it cannot reasonably be expected that he could make a book under such circumstances which would not be sadly defective as to style and composition. However, it can justly lay claim to at least one merit, not often found in similar publications it is not a compilation of imaginary scenes and incidents, concocted in the brain of one who never was beyond the sound of a dinner-bell in his life, but a plain, unvarnished story of the " 'scapes and scrapes" of Big-Foot Wallace, the Texas Ranger and Hunter, written out from notes furnished by himself, and told, as well as my memory serves me, in his own language.

Big-Foot Wallace" is, perhaps, better known throughout Texas as an Indian-fighter, hunter, and ranger, than any one now living in the State ; which is saying a good deal, when the great number who have acquired more or less notoriety in that way is taken into consideration. Few men now living, I am confident, have witnessed as many stirring incidents, had more "hair-breadth escapes," or gone through more of the hardships and perils of a border life. He was a participant in almost every fight, foray, and "scrimmage" with the Mexicans and Indians that took place > in Texas after he first landed on her shores in 1836.

Pioneers, or frontiersmen, are a class of men peculiar to our country, and seem to have been designed especially to meet the exigencies of the occasion. With their "iron nerves," great powers of endurance, and indomitable "go-a-headativeness," they have been essentially useful in clearing the way through the wilderness from such obstacles as would have been perhaps insurmountable to those coming after them. Their mission has been very nearly accomplished. Like the flatboat-men of the Mississippi, who have entirely disappeared as a classsince the introduction of steamboats on that river and its tributaries, their numbers are steadily decreasing before the extension of railroads and the area of civilization.

Only here and there one is still found in our midst, whom disease, wounds, or old age have rendered incapable of further contests with the Indians and other denizens of the forests and plains, and of enduring the hardships and exposure of a life in the wilderness. As a class, frontiersmen are observant and knowing in all that pertains to their peculiar mode of life, and as deeply versed in all the mysteries of woodcraft as the wily savage himself; but they are guileless and unsuspicious as a child, and whenever they come in conflict with the shrewd, calculating man of business, they are as helpless as a "stranded whale." For this reason, they seldom accumulate property, and those who follow after them generally reap the reward of all their perils, toils, and hardships.

Wallace is no exception to this rule, and the best days of his life were freely given to the service of his country.

THE AUTHOR.

Sketch of Wallace's Life

WILLIAM A. WALLACE was born in Lexington, Rockbridge. County, Virginia, in the year 1816. He went to Texas in 1836, a few months after the battle of San Jacinto, for the purpose, he says, of taking pay out of the Mexicans for the *murder of his brother and his cousin*, Major Wallace, both of whom fell at "Fannin's Massacre." He says he believes accounts with them are now about square.

He landed first at Galveston, which consisted then of six groceries and an old stranded hulk of a steamboat, used as a hotel, and for a berth in which he paid at the rate of three dollars per day. From Galveston, Wallace went on to La Grange, then a frontier village, where he resided until the spring of 1839, when he moved to Austin, just before the seat of government was established at that place. He remained at Austin until the spring of 1840, when finding that the country was settling up around him too fast to suit his notions, he went over to San Antonio, where he resided until he entered the service.

He was at the battle of the Salado, in the fall of 1842, when General Woll came in and captured San Antonio. The fight began about 11 o'clock in the day, and lasted until night. General Woll had fourteen hundred men, and the Texans one hundred and ninety-seven, under Caldwell (commonly known as "Old Paint") Between eighty and one hundred Mexicans were killed, while the Texans lost only one man (Jett). Forty men, however, from La Grange, under Captain Dawson, who were endeavoring to form a junction with them, were surrounded and captured by the Mexicans, who massacred them all as soon as they had surrendered their arms.

In the fall of 1842, he volunteered in the "Mier Expedition," an account of which appears in this volume. After his return from Mexico, he joined Colonel Jack Hays's Ranging Company, the first ever regularly enlisted in the service of the "old Republic," and was with it in many of those desperate encounters with the Comanches and other Indians, in which Hays, Walker, McCulloch, and Chevalier gained their reputations as successful Indian-fighters.

When the Mexican war broke out in 1846, Wallace joined Colonel Hays's regiment of mounted volunteers, and was with it at the storming of Monterey, where he says he took "full toll" out of the Mexicans for killing his brother and cousin at Goliad in 1836.

After the Mexican war ended, he had command of a Ranging Company for some time, and did good service in protecting the frontiers of the State from the incursions of the savages. Subsequently he had charge of the mail from San Antonio to El Paso, and, though often waylaid and attacked by the Indians, he always brought it through in safety.

The Meir Expedition

During the fall of 1842, the "Mier Expedition," as it was called, was set on foot by the authorities of Texas, in retaliation, I suppose, for the then recent invasion of the Republic by General Woll, and the capture of the city of San Antonio. Of course, I was "on hand," as usual, and volunteered my services. Young men are always ready to volunteer on "wild goose" expeditions, and I was no exception to the rule; but as we grow older, we learn a thing or two, look some time before we leap, and don't "fly off the handle" quite so easily.

The place of rendezvous was the city of San Antonio, and the volunteers that assembled there for the expedition were placed under the command of Brigadier-General Somerville, an officer of good standing and considerable reputation. A motley, mixed-up crowd we were, you may be certain broken down politicians from the "old States," that somehow had got on the wrong side of the fence, and been left out in the cold; renegades and refugees from justice, that had "left their country for their country's good," and adventurers of all sorts, ready for anything or any enterprise that afforded a reasonable prospect of excitement and plunder.

Dare-devils they were all, and afraid of nothing under the sun (except a due-bill or a bailiff), and if they had been managed with skill and judgment, they would undoubtedly have accomplished all that was expected from the expedition; but dissension, that bane of raw troops, which has so often brought to grief expeditions of this kind, prevailed among our leaders, and in a short time after we had marched on and taken possession of Laredo, a little village on the eastern shore of the Rio Grande, the greater portion of our men became dssatisfied with the way in which matters were managed, and returned home.

We found no troops in Laredo, and no attempt was made by the inhabitants to defend it. By the time we reached the place we had run short of provisions, and a requisition was made upon the citizens for a supply, and after obtaining a small quantity in this way, we marched out of town, and encamped for the night about three miles below it. On our way to camp, a man named De Boyce was accidentally killed by the discharge of a gun in the hands of a messmate.

We remained at this camp all the next day, and the day after we took up the line of march down the left bank of the Rio Grande, keeping well to the eastward of the settlements on the river, so as to conceal our advance as much as possible from the enemy's scouts. We continued on southwardly for several days,through thickets of mesquite and other thorny shrubs, with which this country is covered, and then turning due west, soon struck the Rio Grande, about ten miles above Guerrero.

Here we found two or three small boats, and at once proceeded to cross our whole force, now less than five hundred men, as Colonel Bennett, with all the "drafted men," had left us previously, and taken the "back track" toward home. The boats were quite small, and but few men could go in them at a time, and the day was pretty well gone before one-half of our force had landed on the opposite side.

About this time, General Green, who had been sent forward with a small force to reconnoitre in the direction of Guerrero, came back, closely pursued by a considerable body of Mexican cavalry. The men that had landed were at once formed in line of battle by Colonel Cook, and awaited the anticipated attack of the Mexicans; but, after circling round us and watch ing our movements from a respectful distance, they galloped off, and we saw them no more that day.

The next morning the balance of our men were crossed over to the western side of the river, and we marched at once upon Guerrero. When within half a mile or so of the town, the alcalde came out to meet us, and beggd us not to enter the place, as the citizens had been informed we intended to burn it and put them all to the sword. His request was complied with, and we encamped on the Salado, a small stream about a mile from the village.When we were all anticipating every moment the order to advance upon the town, we were astonished by receiving an order from our commander to "prepare to retreat, and to recross the river as soon as possible." We supposed, however, that of course our commander had received some information of the movements of the enemy that rendered this sudden retreat necessary, and we were soon in motion, and never halted until we were all safely landed again on the eastern bank of the Rio Grande, where we encamped for the night.

Rumors were rife among the men as to the causes of this retrograde movement, but nothing definite in regard to it seemed to be known to any one. Some said that our commander, General Somerville, had received orders from the President of Texas to abandon the expedition, but whether this was so or not I never knew; at an rate, he left us the next morning, and took back home with him a considerable number of the men. I had good reason afterward to regret that I had not continued my retreat with this crowd, for verily "discretion is the better part of valor." Those of us that remained behind (I think, about two hundred and twenty-four, all told) determined, as far as possible, to accomplish the objects of the expedition, and Colonel Fisher, an accomplished and tried officer, was selected for our "commander-in-chief."

The morning after our new organization we recrossed the Rio Grande again, eight miles from Mier,and marched at once upon the town, which we took possession of without any opposition. We found no troops there. The inhabitants were quite friendly apparently, and readily furnished us with such supplies as we needed; but had I known the Mexican character as well then as I do now, I should have suspected their sincerity precisely in proportion to the friendly manner in which they welcomed us.

When I was at school, I learnt a Latin phrase," Timco Danaos et dona ferentes" (I fear the Greeks and the gifts they bring). And beware of the Mexicans, when they press you to hot coffee and "tortillas." Put fresh caps on your revolver, and see that your "shooting-irons" are all in order, for you will probably need them before long. They are a great deal more treacherous than Indians. If you can manage to get into an Indian's camp before he kills you, and can surprise him into offering you any little hospitality, even a drink of water, you are safe from him and his clan as long as you are with them. No temptation would induce him to "lift your hair."

But not so with the Mexican. He will feed you on his best, "sefior" you, and u muchas gracias" you, and bow to you like a French dancing-master, and wind it all up by slipping a knife under your left shoulder- blade ! And that's one reason I hate them so. I have respect for a bold and open enemy, but I despise your sneaking sort, who are forever hoisting "white flags" for the purpose of throwing you off your guard.

When we had got all the supplies we needed from the citizens of Mier, we re-crossed the river once more and pitched our camp about four miles east of the town. No plundering was permitted while we were in Mier, and everything we took from the inhabitants was duly paid for according to our own estimate of its value, and of course the prices were quite reasonable.

The next day after our return to the east side of the river, some of the scouts we had left on the west side to watch the motions of the enemy, came into camp and reported that a large body of Mexican troops were marching into Mier. This we regarded as a banter for a fight; so we struck tents, crossed the river once more and for the last time, and marched on the city, which, as our spies had truly reported, we now found strongly garrisoned by a considerable Mexican force.

Before we crossed the river, a certain number of our men were detailed to remain at camp as a guard for our horses and baggage, and it was my luck to be chosen one of them. But I did not fancy that sort of business, and resolved, by some means or other, to make one of the detachment that was to advance on Mier. As it happened, one of my messmates, who had as little inclination for the front as I had for guarding horses in the rear, proffered to exchange places with me, to which I, of course, readily consented, and at the time we were both, no doubt, well satisfied with the arrangement; but it wasn't a great while before I rued the trade I had made with him, and would gladly have swapped back again and given considerable "boot."

Between 3 and 4 o'clock in the evening, our whole detachment had crossed the river, and we at once took up the line of march for Mier. We saw no signs of the enemy, except a few mounted scouts, who retreated expeditiously as soon as we came in sight, until we came to the Alcantero, a little stream a few hundred yards from the town. Here we made all necessary preparations and advanced to the attack; but in place of hot coffee and tortillas, as in the former instance, we were received by the Mexicans with shouts of defiance, and heavy discharges from their "escopetas."

These "escopetas" are a short bell-mouth, bulldoggish looking musket, carrying a very heavy ball,which is "death by the law" when it hits, but that is seldom, for they shoot with little accuracy. They are good for nothing, except to make a noise, and a volley from them always put me in mind of the old saying about shearing hogs "Great cry and little wool." I never fired one of them but once, and that was at the battle of the Salado, near San Antonio.

During the fight, I came across a dead Mexican with one in his hands, and as my rifle was empty at the time, I hastily caught it up, placed the breech against my shoulder, as I was in the habit of doing with my own gun, and fired at a party of the enemy who were retreating from the field. My first impression was that I had been struck with a nine-pound cannon-ball. It kicked me heels over head, and I suppose kept on kicking me after I was down, for when I "came to"

I found that my nose was unjointed and two of my ribs stove in. I have since found that the Mexicans never place them to the shoulder, but hold them with both hands above their heads and fire at random, which accounts in a great measure for the little execution done by them. But to come back to my story. The Mexicans received us, as I said, with heavy discharges from these escopetas, and after some sharp skirmishing we got possession of a portion of the town, and the fighting began in earnest.

Among us there were some of the best marksmen in the world, backwoodsmen from Kentucky, Tennessee, and Arkansas, and every "greaser" that ventured to peep at us above the parapets of the houses, and round the corners of the streets, was sure to get a bullet through his head. In the meantime, with crowbars and picks, some of us were busily engaged in breaking through the stone walls of the buildings, and in this way we were rapidly advancing toward the "square," in the centre of the town. Night, however, came on and put an end, for the time, to the contest. So far, we had lost but one man killed (Major Jones, former Postmaster-General of the Republic). The Mexican loss must have been considerable, but we had no means of ascertaining the extent of it.

Just as the fight ended, two or three of us had picked our way into a room, where we found a table well covered with various sorts of eatables "chili con carne" "tortillas" etc., several bottles of "pulque" and a box full of fine "puros" or Spanish cigars. At first, we were a little suspicious of these things, thinking the Mexicans might have left them on purpose, for the benefit of the "gringos" and we touched them sparingly. But hunger at length got the better of our suspicions, for we had not eaten a bite all day, and we pitched into them, regardless of consequences, and made a jolly night of it.

At daylight the next morning the Mexicans began to blow their bugles and beat their drums, and to make a great to do generally; but Tom Hancock, a messmate of mine, who had been among the Mexicans a long time, and knew their character well, told us not to be alarmed, as they were merely playing a "game of bluff," and that he had always noticed that the more noise they made the less stomach they had for fighting.

But it seems that the garrison had been greatly reinforced during the night, and, confident in their numbers, they charged us in the position we occupied with more spirit than we had given them credit for. The fire from our rifles, however, was so rapid and deadly that they at length fell back in confusion, leaving the streets and plaza strewed with their dead and wounded.

For some time after this repulse they contented themselves with firing upon us at "long taw," from their port-holes and the flat roofs of their houses, but gradually growing bolder, as reinforcements came in, they charged us again and again with great impetuosity and courage, but each time with the same result.The last time, they pressed us so closely that our men, in many instances, after discharging their guns, fought them with rocks, which they tore up from the streets,and from the walls of the surrounding buildings.

Captain Cameron, a man of extraordinary strength, having just discharged his rifle, was observed by several of us to seize a rock and dash out the brains of a Mexican soldier who was in the act of charging upon him with his bayonet. Some of the Mexican soldiers told us subsequently that they were all drunk, having been furnished by their officers with as much "pulque" as they could drink, in order to stimulate their courage, which will account for their unusual intrepidity and daring.

I had a very narrow escape myself in this fight.After one of their charges upon us, in the excitement of the moment, I followed too far a party of retreating soldiers, when they suddenly turned upon me, and before I was aware of my danger, I was surrounded, and escape was apparently impossible. I was determined, however, to retreat at all hazards, and turned and dashed through their line. One fellow, as I passed, made a lunge at me with his bayonet, slightly wounding me in the left arm; but I made good my escape and rejoined my comrades, who had given me up for lost.

The adventure had the effect of cooling my courage considerably, and during the rest of the fight I kept within supporting distance of my comrades. The battle continued with great obstinacy on both sides, until perhaps one or two o'clock in the evening, when a temporary cessation took place of an hour or so, apparently by the tacit consent of both parties. And then affairs took a turn I have never been able to account for satisfactorily to this day.

Everything seemed to promise us a complete victory, for we had evidently been getting the best of the fight all day long, when the Mexicans hoisted a "white flag," and sent it into our lines by a Dr. Sennickson, one of our men that had been previously captured by them. Sennickson was accompanied by two Mexican officers, Algereto and Carasco. They bore a message from General Ampudia, requesting an entire cessation of hostilities for an hour longer.

This request was readily acceeded to by our commander, for no doubt he was fully under the impression that the Mexicans intended surrendering the town to us. You can form some idea of our astonishment, then, when in a little while another officer came into our lines, with a message from General Ampudia to the effect "that it was a useless waste of life for us to contend longer against his greatly superior force; that he had received large reinforcements, and had us completely in his power; and if we would surrender at discretion, he pledged us his word we should be treated with liberality and clemency, but that if any further resistance was made, every one of us should be put to the sword without mercy."

I had but little knowledge at that time of the Mexican character, but I have since learned whenever they hoisted the white flag and succeeded in persuading the Americans into a "parley," they invariably got the better of them in some way or other. It was so at the storming of Monterey, during the Mexican war. There, after three days' hard fighting, and just when we got the whole Mexican army completely in our power, they hoisted that same white flag, and "bamboozled" General Taylor into a "parley," and then into an "armistice" of thirty days, and finally into a "capitulation," which enabled them to march out with the "honors of war," and with all their arms and equipments; and he had to fight them over again at Buena Vista under the most disadvantageous circumstances.

A long "talk" followed between our officers and the messengers of General Ampudia, going and coming. Some of our officers were in favor of fighting to the last, but the "white flag" had produced its usual effect upon the majority, who were in favor of surrendering, providing General Ampudia would grant us such terms as they thought honorable and reasonable. These terms were at length decided upon, written out on paper, and forwarded to General Ampudia. After some further parleying, General Ampudia finally consented to ratify the capitulation upon the terms demanded. It was at this stage of the proceedings that General Green stepped out of the ranks, and called for a hundred volunteers to go with him and cut their way through the Mexicans. But the attempt seemed such a hopeless one, that no one responded to his call, and mortified, the gallant soldier dashed his rifle to pieces against the ground, and resigned himself to his fate.

We delivered up all our arms to the Mexicans, who marched us off in double file to our quarters in some deserted stone buildings. Never shall I forget the humiliation of my feelings, when we were stripped of all our arms and equipments, and led off ignominiously by a numerous guard of swarthy, bandy-legged, contemptible "greasers." There we were, two hundred good men and true as ever shouldered a musket (for we had lost only about thirty men killed and wounded in the fight), for no earthly reason that I could see, bound hand and foot, and delivered over to the tender mercies of these pumpkin-colored Philistines, and all through the workings of that miserable little "white flag." I could have cried with a right good will if I hadn't been so mad.

The force opposed to us in Mier, at the time of our surrender, as stated by the Mexicans themselves, amounted to something over three thousand men. Of these, upwards of five hundred, according to their own estimate, were killed in the fight. How many were wounded we never knew. In no other battle with the Americans, before or since, have the Mexicans ever displayed as much intrepidity and daring, and I have no doubt myself, that, as some of them told us, the greater portion of their soldiers were drunk on "pulque," with which they were liberally supplied by their officers during the fight. The guard we had left on the east side of the river, in charge of our horses and camp equipage, it seems got information of our surrender, by some means, in time to effect their escape. They made their way safely back to Texas.

As I have before stated, as soon as we had surrendered, they fastened us up in some deserted stone buildings, like so many pigs, where we were kept for five or six days with nothing to eat except a little dried beef, which was so tough I gave one-half of my rations to a messmate, who had a remarkably strong set of teeth, to chew the other half for me; and, to wash this down, we were furnished with a limited supply of muddy water from the Rio Grande. However, there was no use to complain; we knew we were "in for it," and principally through our own stupidity and folly, and we resolved to make the best of the worst situation in which we might be placed. As for myself, I was somewhat better prepared for this course of dieting than most of the men, having had, as I have already mentioned, a good meal the night before we surrendered.

Besides, on the morning of the second day after we were imprisoned, while I was sitting in front of a small grated window, looking out ruefully and hungrily upon the passers-by, a little Mexican maiden bless her little tawny hide came tripping alone, and suspecting, from my woe-begone visage, the empty condition of my stomach, made signs to me to know if I didn't want something to eat. I could not speak a word of Spanish at that time, but I easily made her understand that she had guessed how matters were with me precisely, and she forthwith tripped off, and soon returned with a batch of the inevitable tortillas, some red peppers, and a considerable chunk of roast kid-meat, which she handed to me through the little window.

I made her a low bow, pulled my forelock, and smiled as sweetly and as amiably as I could with my powder-burnt and dirt-begrimed countenance. She went off laughing at my grimaces, and turning a corner, I lost sight of my little pumpkin-colored angel forever. The tortillas were cold and tougher than army "flap-jacks," and the red pepper was as hot as mesquite coals; but I was as sharp-set as a new saw, and of course not inclined to "look a gift horse in the mouth." I felt just as grateful to the little saffron colored maiden as if she had feasted me on roast turkey and plum pudding, and I hope her lot has been a happy one tortillas, Chili pepper, and black-eyed papooses in abundance.

For five or six days we were kept closely guarded and watched in these miserable quarters, until the morning of the 28th of December, when a pompous little Mexican official came into our prison and told us to get ready immediately for a march, as we were to start at 10 o'clock that day for Camargo. But this intimation was hardly necessary, for we were prepared to start at a moment's warning, as everything we had in the world was upon our backs. The extra wardrobe of our whole force could easily have been packed in a lady's bonnet-box.

At the hour designated by the little official, our Mexican guard made its appearance. It was commanded by General Ampudia in person, and consisted of about six hundred infantry and a considerable body of cavalry, and a small company of artillery, with two six-pound fieldpieces. Certainly a most ample guard for two hundred half-starved, unarmed prisoners. After a fatiguing march of twenty-five or thirty miles, we reached the town of Camargo, and encamped in the vicinity, in commodious and well-ventilated quarters, at a corral or cattle-pen. Camargo differs in nothing from every other Mexican town I have seen the inevitable square in the centre, enclosed by lines of low, flat-roofed houses, with wretched "jacals" [pronounced "hakals"] built of sticks and mud, scattered promiscuously about the suburbs.

Nothing that I remember now worthy of note occurred on our route to this place. Sometimes we were very roughly treated by the guard. Whenever a poor fellow lagged behind the column for an instant, they seemed to take an especial pleasure in accelerating his speed by the vigorous application of a bayonet. A bayonet is undoubtedly a powerful persuader. I have seen men when broken down and u beat out" by a hard day's march, wake up to new life and energy on the receipt of some welcome news, and, under like circumstances, I have seen renewed vigor instilled into them by the spirited strains of a fine band of music; but nothing is so effective in this way as one or two inches of cold steel in the body. I know this is so, for I speak from sad and personal experience of the fact.

In the morning, after our guard had paraded us several times around the public square, to give the good people of Camargo a chance to look at the wild Texans," we bid them farewell, and again took the road down the river. Occasionally we were halted to rest for an hour or so at the "haciendas" and "ranchos" on the way.

In some places the inhabitants, and especially the women, seemed to compassionate the miserable condition of the "Gringos," as they called us, and gave us water to drink, and sometimes more substantial refreshments. In others, we were hooted at by the mob, that was sure to collect around us whenever we stopped for a few moments, who would call us by all sorts of hard names, and pelt us with stones and clods of earth, and stale eggs.

Foot-sore and weary, we at length reached Reinosa, a town about fifty miles below Camargo. Here we were received with discharges of musketry, and a general ringing of cracked bells, and as we were marched "triumphantly" into the public square, where banners of all kinds were flaunting in the air, the flat roofs of the houses and the porches and the balconies were thronged with women and children, anxious to get a peep at the "terrible Gringos." It made us feel quite proud, to think that we were the cause of all this noise and bustle. When they had paraded us around the square often enough to satisfy the curiosity of the good people of Reinosa, we were marched off to our quarters, in some old buildings just outside of the town.

We remained at Reinosa a day or two, and then took the road again for Matamoros. We never halted except at night, when we were "corralled," like so many cattle, on the bleak prairies, without any shelter, and scarcely food enough to keep body and soul together, until we reached the "Heroic City."

Is it any wonder, with the recollection of such treatment still fresh in their memory, that in the war which subsequently took place between Mexico and the United States, the Texans should have sent many a “greaser" "up the spout," without the formality of a court-martial to decide upon his guilt or innocence? However, I can say for myself that I never killed one in cold blood. I always turned them loose first and gave them a chance for their life; nevertheless, very few of them ever were heard of again, as in those days I was hard to beat in a "foot-race."

As we approached the city of Matamoros, a great crowd of men, women, and children came out to meet us, who were so anxious to get a look at the "wild Texans," that they couldn't await our arrival. Some were in vehicles of various kinds, some mounted on mustang ponies, and not a few upon the backs of the little "burros," or jacks, that were hardly as tall as a good-sized Newfoundland dog. Escorted in this way, we entered the city of Matamoros, as at Reinosa, amid the firing of muskets, the waving of flags, and the clanging of bells, and after parading us, as usual, several times around the public square, we were marched off to our quarters in the outskirts of the city.

Our men were so worn down by fatigue and "short commons" that it was found absolutely necessary to remain in the city for a few days, to give them a chance to recruit. But our stay there could hardly have been considered an entire respite from the hardships and discomforts of the road, for at that time the lower classes of the citizens of the "heroic city" were exceedingly bigoted and intolerant toward all "heretics," and especially the Texans, and, consequently, whenever an opportunity offered, they never failed to render our situation as disagreeable as possible. Some among the better class of inhabitants, however, were very kind to us, as were many of the foreign residents, and furnished us with occasional supplies of food and clothing; indeed, if it had not been for these charitable individuals, we might have suffered considerably during our stay in the city, as the rations furnished us by the "commissariat" were remarkable neither for quantity nor quality.

One morning, a little bow-legged chap, dressed in uniform, and covered with stars and spangles, came into our quarters, and made us a regular set speech, which was interpreted to us by an attendant. What was his purpose I could never understand, unless, like some of our politicians, he was resolved to neglect no opportunity of bringing himself prominently before the public.

He told us, among other things, what a great and magnanimous people the Mexicans were; that but for this fact, we would all have been taken out and executed as soon as captured, etc. and he finally wound up by saying that if the United States did not cease giving "aid and comfort" to the rebels of Texas, this great and magnanimous nation would collect an irresistible army, that would march from one victory to another, until the Mexican flag would float proudly from the dome of the Capitol at Washington City!

Here some fellow in the crowd sang out, "Oh ,spare the women and children!" and another, You had better whip Texas first, before you tackle Uncle Sam!" But the little hero paid no attention to them. He continued in the same strain for some time, and then with a "grand flourish" turned on his heel, and stalked away majestically, his spurs clanking and his little sword trailing on the pavement behind him. I wonder what he thought of the "magnanimous nation" when, a few years afterward, General Scott planted the Stars and Stripes on the "Halls of the Montezumas," with only ten thousand raw troops to back him !

Before we left Matamoros, a rumor reached the city in some way that General Rusk, with a considerable force of Texans, was on his way to attack it. In consequence of this report, General Ampudia and the detachment of troops that had guarded us from Mier were ordered away, and another guard of raw recuits, under General Canales, was substituted for it. This substitution of new levies for the regular soldiers, of which our previous escort was composed, emboldened us afterward to concert a plan to surprise them, and to make our escape into Texas. The officer in command of our guard (General Canales) was well known to many of us by reputation as a cruel, cowardly tyrant, and we knew very well we had no favors to expect from him, or from the ignorant, undisciplined recruits of which his force was composed.

We were detained at Matamoros for six days,and on the morning of the seventh took up the line of march again for Monterey. With all its hardships and discomforts, I much preferred being on the road to confinement in the filthy prisons and quarters of the towns and villages through which we passed. While marching, I could, at any rate, breathe the pure, fresh air of heaven without being hooted at and reviled by the mob or rabble that always collected around us whenever we were halted on the way.

On the first day's march I met with a serious misfortune in the loss of my "fine-tooth comb," which I had safely kept until then, in spite of our thievish Mexican guard, and which was my only "ark of safety" against the swarms of vermin with which they were infested. I have since lost articles of a thousand times more intrinsic value than that little horn comb, but never anything the want of which I felt so sensibly. The want of many things may be supplied by tolerable substitutes, but nothing will answer in place of a "ridding comb," under such circumstances.

Neither is there any substitute for tobacco. I have tried all sorts of leaves, herbs, and roots, and have never yet found anything that would take its place. Ever since we were captured we had suffered as much for the want of tobacco as for lack of sufficient and healthy food. The miserable little "cigaritos," with which we were occasionally supplied by the Mexicans (each containing about a good "pinch" for a snuff- taker) only served to tantalize us.

No one except an habitual consumer of the weed can appreciate or understand the soothing effects, after a hard day's march, of a pipe full of the genuine Virgina cut and dried." It will even mitigate, to a considerable extent, one's grief for the recent loss of dear friends. But these wretched little cigaritos always reminded me of the old saying of "feeding a hungry man on soup with a fork." I never got a chance at a good old solid plug again until after we were liberated and landed at New Orleans, on our way home.

We encamped, the first night out from Matamoros, near a forlorn-looking little Mexican "ranch," where nothing was to be had to eat except a few scraggy goats, and naked, sore-eyed dogs. Even the inevitable tortillas could not be had for love or money.Some of the boys pressed one of the little naked dogs into service for supper, but according to their report (for I did not taste it myself), it must have been a poor substitute for roast pig.

They said it was exceedingly tough, and when cooked, that "it smelt worse than a wet dog." The mess, however, to which I belonged was fortunate enough to kidnap one of the goats, which furnished us with a tolerable meal. If that goat has ever been paid for yet, I am not aware of the fact. The inhabitants of this "ranch" were certainly the most wretched-looking specimens of humanity that I had ever seen, except perhaps the Digger Indians of California. Men, women, and children were squalid and filthy beyond description, and one would have supposed that the very height of their ambition, in a worldly point of view, was to make shuck cigaritos and smoke them, and loll all day upon a beef-hide, eating dried meat and red pepper.

But this will apply equally as well to all the lower classes of the Mexican people as to the inmates of this ranch. If you wish to bring a Mexican to the lowest depths of despair, cut him off from cigaritos, red pepper, and tortillas. He might sustain existence for a little while under such distressing circumstances by the stimulus of gambling, but even that would afford him but a temporary respite.

If I were fishing for men, and wanted to catch an Englishman, I would bait my hook with the "TIMES," and a bottle of "Brown Stout;" if for an Irishman, I should use a small bit of the "blarney stone," and a "drop of the crathur;" if for a Frenchman, the customary frog and a little "eau sucre;" and for a Yankee I should put on a new copper cent and a "Congress penknife" for whittling; but if I wanted a "greaser," a cigarito, a pod of red pepper, and a tortilla would insure a bite at any stage of the tide.

The last chapter left us on the road to Monterey. One day's march was pretty much the counterpart of another, ending at night by our being driven into some corral or other enclosure, like so many cattle, where we were left to recruit our exhausted energies as best we might, without any protection from the weather, and with barely enough food to keep us from an absolute state of starvation.

On one occasion, while passing through a little village on the way, the customary crowd of men, old women,and boys flocked around us, shouting u death to the ladrones" (robbers), and "down with the hereticos" (heretics) . One little scamp, in the excitement of the moment, ventured within reach of my foot, which you see is a No. 12, and I gave him a kick which would have done credit to a vicious mule, and of which, no doubt, he has a distinct recollection to this day. He went off howling like a full-grown cayote, but, fortunately for me, none of the guard noticed this little by play, as otherwise I should certainly have been punished for it with a thrust from a bayonet, or cut from a broadsword.

When shivering around our miserable camp-fire at night, exposed to the cold winds or pitiless rains, we frequently debated among ourselves the possibility of surprising and capturing our guard, and, with the arms and ammunition we expected to secure in this way, of effecting our retreat into Texas before a sufficient force to retake us could be collected together.

We at length agreed that the attempt should be made at a little place called Sacata, one day's march ahead of us, and where we expected to be quartered for the night. Captain Cameron was chosen for our leader in the perilous undertaking, and it was understood, when the propitious moment for the attack should come, that he was to notify us of the fact by the command "draw," the usual order given when we were formed in line to receive our rations.

On our arrival at Sacata, we were turned into the corral as usual, and the guard took up their usual positions on the outside around us. Our plan for the attack was all arranged, and late in the evening, when our rations were brought in, every one fixed his eyes upon Captain Cameron, momentarily expecting him to give the wished for signal, but when he stepped out and said, "Draw your rations first," the men concluded that something had gone wrong, and every one quietly kept his place. The rations were then distributed to us, and the attack postponed for a time.

I never understood what reason Captain Cameron had for giving up the attempt of surprising the guard at Sacata, but I have no doubt it was a good one, for he was as reckless of danger as any man that ever lived. The next day we continued our march, and at length, wearied and worn down with the fatigues and hardships of our tramp, we reached the city of Monterey. A great crowd of persons, as usual, came out to meet us as we approached, by whom we were escorted into the city, but though they evidently regarded us as a species of wild cannibals, they treated us with more courtesy and respect than we had met with elsewhere, and on our arrival in the city we were at once conducted to our quarters.

They were much more cleanly and comfortable than those that had been allotted to us in other places. Still they were not such quarters as one gets, for instance, at the St. Charles Hotel, in New Orleans. No sofas, no lounges, no chairs, no tables, nor even a common stool to
dignify with the name of furniture nothing, in fact, but bare stone walls, marked and scribbled over with pieces of charcoal by former occupants of these luxurious abodes.

Here one of our guard was bitten the first night of our arrival by a tarantula, a sort of large venomous spider, and although the Mexicans tried many kinds of "remedies" to relieve him, they all failed, and he died in a few hours. I can't say I lost much sleep from grief on account of his death, for he was a noted tyrant, and treated our men most cruelly whenever he could do so with impunity.

Only a day or two before, I saw him cut one of our men severely with his sword, merely because he had loitered a moment behind the rest to tie his shoestring. On another occasion, when some of us were complaining of the scantiness of our fare, he went out and collected from the back-yard a quantity of old bones and other filthy offal, and returning, threw them on the ground before us, bidding us eat, as that was good enough for such abominable heretics as we were. Before he died, the priest was sent for to "confess" him, and if he only owned up to a tithe of the rascalities and crimes he had committed, I am sure the long list must have astonished even a Mexican padre.

Monterey was the handsomest Mexican town we had yet seen. It is situated in a rich and fertile valley, watered by clear cold streams that take their rise in the adjacent mountains. In the suburbs I noticed a number of handsome private residences, with beautiful grounds and gardens attached, in which the orange, and the lemon, and the fig, and many other tropical fruits and plants were growing luxuriantly. The city seemed to be well fortified, and some years subsequently, when General Taylor attacked the place with his forces, I had the satisfaction of seeing them pretty well battered by our artillery. They would, no doubt, have been demolished if it had not been for that "same white flag" which the Mexicans always hoist when in a tight place,and with which they "bamboozled" General Taylor on that occasion in the same way they had so often done the Texans before him.

We were much more kindly treated while in Monterey than we had been in Camargo, Matamoros, and other places on our route. We were indebted particularly to many of the foreign residents in the city, who took pity on our forlorn condition, and supplied us with provisions and such articles of clothing as we stood most in need of. Among our men there were a few Roman Catholics, and we noticed, as a general thing, that they fared better and were allowed more privileges by the guard than the rest of us. Several sudden conversions to the Catholic creed were the result of this, but I remained constant to my "heretical opinions;" for I did not think it right to assume the garb of religion for such purposes.

Before leaving Monterey, to our great joy, Colonel Barragan, a kind and humane man, was placed in command of our guard, and the cowardly tyrant Canales assigned to some other duty. We remained in Monterey three or four days, and then took up the line of march once more this time for Rinconada, a large hacienda about thirty miles distant, situated near the river San Juan. On the route, the question of attacking our guard was discussed, and we determined to make the attempt the first good opportunity that might present itself. We knew that the farther we penetrated into the country the poorer would be our chances of making our way back safely to Texas after we had effected our liberation.

On our arrival at Rinconada we were quartered in an old building, and our guard in two adjoining ones, one on each side. After supper, our project of attacking the guard was talked over again, and it was resolved to carry it into execution at that place. The attack was to be made at daylight in the morning. At the appointed time we were all up and ready for the fray, but just then we heard the rolling of the Mexican drums, and, looking through the chinks in our quarters, discovered the whole Mexican force filing out from the rooms they had occupied, and in a few moments we were ordered to fall into line and prepare to march.

There was no doubt in our minds that our plans had been divulged to the Mexicans by some one, but who it was we never knew to a certainty. What could have been his motive for betraying us it is hard to say, unless he thought by so doing he would secure better treatment from the Mexicans for himself. Had not our secret been divulged, it is more than probable we would have effected our intended surprise of the guard at Rinconada, from whence the chances are we could have made good our retreat into Texas, before a sufficient force of regular soldiers could have been collected to stop our advance. What a number of lives might have been saved, and what an amount of suffering avoided, if we had not been thus basely betrayed.

We at length reached Saltillo, without anything of particular interest transpiring on the way. Saltillo is a city of considerable importance, containing, I suppose, some ten or twelve thousand inhabitants. Here our guard was considerably reinforced, which satisfied us all the more that our intention of attacking them had been divulged by some one. We remained at Saltillo but a day or two, and then continued our route through a desolate and barren country, toward the city of San Luis Potosi. Our guards were so wary and vigilant that we had no chance of surprising them, until we came to Salado, a little place about one hundred and twenty-five or thirty miles from Saltillo, near which we were quartered for the night. We were confined in some old adobe buildings, and our guards, as usual, were posted in front of us. Here we were resolved to carry our plan of attack into operation, no matter what might be the result.

Next morning about sunrise, observing that the guard had stacked their gunswhile engaged in cooking their breakfast, we held a hurried consultation as to what was best to be done, and proceeded at once to put it into execution. Our plan was for Captain Cameron to "raise the yell" when he thought the proper moment for the attack had arrived, and at this signal our men were to rush out suddenly, knock down the sentinels stationed at the doors, and hasten to seize the guns that were stacked in front of the Mexican camps all of which we hoped to accomplish before the Mexicans could recover from the disorder into which we thought the suddenness of our attack would throw them.

Captain Cameron took his seat near the open door, for the purpose of watching the movements of the Mexicans, who were scattered about here and there, busily engaged in cooking their breakfast, and totally unsuspicious of our designs. We anxiously waited for the concerted signal from him, and when it was given a "yell was raised" that might have been heard for miles, and out we poured from our dens like a pack of ravenous wolves. In an instant, the sentinels who were stationed at the doors were knocked down and trampled under foot, and we dashed forward as rapidly as possible to where the guns were stacked.

The Mexican soldiers made a rush for them at the same moment, and a fierce struggle took place for their possession. But the Americans generally had the advantage over their foes in strength and weight, and the contest was of short duration. As soon as we had secured the guns, the Mexicans fled in the wildest confusion, leaving ten of their number dead upon the ground. Our loss was five killed Dr. Brennan, Fitzgerald, Rice, Lyons, and Haggerty; wounded Hancock, Captain Baker, Sansbury, Harvey, and another, whose name I have now forgotten. Colonel Barragan made every effort to rally his men after they were thrown into confusion by the suddenness of our attack.

He exerted himself manfully to retrieve the fortunes of the day, but without success, for the Mexican soldiers were thoroughly "stampeded," and nothing could stop their precipitate flight. At length he succeeded in rallying a small party of the fugitives, and with these he returned. Advancing fearlessly up to our lines, he addressed himself to Captain Cameron, telling him that if we would surrender ourselves to him, he pledged his word that we should be treated kindly, and not be punished for anything we had done. He said a good deal more to the same purpose, but we were deaf to all his arguments and he finally rode off and left us.

Before Colonel Barragan left, in consideration of his bravery and of his kind and humane treatment when in command of our guard, Captain Cameron ordered his horse, sword, and other equipments, which we had captured, to be returned to him. Captain Cameron then recommended our wounded men to his care, and he promised us he would take charge of them and protect them to the extent of his power, and this promise we afterward learned he faithfully kept.

And I here take great pleasure in giving my testimony of the fact that this Colonel Barragan was a clever fellow and a gentleman, although he had the misfortune to be born a Mexican. He was another example of the truth of the old saying, that bravery is always an ingredient of a magnanimous disposition. The domineering, tyrannical man is generally, though not always, a coward at heart. I have seen butfew exceptions to the rule.

We captured all the guns the Mexicans had, and ninety-seven horses and mules, and all their baggage and equipments. Having selected the best guns for our own use, we broke up the balance. I think I destroyed nearly a cart-load myself. I took an especial pleasure in demolishing the "scopets," remembering the kicking I got from one of them on a former occasion. After destroying the arms, and burning up all the tents and camp-equipage belonging to the guard, we put out from Salado with as little delay as possible, taking the road to Agua Nueva, and riding and walking by turns, as we had only one animal for every two men.

About midnight we stopped for a few moments at a "ranch" on the road, to get something to eat, and then continued our route and traveled until about sunrise the next morning, when we halted for several hours to rest ourselves and our jaded animals, having made upward of seventy miles from Salado, our starting point. On the road we went at a full trot the whole time, those on foot keeping up with the horsemen at that gait until exhausted, when they would take their turn in riding.

The Mexicans, who are hard travelers themselves, particularly when an enemy is behind them, could scarcely believe we made so great a distance in such a short time, and under such circumstances. Our object, of course, in making this forced march, was to get so far ahead on the road as to render it next to impossible for a Mexican force to pursue us with any chance of overtaking and recapturing us ; and had we followed the road from this on,as we ought to have done, there is every probability we would have succeeded in making our way back to Texas; but, as will be seen hereafter, rash counsels prevailed among our readers, and we were persuaded to leave the high road and take our course through an unknown and barren country, in which neither food nor water could be found.

We remained but a short time at our halting place,and then took up the line of march again. After traveling some eighteen or twenty miles, we quit the road, and turning to the left, entered a deep ravine that ran up toward the mountains. Passing for some distance along this, we at length came out upon an open plain, in which we discovered a ranch in the direction we were traveling.

We hurried forward toward it, hoping to get a supply of water, but as we approached it more nearly, we saw that the housetops and windows were crowded with soldiers. We concluded it would be paying too dearly for a little water to storm such a strong position to obtain it, so we passed on.

After crossing a spur of mountain we came to a little pool of muddy water, from which we procured enough to quench our thirst. Some miles further on we came to a ranch, and finding out from the people living there that we were considerably to the east of the course we wished to pursue, we turned more to the west and traveled on till night overtook us, when we encamped at the foot of a mountain.

The next morning we were on the way again by sunrise, and after going some distance we came to the road running to Monclova. This we followed for a few miles, when we met an Englishman (an old acquaintance of Captain Cameron), who advised us to go with him to a ranch he owned near by, where he told us he could furnish us with provisions, of which by this time we were nearly destitute.

He urgently advised us not to leave the road and go off into the mountains, as the majority of our men and officers were in favor of doing, and told us there were no troops along the road to intercept us, and that if we abandoned it and turned off through the desolate and mountainous country that lay in the direction we proposed traveling, we would inevitably perish for want of food and water.

Well would it have been for us if we had done as this Englishman advised, for I am satisfied now that he was truly anxious to aid us in effecting our escape into Texas; but somehow the majority of us came to the conclusion that he was not to be trusted, and that he was merely seeking to delay us on the way until a sufficient force should have time to pursue and overtake us.So we declined accepting the friendly offers and advice of the Englishman, and continued our route until late at night, when we came to a little place called San Felipe, where we left the road and encamped at the foot of the mountains.

The next morning, we continued our course over the dreary-looking mountains that rose up before us, and their barren and desolate appearance disheartened even those who had been the warmest advocates for seeking the protection of their solitudes. All day long we toiled through this sterile region, until night overtook us again, when we encamped at the bottom of a "barranca" or deep canon. Several times during the day our progress was impeded by deep gulches and almost perpendicular walls of rock, through and up which we found it nearly impossible to force our jaded and broken-down hacks. Not a single drop of water had been seen upon the whole route, and, thirsty and dispirited, we wrapped our scanty covering around us and lay down upon the cold ground, and endeavored to forget our troubles in sleep.

Some time during the night, a small party, that had been sent in advance of the main body to reconnoitre,came into camp and reported an abundance of water within two or three miles of us. This was joyful news to us, and all hands were at once roused up, and without attempting to preserve any order in our line of march, we stumbled along in the darkness among the gulches and ravines, each one for himself, until we all safely reached the desired locality, and, after slaking our thirst with hearty draughts of the grateful fluid, we lay down once more and slept soundly until morning. We then went back to our first halting-place, gathered up our half-starved, broken-down horses, and such of our equipments as had been left behind, and returned with them to our "water-hole."

Here we concluded to remain and recruit ourselves for the balance of the day, and as it was deemed impossible to take our poor jaded hacks any farther through the rugged and mountainous country ahead of us, and as there was not a blade of grass in all that barren region, we determined, as a matter of humanity, to kill them all, and thus save them from the miseries of a prolonged death from starvation. This we proceeded at once to do, cutting the throats of the poor animals with our butcher-knives, in order to save our ammunition, of which we had but a limited quantity. This was a job I didn't fancy at all, and when I came to cut the throat of the wretched "Rosinante" that had borne my comrade and myself so faithfully and well ever since our escape from the guard, he looked at me so knowingly and pleadingly out of his sunken, hollow eyes, that my heart failed me entirely, and my comrade, who was not so "squeamish," had to play the part of executioner.

When the horses had all been killed, we selected a few of them that were in the best condition (and a decent Mexican buzzard would have disdained to whet his bill upon any one of them), skinned and cut them into small strips, which we "jerked" over fires made of a sort of weed, the only fuel to be found in many places in these barren mountains. When we had prepared as much of the meat in this way as the men could conveniently carry in their knapsacks, it was divided out among them, and at daylight the next morning we were again on the march.

Little did we think, when we left this camp, that we had tasted the last drop of water that was to moisten our parched lips for six weary days and nights; but so it was. Oh, those wretched, wretched days and nights! I shall never forget them as long as I live. After toiling all day over broken, barren, rugged mountains, suffering the agonies of torturing, excruciating thirst, we laid down at night upon our hard rocky beds, to dream in broken and disturbed slumbers of bubbling springs and rippling brooks, which somehow always mysteriously disappeared when we were just in the act of quaffing their sparkling waters.

With much difficulty, even on foot, we at length forced our way up the almost perpendicular walls of the mountain that rose up in the direction we proposed traveling. When, at length, we had reached the top, we were appalled by the dreary prospect that presented itself to our view. Behind us lay the dismal valley from which we had just emerged, over the surface of which, scattered here and there as they had fallen, lay the dead carcasses of our animals, looking like little specks in the distance; and the thin columns of smoke, rising up from some smouldering fires we had left burning, and the naked rocky hills that encircled it, gave it a remarkable resemblance to the crater of some slumbering volcano. Before us, as far as the eye could reach, mountain after mountain, rose up, rough, rugged and broken, and the total absence of everything like vegetation too surely indicated that no water was to be found in the parched-up valleys that lay between them; and over all this dismal prospect a burning sun poured down its scorching rays from a cloudless sky, with an intensity that greatly aggravated the thirst from which we had already begun to suffer excessively, before the toils of this weary day had ended.

We were much depressed in spirits by the barren and desolate appearance of the country ahead of us; still there was no alternative but to go forward, for we knew if we retraced our steps we would inevitably be recaptured by our vigilant foes, and anything, we thought, was preferable to that. And so we went on all through that weary day, up one rugged, rocky mountain, and down, and up another, still hoping and trusting that in some of the ravines or canons that lay between them (for they could not be called valleys), we would, at length, come across a stream or pool of water sufficient to slake our thirst.

But everywhere we were continually disappointed. No murmuring streams rippled over the sandy beds of the gulches and canons we passed all were parched up, and as dry as if a drop of rain had not fallen in that dreary region since the days of the Flood. And thus we struggled on till darkness closed over the scene, when ,abandoning all hope of finding water that day, we encamped for the night among the broken rocks and debris at the bottom of a deep canon.

In the morning, dispirited and unrefreshed by our comfortless bivouac in the canon, we again took our toilsome way over gulches and ravines, and up and down the steep and scraggy sides of the mountains, that everywhere impeded our progress. And thus we staggered on for several days in succession, through this desolate wilderness, still hoping for, but never finding, the cooling stream or pool of water, in which we longed to slake the intolerable agonies of thirst that was consuming our very vitals.

On the morning of the fifth day we made an early start, in order to get over as much ground as possible before the sun should acquire its full power. But, by the time it had risen above the tops of the mountains to the eastward of us, the suffering of the men became so intolerable that many of them, to relieve themselves of all superfluous weight, threw away their guns and equipments, and what remained of their rations of jerked meat for hunger was not felt or feared our whole craving was for water! water!

Many of the men gave out entirely, and laid down on the wayside to die, but no one paid any attention to them, for great suffering, such as we were enduring, is apt to render men callous and unfeeling toward each other. Still the rest of us struggled on, hoping that our strength might hold out until we came to water; but we toiled up one rugged, barren mountain, only to see another as rugged and barren rise up before us.

The way appeared interminable, and no change of scenery varied the tiresome monotony, or gave us any indication of our approach to the long-wished-for stream or pool of water. Not a tree, nor a sprig of grass was to be seen anywhere. Nothing like vegetation, except in a few localities, where a species of leafless, withered weed managed, in some way, to draw a precarious sustenance from the parched and gravelly soil. All was barren, desolate, and scorched up by the long-continued drought that had prevailed in that country.

Not a living animal was to be seen, nor was the song of a bird, or even the chirping of a cricket ever heard during all our wandering in this wilderness which might appropriately have been termed anything but "howling;" for even a "cayote" could not, by any possibility, have existed there for many days. Night at length overtook us again, and, worn out, despairing, and suffering indescribable tortures from thirst, we threw ourselves upon the ground to pass away, as best we might, the wretched hours till morning.

All night long I could hear the men moaning in their uneasy slumbers, and crying out for water! water! I had heard that by chewing a leaden bullet thirst could be partially alleviated, and I tried it upon this occasion, but without success. It may afford some relief when the thirst is not excessive, but in the extremity to which we were reduced for the want of water, it did not seem to have the least effect.

The suffering caused by hunger is not comparable to that resulting from long-continued thirst. I had rather starve a week than go two days without water in warm weather, especially when compelled to travel on foot. Hunger comes by "fits and starts." There are intervals, even after starving for days, when one is comparatively free from pain or suffering, and in sleep one can occasionally find temporary relief from its pangs. But with thirst it is very different.

There is not a moment's cessation of the suffering, but, on the contrary, it increases regularly in intensity, until the tongue becomes black and swollen and protrudes from the mouth, and the eyes, bloodshot and bleared, seem as if they are about to start from their sockets. Even sleep affords no respite whatever from this terrible anguish. Horrible shapes gibber and moan around the wretched slumberer, and frighten him away from the gushing spring or rippling brook that forever haunts his mind, waking or sleeping.

One spring in particular haunted me in my dreams one that I had often frequented in my boyhood's days, in one of the gorges of the Blue Ridge in Virginia. It poured out from the mouth of a small cave, at the foot of a high bluff, in a bold rivulet larger than my arm, which ran off through an old mossy trough six or eight feet long to the edge of a ravine, down which it pattered and splashed with a musical sound very agreeable to the ear of the passer-by on a hot summer's day. During my disturbed slumbers that night, if I attempted to take a drink at that spring once, I tried it a thousand times, but something always prevented me, just as I was on the eve of placing my parched lips to the mossy spout.

On the morning of the seventh day after our entrance into these dismal solitudes, we resumed our cheerless march again, or rather, I should say, those of us that still remained together under the leadership of Captain

Cameron; for by this time many had fallen by the way from exhaustion, and others singly or in small squads had wandered off in various directions in their search for water. Mechanically as it were, and scarcely knowing or caring whither we went, we crawled up the side of the sterile mountain ahead of us, frequently pausing on the way to recover our breath and strength. At last, when with much labor we had gained the top of the mountain, a joyful prospect presented itself to our view, and instilled new life and vigor into our worn-down frames. Below us lay an extensive valley, covered in places with thick chaparral, and in others with clumps and groves of trees, indicating certainly, as we thought, the proximity of water.

Resting for a few moments, we hastened down the mountain as rapidly as we could, toward this "land of promise," but the distance to it, and the difficulties of the way were much greater than we had at first supposed, and the day was pretty well advanced before we reached the level plain. But even then, in vain did we search every gully and ravine for the long-wished- for element. Not one drop of water could we find. Some of the stoutest and strongest men now "gave in" entirely, and dropped down by the way, either exhausted by the fatigue they had undergone, or else utterly overcome by the sufferings they endured from the intense thirst that preyed upon them. Others wandered off into the dense "chaparral" in their frantic search for water, and were lost.

When the sun went down, on this wretched day, only fifty or sixty men still dragged their weary limbs after our gallant leader, Captain Cameron, of which remnant I was one. But so terrible were our agonies from the internal fires that consumed our vitals, that we never thought of halting as usual when night overtook us, but staggered on, like drunken men, scarcely conscious of what we were doing. Some wandered in their minds, and prayed, or cursed, or sang, as their different dispositions prompted.
I noticed one poor fellow, as he stumbled along in the dark over the stones and bushes that beset our path, cheerily singing, in a very faint voice, "And who'll be king but Charlie." He was as crazy as a bed-bug, of course.

About 10 o'clock at night, we discovered some fires burning ahead of us, and supposed that they were the signal fires of some one of our straggling parties, to let us know that water had been found. We pushed on toward them as fast as we could, and scarcely had the foremost men reached them, when a large Mexican force rose up on all sides of us, and enclosed us in their lines. No attempt at resistance was made; indeed, we would have been utterly incapable of anything of the sort, even if we had been armed, but as well as I remember now not more than ten or a dozen of our men still carried their muskets with them the rest having thrown them away while in the mountains. The Mexicans seized us at once, tied us in pairs together, and laid us on the ground. We begged and implored them, in the most piteous terms, to give us some water, and they measured out from the gourds they had brought along with them a very small drink for each one of us, not venturing to trust them in our hands for fear we should take too much at one time.

Though not the twentieth part of what would have been sufficient to quench our thirst, I shall never forget as long as I live the delicious taste and coolness of that drink of water! If you want to know what that "nectar" is which is said to be imbibed by the gods alone, travel for six days under a burning sun, without a drop of water to cool your "coppers," and then take a long "swig" at a Mexican gourd, filled to the brim with the pure element. Then, and not till then, can you fully appreciate its great superiority over all other drinks.

At early dawn the next morning, the Mexicans untied us, and we all started for the nearest water. From the place where the Mexicans recaptured us, as well as I remember, it was fifteen or twenty miles distant, and without their guidance and assistance we could never have made half the distance, even if we had known in what direction to travel. But such of us as were the weakest and least able to walk were mounted on horseback, and in this way we reached the "water-hole" by 12 or I o'clock in the day.

On the route, the Mexicans gave us two or three times a small drink of water from the gourds they had brought along with them, but this only seemed to aggravate our thirst the more, though I have no doubt it was better for us than it would have been if we had been supplied with all we wanted at one time. I am inclined to think so from the fact that, soon after we were recaptured by the Mexicans, one of them carelessly left his gourd, filled with water, within reach of one of our men, who instantly snatched it up and drained it dry before he took it from his lips. He was seized with terrible pains a little while afterward, and for several hours we thought he would die; but he eventually recovered.

When at length we came to the water (which was contained in a sort of artificial tank or reservoir), we were led down to it under guard, and only permitted to drink for a few moments; but before they succeeded in "horning me off," I am confident I secured at least a gallon "under my belt." But I experienced no bad results from having drank so freely; on the contrary, my strength was rapidly restored to me from that moment.

Here we found encamped the main body of the Mexican cavalry that had been sent in pursuit of us. The wretched appearance we presented seemed to touch, to some extent, even the callous hearts of our enemies, and with the exception of again tying us together in pairs, they offered us no indignity. In truth, we were as woe-begone a looking set of "scarecrows" as were ever congregated together, I suppose, at one time. Some were without hats, some without shoes, and one could scarcely tell from the shreds of clothing that still hung about us to what garments they originally belonged.

I noticed one fellow who was hatless, and with but one shoe on, with one leg of his pantaloons torn entirely off, and nothing left of his coat except the collar and sleeves, and a few little strips of the lining still dangling behind his back. Our beards were rough and unshaven, and our matted and uncombed locks hung down in "swads" around our faces, pinched and sharpened by long abstinence from food and water, from out of which our sunken hollow eyes glared with a wild and demoniac expression not at all pleasant or assuring to the beholder.

We remained at this tank, or water-hole, for three days, during which time most of the men that had wandered off from us in the mountains, singly or in squads, were hunted up by the Mexicans and brought into camp. Only thirteen, I think, were missing, and as nothing was ever afterward heard from them, it is reasonable to suppose they ultimately perished from thirst, and no doubt their bones are bleaching to this day in some of the dreary gulches and ravines of those inhospitable mountains, from which we had made such a narrow escape ourselves.

Early on the morning of the fourth day, we were told by our guard to prepare for the march, and in a few moments they started off with us on the road toward Saltillo. To prevent all chance of another "uprising' on our part, of which they seemed to be continually in dread, they tied our hands securely behind us with raw-hide thongs, and thus "hampered" we had to march all the way back to Saltillo. I never knew before how necessary a free use of the arms was, to enable one to walk with ease and celerity. A twenty miles' march, with our arms pinioned down in this way, fatigued us as much as twice the distance would have done if they had been unfettered.

We were several days on the road, during which time nothing of interest that I remember now occurred. On our arrival at Saltillo, we were taken again to our old quarters, in which we were once more securely fastened up and closely guarded. I had been but a day or so in these dirty barracks when I fully appreciated the extent of the misfortune I had met with in the loss of my fine-tooth comb. And here I may as well say to over-sensitive readers, that perhaps it would be as well for them to skip to the bottom of this chapter, as it is not my wish or intention to say anything offensive to "ears polite "but I have started out with the determination of telling my story my own way, and I must do it, or abandon the attempt altogether. With this fair warning, I shall resume the story of what I endured from the loss of my fine-tooth comb.

Vermin swarmed in countless numbers in the miserable quarters in which we were confined. Even the bare floors at times were thickly covered with them. Our Mexican guard did not seem to mind them much; in fact, I rather think they liked them, and that, in some way or other, they were absolutely necessary to their health and comfort. I verily believe, if one of them had been suddenly freed from all sorts of the vermin with which they were infested, that he would not have slept soundly for a week afterward. They never use a comb, and of course it is only the larger and overgrown fellows they succeed in capturing by the primitive method of "looking each other's heads.' 1 These they "crack" between their teeth, apparently with much gusto and relish, by way, I suppose, of retaliation "bite for bite."

For want of a comb myself, I was compelled to have my hair cut off short and permit my finger nails to grow untrimmed. With these I became so expert, after long practice, that I could rake out a fellow above a certain size with unerring certainty, whenever by biting or crawling he designated sufficiently his exact locality. By this primitive method, I kept myself pretty well rid of all the full-grown chaps, but the "small fry" dodged the question entirely, and unfortunately it is the nature of the "brute" to commence the propagation of his species the moment he is hatched. Often when a fellow, by hard raking and combing, if lucky enough to be the possessor of a fine-tooth comb, has come to the conclusion that at last he has entirely freed his head from these disgusting occupants, he wakes up the next morning to find it as densely populated as ever. I speak advisedly on this subject, for I had many opportunities, while imprisoned in Mexico, of studying the "habits and customs" of all kinds of "vermin."

The evil was aggravated by our want of a change of clothing and the scanty supply of water furnished us for our ablutions. As for myself, I had worn from necessity the same suit of clothes I had on when we made our escape from the guard, and after traveling in them all this time over dusty roads, and sleeping in them at night upon the ground, it can easily be imagined that my costume was not exactly a suitable one for a ball-room or a fashionable assembly. But little was left of my shirt. My hat had long since gone by the board, and in place of it my head was partially protected from the sun by a red cotton handkerchief, wrapped around it somewhat in the fashion of a Turkish turbin.

I had but one shoe left, which was in a very dilapidated condition, and in lieu of the other a raw-hide sandal was strapped on my foot with leathern thongs. My coat was tattered and torn by thorns, and like Joseph's, from frequent mending with all sorts of materials, was of many colors, but the dirt or drab color predominated over all. It is impossible to describe the remnant of my pantaloons. They hung upon me in shreds that were inextricably bound together by thongs and strings, and upheld by a system of "tackling" as complicated as that of a seventy-four gun ship, to compare small things with great; and from out of these habiliments a countenance presented itself that had been guiltless of a thorough cleansing for I am ashamed to say how long. "Such a beauty I did grow!" If my old sweetheart, Jenny Foster, could have seen me then, I am sure her heart would have relented, and that she would have reversed that cruel decision that sent me "packing off" to Texas some years before.

Everything that grows in Mexico, or at least in that portion of it in which I had traveled, has thorns upon it, which will account satisfactorily for the dilapidation of my clothing in such a short space of time. Even the very grass has thorns or spurs upon it, as we frequently found to our cost, whenever, forgetful of the fact, we seated ourselves upon it for a moment's rest or repose. I remember once seeing a shrub in Mexico which at a little distance appeared to be covered with dense green foliage, but upon a nearer approach I discovered that it was leafless, and simply a mass of thorns, thorns growing out of thorns.

The prickly pear, or cactus, the varieties of which comprise three-fifths of the vegetable kingdom in that part of Mexico, are all armed with spears, pikes, and thorns, of all lengths and sizes, from the minute prickle, so small that it is impossible to ascertain its locality when fixed in the flesh, except by the irritation it causes, to pikes and spears long enough for a buffalo to impale himself upon. Everything in this wretched country seems to be constituted for a state of continual warfare, as if nature herself sympathized in some way with the normal political condition of its inhabitants. The cattle have the longest horns; the snakes all have fangs; every insect you touch stings or bites you, or both at the same time ; and, as I have said before, the trees, shrubs, and grasses are all thorny : even the frogs and toads, so soft, moist and flabby in other countries, are here protected by hard, dry, scaly hides, and horns upon the head.

There was one species of shrub, with long, crooked, cat-like talons, which was dubbed by our men the "tiger thorn," from the fact that its slender elastic branches were frequently held in a constrained position by the twigs and foliage of other shrubs, and when loosened by the passer-by, they would spring up voluntarily, as it seemed, and seize him with their crooked claws, very much as the tiger springs from the jungle and seizes upon its prey. I have left many a rag fluttering in the breeze upon the branches of this terrible "tiger thorn," proud trophies of the numerous contests we had had, and showing that I had always retreated from the field with the loss of part of my baggage and equipments.

We remained at Saltillo several weeks, awaiting, it was said, orders from Santa Ana, the President of Mexico, as to our further disposition. During this time many and conflicting rumors reached our ears as to what was to be our ultimate fate. For some days it was generally believed that we were to be sent on to the City of Mexico, where we were to be immediately released or paroled and shipped back home by the way of New Orleans. Then again it was reported among us, that a ispatch had been received from Santa Ana ordering the immediate execution of every one of us, but that General Mexier had refused to carry out the barbarous mandate.

I am inclined to think there must have been some truth in this report, as General Mexier, before we left Saltillo, resigned his joint ommission of Commandante and Governor of the State. Besides, I learned afterward, when a prisoner in the City of Mexico, that all the foreign ministers resident there, as soon as they heard of this order, remonstrated against it as barbarous and inhuman, and Santa Ana revoked the order, substituting in its place the one requiring the "decimation" or execution of every tenth man. We had no intimation, however, that any of these orders had been actually passed until both of the latter were read to us subsequently at Rancho Salado, where the "decimation" took place.

The Saltillo Lottery

After our long sojourn at Saltillo, we were one morning roused up by our guard, and told to get ready to march, as we were to start that day to the City of Mexico. A few moments afterward the guard paraded in front of our quarters. We were taken out and formed into line, and marched off on the road back toward Rancho Salado, where, some weeks previously, we had risen upon and surprised the guard under the command of Colonel Barragan. Handcuffed and bound together in pairs, to cut off all chance of our escaping or making another attack upon the guard, we were driven along the road at a gait that would have been "killing" even to men that were not fettered as we were.

On the evening of the fourth day, I think it was, after leaving Saltillo, we came in sight, once more, of the lonely, desolate Rancho Salado. The officer now in command of the guard, Colonel Ortez, had spoken kindly to us frequently during the day, telling us to "be cheerful and walk up fast, for that the sooner we arrived at the City of Mexico, the sooner we would be liberated and sent back home." Notwithstanding these assurances, from the first moment the men caught sight of the dismal old ranch, whether it was the dreariness of the locality, or the recollection of what had happened there when we rose on the guard, and of the sufferings and disasters that followed in the wake of that event, or whether it was some dim foreboding of the "bloody scene" that was to be enacted there again so soon, that weighed upon the minds of the men, I know not; but not a word was uttered by any one, as we trudged along silent and depressed, until we reached the hated spot, and were once more securely fastened up in the same corral we had occupied before.

But a few moments elapsed before an officer, accompanied by an interpreter, entered the corral, and calling our attention, proceeded to read to us from a paper he held in his hand, a mandate from the "Supreme Government of Mexico," ordering the instant execution of every tenth man. Some of the more sanguine among us fully thought that the paper contained an order for our release, and eagerly crowded around the interpreter to hear the joyful news; but when the purport of the writing was explained to us by the interpreter, this barbarous decimation of our number came upon us so unexpectedly that we stood for a moment stunned and confused by the suddenness of the shock. Then a reaction took place, and if our hands had only been unshackled, unarmed as we were, the old Rancho Salado would have witnessed another up-rising, ten times as bloody as the first; but when we looked upon our manacled limbs, and the serried ranks and glittering bayonets of the large guard drawn up around us, we saw at once that any attempt at resistance would be utter folly, and we quietly submitted to our fate.

It was determined that *the seventeen men to be executed should be selected by lottery*, and in a little while a squad of Mexican officers came into the corral, preceded by a soldier bearing an earthen vessel, which he placed upon a low stone wall bounding the farther side of the corral, and which was intended to hold a number of white and black beans, corresponding to the number of men and officers in our command. The Mexican officers stationed themselves near the earthen pot, to overlook and superintend the lottery, and see that every one had a fair chance for his life. One of them then proceeded to count out so many white beans, which he poured into the vessel, and then dropped in the fatal seventeen black ones on top of them, covering the whole with a thick napkin or cloth. We were then formed into line and drawn up in front of the low wall on which the earthen pot had been placed. Before the drawing began, they informed us that if any man drew out more than one bean, and either of them should prove a black one, he would be regarded as having drawn a black one solely, and be shot accordingly.

Our commissioned officers *were ordered to draw first*. Captain Cameron stepped forward, and without the slightest visible trepidation put his hand under the cloth and drew out a white bean. He had observed, when the Mexican officer put the beans in the pot, that he poured the white in first and the black ones on top of them, and then set it down without shaking, possibly with the intention of forcing as large a number as possible of the black beans upon our commissioned officers, who were to have the first drawing.

When he returned to his place in the line, he whispered to those nearest him, "*Dip deep, boys*," and by following his advice all the officers drew white beans except Captain Eastland. After the officers had all drawn, the "muster-rolls" of the men were produced, and we were called forward as our names appeared upon them. Some of the Mexican officers present were evidently much affected by the courage and nonchalance manifested by the men in this fiery trial; others, on the contrary, seemed to enjoy the whole proceedings hugely, particularly one little swarthy baboon-visaged chap, that looked as if he had subsisted all his life on a short allowance of red pepper and cigaritos. He appeared to take an especial delight in the hesitation of some of the men when they put their hands into the vessel, for even the bravest felt some reluctance to draw when he knew that life or certain death depended upon the color of the bean he might select.

Whenever there was the slightest hesitation, this officer would say, in apparently the most commisserating tone: "Take your time, mi nino (my child); don't hurry yourself, mi muchaco (my boy) ; be careful, mi pobrecito (poor fellow) ; you know if you get a black bean you will be taken out and shot in ten minutes" a fact we had already been fully apprised of. "Ah! that's unfortunate," he would say, when a poor fellow drew a black bean; "but better luck to you the next time." Yet, all the while he was talking in this way, in the kindest accents, a devilish grin on his baboon-face indicated the great pleasure he took in the anxiety and distress of these "poor fellows."

I am not of a revengeful disposition, but if that Mexican had ever fallen into my power, his chances of living to a "good old age" would have been miserably slim, and I could have recognized him among ten thousand, for his weazen features and his diabolical grin were indelibly impressed upon my memory. I'll tell you how I would have served him. I would have bought a bushel of black beans, cooked them about half done in a big pot, and made him sit down upon it and eat until he bursted. I'd have given him a dose that would have stretched his little tawny hide as tight as a bass-drum. He should have had his fill of black beans for once, to a certainty.

Those who drew black beans seemed to care very little about it. Occasionally one would remark, as he drew out the fatal color "Well, boys, the jig is up with me;" or, "They have taken my sign in at last;" or something of a similar character, and then give way to the next, apparently as unconcerned as if he had no interest whatever in what was going on around him.

There was but a single exception to this. One poor fellow, a messmate of mine, too, appeared to be completely overcome by his apprehensions of drawing a black bean. He stood until his own time to draw came round, wringing his hands and moaning audibly, and continually telling those near him that he knew he should draw a black bean; that he had a presentiment such would be his fate. When his turn came, he hung back, and absolutely refused to go up at all until a file of Mexican soldiers forced him forward at the points of their bayonets. He hesitated so long after he put his hand into the vessel containing the beans that a Mexican officer near him pricked him severely with his sword to make him withdraw it.

All this, of course, was immensely gratifying to the little baboon-faced official, who "ninoed" and "pobrecitoed" him in his kindest tones, all the while, though, evidently snickering and laughing in his sleeve at the fears exhibited by the "pobrecito." At last the poor fellow was forced to withdraw his hand, and his presentiment proved too true, for in it he held the fatal black bean. He turned deadly pale as his eyes rested upon it, but apparently he soon resigned himself to his inevitable fate, for he never uttered a word of complaint afterward. I pitied him from the bottom of my heart.

My name beginning with W, was, of course, among the last on the roll, and when it came to my turn to draw, so many more white beans than black had been drawn out in proportion, that there could have been no great difference in the number of each. I observed twenty-four white beans drawn out in succession. The chances of life and death for me were, therefore, not so very unequal. I will frankly confess, when I put my hand into the pot and this fact recurred to my mind, a spasm of fear or dread sent a momentary chill to my heart, but I mastered it quickly, and before even the lynx-eye of the little baboon-faced official detected any sign of such weakness. At any rate, he bestowed none of his endearing epithets upon me.

All the time the drawing had been going on I stood pretty close to the scene of operations, and I thought I could perceive a slight difference in the size of the black and white beans that the former were a shade larger than the latter. This difference, I know, may have been purely imaginary, but at any rate, I was eventually decided by it in my choice of a bean. When I first put my hand in the pot I took up several beans at once in my fingers, and endeavored to distinguish their color by the touch, but they all felt precisely alike. I then dropped them and picked up two more, and after fingering them carefully for an instant, I thought that one of them seemed a little larger than the other. I dropped that one like a hot potato, and drew out the one left. It was a white one, of course, or I should not now be here to tell my story but not a very white one, and when I cast my eyes upon it, it looked to me as "black as the ace of spades."

I felt certain for a moment that my fate was sealed, but when I handed it to the Mexican officer who received them as they were drawn out, I saw that he put it on the wall with the white beans, and not into his waistcoat pocket, as he had done the black ones. I knew then that I was safe, and the revulsion of feeling was so great and rapid that I can compare it to nothing except the sudden lifting of an immense weight from off one's shoulders. I felt as light as a feather, though I weighed at least one hundred and seventy pounds net (after all my hardship and starvation), exclusive of the remnant of clothing I had on me.

Among the last to draw was an Irishman, by the name of W, a fellow noted for his wit and humor, as well as for his reckless, dare-devil character. He put his hand into the pot, and feeling around, discovered that there were but few beans left in it. "Arrah now, my hinnies !" he said, "and is this the way you would try to desave an innocent man to his destruction? Faith, and there's not a dozen beans left in the pot, and I'll not draw one at all." He was peremptorily ordered to take one out immediately. "Oh, it's for murtherin' me ye are, ye bloody spalpeens," said Pat, "widout judge or jury. I see that as plain as the nose on my face. Yees let the ither men pick and choose as it suited 'em, out of scores of beans, and now there isn't a dozen left in the pot, and I've no choice scarcely at all. Divil take such a lotthery, say I. But I suppose there's no help for it, so here goes."

And Pat drew forth his bean, and everybody, even the Mexican officers themselves, I believe, were rejoiced when it proved to be a white one. "There, you bloody nagurs," said Pat, handing the bean to the officer, "it was a black one, but I offered up a short prayer to Saint Pathrick, you see, and in the twinkling of an eye, he convarted it into a white one ! Hooray for Saint Pathrick and Ould Ireland forever." When the drawing was completed, the white and black beans were carefully counted over again, and the number found to tally with that of the men. *Those that had drawn black beans were kept separate from the rest of us*, and, in a few moments after the drawing was concluded, *they were marched off in two squads*, and shortly afterward repeated volleys of musketry were heard, and we knew that their cares and troubles were forever ended in this world.

One of them, however, a man by the name of Shepperd, as we learned subsequently, made a most miraculous escape for the time being. When they were fired upon by the guard, Shepperd fell and pretended to be dead, though, in fact, he was only slightly wounded. He was left on the ground with the dead bodies of his companions, and when night came he got up and went off without being observed. The next morning, when the Mexicans examined the bodies again, they were greatly astonished to find that one was missing, and could not be accounted for satisfactorily in any way. Shepperd wandered around for several weeks without being recaptured, but at length he was discovered, taken back to Saltillo, and shot to death in the public square, and his body carried out and left unburied on the commons.

CYNTHIA ANN PARKER – INDIAN CAPTIVE

Cynthia Ann Parker, or *Naduah* (also sometimes spelled "Nadua" and "Nauta" means "Someone Found"), (ca 1827–1870) was an Anglo-Texas woman of Scots-Irish descent who was kidnapped at the age of nine by a Native American raiding party. Cynthia Ann was a member of the large Parker frontier family that settled in east Texas in the 1830s. She was captured in 1836 by Comanches during the raid of Fort Parker near present-day Groesbeck, Texas. She was later liberated at age 34 by the Texas Rangers.

Cynthia Ann Parker was born to Silas M. Parker and Lucy (Duty) Parker in Crawford County, Illinois. There is considerable dispute about her age, as according to the 1870 census of Anderson County, Texas, she would have been born between June 2, 1824, and May 31, 1825. When she was nine years old, her family moved to Central Texas and built Fort Parker, a log fort, on the headwaters of the Navasota River in what is now Limestone County. Her brother James was killed on the way from Illinois to Texas, when the wagon lost a wheel and he was struck through the chest with a piece of splintered wood.

Her Grandfather, Elder John Parker, the Patriarch of the family, had negotiated treaties with the local Indians, and historians conjecture that he believed those treaties would bind all Native Americans, and that his family was safe from attack. If so, this was a tragic error.

On May 19, 1836, a huge force of Comanche warriors approximately 500 strong, accompanied by Kiowa and Kichai allies, attacked the fort and killed a number of its inhabitants. During the attack, the Comanches seized five captives, including Cynthia Ann Parker The other four were released after the typical ransom was paid, but Cynthia remained with the Indians for nearly twenty-five years. She completely forgot whatever she had known as a white child, and became a Comanche[1].

---

[1] It is said that in the mid-1840s her brother, John Parker, who had been captured with her, asked her to return to their white family, but she refused, explaining that she loved her husband and children too much to leave them. She is also said to have rejected Indian trader Victor Rose's invitation to accompany him back to

Cynthia Ann Parker and Peta Nocona

Peta Nocona was one of the war chiefs present at the Fort Parker massacre, and had formed his own band of the Comanche called the Noconi or Nokoni, afterwards. He became the husband of Cynthia Ann Parker. A great tribute to his affection to her was that he never took another wife, though it was common among the Comanche for such a successful war chief to do so. The couple had three children, famed Comanche chief Quanah Parker, another son named Pecos ("Pecan") and a daughter named Topsanna ("Prairie Flower").

Recapture by Texas Rangers at Pease River

In December 1860, Cynthia Ann[2] and her 2-year-old daughter, Toh-Tsee-Ah (*Topsannah*), or "*Prairie Flower*", were among a Native American party captured at the Battle of Pease River by Texas Rangers led by Lawrence Sullivan "Sul" Ross.

After fierce fighting, the Comanche realized they were losing and fled. Ross and several of his men pursued the chief who had been giving orders. The chief was fleeing alongside another rider. As Ross and his men neared, the other rider held a child over her head; the men did not shoot, but instead surrounded and stopped her. Ross continued to follow the chief, eventually shooting him three times. Although the chief fell from his horse, he was still alive, and refused to surrender. Ross's cook, Antonio Martinez, who had been taken captive in Mexico after Nocona killed his

---

white settlements a few years later, though the story of the invitation may be apocryphal. A newspaper account of April 29, 1846, describes an encounter of Col. Leonard G. Williams's trading party with Cynthia, who was camped with Comanches on the Canadian River. Despite Williams's ransom offers, tribal elders refused to release her. Later federal officials P. M. Butler and M. G. Lewis encountered Cynthia Ann with the Yamparika Comanches on the Washita River; by then she was a full-fledged member of the tribe and married to a Comanche warrior. She never voluntarily returned to white society. Indian agent Robert S. Neighbors learned, probably in 1848, that she was among the Tenawa Comanches.

[2] On December 18, 1860, Texas Rangers under Lawrence Sullivan Ross attacked a Comanche hunting camp at Mule Creek, a tributary of the Pease River. During this raid the rangers captured three of the supposed Indians. They were surprised to find that *one of them had blue eyes*; it was *a non-English-speaking white woman with her infant daughter.* Col. Isaac Parker later identified her as his niece, Cynthia Ann. Cynthia accompanied her uncle to Birdville on the condition that military interpreter Horace P. Jones would send along her sons if they were found.

family, identified the captured chief as Nocona. With Ross's permission, Martinez fired the shot that took Nocona's life. There is some dispute whether the man killed was actually Nocona or someone else.

When Ross arrived back at the campground, he discovered that the woman his men had captured had blue eyes. He assured her that no young boys had been killed in the battle, so her sons, Quanah and Pecos were safe. The woman could not speak English, and did not know her American name or where she came from. After much questioning, she remembered a few details of her capture as a child. The details matched what Ross knew of the Fort Parker Massacre of 1836. Though some of the Rangers urged Ross to set her free to return to the Comanches, he considered it best to try to return her to her white family. Ross knew many settlers had lost children to the Indians, and many of them might feel this was their child or relative. Ross sent the woman to Camp Cooper and sent a message to Colonel Isaac Parker, the uncle of a young girl kidnapped in the raid.

Cynthia Ann never adapted to her new life among the whites, and attempted to escape on several occasions. Her brother, Silas Jr., was appointed her guardian in 1862, and took her to his home in Van Zandt County. When Silas was mustered into the Confederate Army, Cynthia Ann went to live with her sister, Orlena. According to some accounts, the Parker family was negotiating to return her to west Texas and her adopted people when the American Civil War broke out. The chief cause of Cynthia Ann's unhappiness was that she missed her sons and never knew what had happened to them. In 1863, her daughter, Toh-Tsee-Ah, caught influenza and died from pneumonia.

In her grief, Cynthia Ann stopped eating. She became sick and died in 1870. She was buried in Fosterville Cemetery in Anderson County near Frankston. Her son, Quanah, moved her body in 1910 to the Post Oak Cemetery near Cache, Oklahoma. He was buried there in 1911. She and her son were moved in 1957 to the Fort Sill military cemetery in Oklahoma.

There is some confusion about the correct birth and death dates for Cynthia Ann Parker. Different sources place her birth from 1825 to 1827 in Coles, Clark or Crawford counties of Illinois, and her death from 1864 to 1871 in Anderson County, Texas. However, her presence in the 1870 Anderson County census makes an earlier death date unlikely. Writing in the *Crowell Index* on October 8, 1909, Tom Champion opined, "...I am convinced that the white people did more harm by keeping her away from them than the Indians did by taking her at first."

Bibliography

James T. DeShields, Cynthia Ann Parker: The Story of Her Capture (St. Louis, 1886; rpts.: The Garland Library of Narratives of North American Indian Captivities, Vol. 95, New York: Garland, 1976; Dallas: Chama Press, 1991).

Jo Ella Powell Exley, Frontier Blood: The Saga of the Parker Family (College Station: Texas A&M University Press, 2001).

Margaret S. Hacker, Cynthia Ann Parker: The Life and the Legend (El Paso: Texas Western Press, 1990).

Grace Jackson, Cynthia Ann Parker (San Antonio: Naylor, 1959). Paul I. Wellman, "Cynthia Ann Parker," Chronicles of Oklahoma 12 (June 1934). Women of Texas (Waco: Texian Press, 1972).

Margaret Schmidt Hacker. Source; *Margaret Schmidt Hacker- Handbook of Texas Online*, s.v.

REBECCA JANE GILLILAND FISHER – INDIAN CAPTIVE

Rebecca Jane Gilleland Fisher (1831–1926), preservationist, was born in Philadelphia on August 31, 1831, the daughter of Mary (Barbour) and Johnson Gilleland. Around 1837 the family arrived in Texas and settled in Refugio County near the Don Carlos Ranch. In 1840 Comanches attacked the home, killed the parents, and captured Rebecca and her brother William.

The children were rescued by Albert Sidney Johnston and a detachment of Texas soldiers and taken to Victoria, where they stayed with William C. Blair until they could be sent to live with Jane Trimble, an aunt in Galveston. Rebecca Gilleland attended Rutersville College from about 1845 to 1848, when she married Orceneth Fisher, a Methodist minister. The couple had six children. In 1855 the Fishers left Texas for the Pacific coast, where for nearly sixteen years Fisher served as a pastor in California and Oregon.

They returned to Texas about 1871 and eventually established a home in Austin, where Fisher died in 1880. Mrs. Fisher was a charter member of the Daughters of the Republic of Texas and served as its state president for eighteen years. She was also president of the Austin chapter. She delivered an oration at the unveiling of the Sam Houston monument at Huntsville and aided Clara Driscoll in saving the Alamo from destruction. For several years she gave the opening prayer when the Texas legislature convened. She was the only woman elected to the Texas Veterans Association and was its last surviving member.

Her portrait was the first of a woman to be hung in the Senate chamber at the Capitol. She died in Austin on March 21, 1926. Her body lay in state in the Senate chamber, where funeral services were held. The Senate unanimously adopted a resolution in her memory and draped her portrait in mourning cloth. Honorary pallbearers included the two United States senators from Texas and four former governors. She was buried in Oakwood Cemetery, Austin.

Bibliography

Elizabeth Brooks, Prominent Women of Texas (Akron, Ohio: Werner, 1896)

Rebecca J. Gilleland Fisher, "Capture and Rescue of Mrs. Rebecca J. Fisher, née Gilleland,"

Quarterly of the Texas State Historical Association 3 (January 1900)

Annie Doom Pickrell, Pioneer Women in Texas (Austin: Steck, 1929).

SOURCE ; Handbook of Texas Online, s.v. http://www.tshaonline.org/handbook/online/articles/FF/ffi22.html .

MATILDA LOCKHART – INDIAN CAPTIVE

Matilda Lockhart, (ca. 1825–?) who as a young girl was taken captive by Comanche Indians, was probably born in Illinois around 1825. Her father, Andrew Lockhart, emigrated with his family from Illinois to Texas in 1828 and settled on the Guadalupe River in Green DeWitt's colony. In the fall of 1838, when Matilda was about thirteen years old, she and four children of Mitchell Putnam were captured by Comanche Indians and carried into the Guadalupe Mountains.

Two unsuccessful excursions were made to free the children, one to the head of the Guadalupe River in late 1838 and one under John H. Moore in 1839 to Spring Creek, a tributary of the San Saba River. Under the terms of a treaty, sixty-five Indians led by the chieftan Muguara (Muk-wah-rah) delivered Matilda to authorities in San Antonio on March 19, 1840.

Mary Ann Maverick, who witnessed the event and helped to bathe and dress the girl after she was returned, later recounted that Matilda had been badly tortured and "was utterly degraded, and could not hold up her head again. Her head, arms, and face were full of bruises, and sores, and her nose actually burnt off to the bone-all the fleshy end gone, and a great scab formed on the on the end of the bone. Both nostrils were wide open and denuded of flesh. She told a piteous tale of how dreadfully the Indians had beaten her, and how they would wake her from her sleep by sticking a chunk of fire to her flesh, especially to her nose...her body had many scars from the fire."

During her two years with the Comanches, Matilda had learned to understand some of the Comanches' language, and she revealed to the Texan authorities in San Antonio that the Indians still held thirteen other captives and that they planned to bring them in one by one and bargain for each in exchange for ammunition, blankets, and other supplies. Her harrowing tale of privation and torture and the failure of the Indians to deliver the Putnam children and other captives resulted in the Council House Fight, which took place the day Matilda was returned. According to Maverick, the girl never recovered from her experience and died two or three years later.

Bibliography

T. R. Fehrenbach, *Lone Star: A History of Texas and the Texans* (New York: Macmillan, 1968). Rena Maverick Green, ed. *Memoirs of Mary A. Maverick* (San Antonio: Alamo Printing, 1921 rpt., Lincoln: University of Nebraska Press, 1989)

Andrew Jackson Sowell; *Early Settlers and Indian Fighters of Southwest Texas* (Austin: Ben C. Jones, 1900; rpt., Austin: State House Press, 1986).

J. W. Wilbarger, *Indian Depredations in Texas* (Austin: Hutchings, 1889; rpt., Austin: State House, 1985).

SOURCE Alice Gray Upchurch - *Handbook of Texas Online*, s.v.

BUFFALO HUMP

BUFFALO HUMP (?–1870). The English name Buffalo Hump is evidently a euphemism for an Indian name represented in written sources in various forms, among them *Pochanaquarhip* and *Ko-cho-naw quoip*. The name had a phallic significance not precisely recorded.

There were more Buffalo Humps than one. The famous one was a celebrated war chief of the Penateka Comanches. He first became prominent after the Council House Fight in San Antonio in March 1840. Outraged at the incident, he carried out a revenge raid across southeastern Texas. In August 1840 he led nearly 400 warriors and an equal number of women and children in raids on Victoria and Linnville, from where he carried off both livestock and human captives. Texas militiamen intercepted Buffalo Hump's party and administered it a stinging defeat in the battle of Plum Creek on August 12, but most of the Comanches and their leader escaped to their camps on the upper Colorado River.

After the Linnville raid and the fight at Plum Creek, Buffalo Hump continued to resist the attempts of Texas settlers upon his people's hunting grounds. In 1844 he met with Sam Houston and demanded that the white men remain east of the Edwards Plateau. Houston tacitly agreed to this proposal, and Texas Indian agents provided Buffalo Hump with gifts to demonstrate their good will. However, the Texas government was unable to stem the flood of settlers onto Comanche lands, and so the Indians resumed their raids. In response to these attacks, Texas Rangers[qv] struck at Penateka camps, but Buffalo Hump managed to hold his own for some time. Finally, in May 1846 he led the Comanche delegation at Council Springs that signed a treaty with the United States.

As war chief of the Penatekas, Buffalo Hump dealt peacefully with American officials throughout the late 1840s and 1850s. In 1849 he guided John S. Ford's expedition part of the way from San Antonio to El Paso, and in 1856 he led his people to the newly established Comanche reservation on the Brazos River. Threats from horse thieves and squatters, coupled with his band's unhappiness over their lack of freedom and food, forced Buffalo Hump to move his band off the reservation in 1858.

While camped in the Wichita Mountains, the Penatekas were attacked by United States troops under the command of Maj. Earl Van Dorn. Unaware that Buffalo Hump's band had recently signed a treaty of peace with military authorities at Fort Arbuckle, Van Dorn and his men killed eighty Comanches.

In 1859 Buffalo Hump settled his remaining followers on the Kiowa-Comanche reservation near Fort Cobb in Indian Territory. There, in spite of his distress at the demise of the Comanches' traditional way of life, he asked for a house and farmland so that he could set an example for his people. He died in 1870.

Bibliography

E. W. Henderson, "Buffalo Hump, a Comanche Diplomat," West Texas Historical Association *Yearbook* 35 (1959).

John Holland Jenkins, *Recollections of Early Texas*, ed. John H. Jenkins III (Austin: University of Texas Press, 1958; rpt. 1973).

Kenneth F. Neighbours, *Robert Simpson Neighbors and the Texas Frontier, 1836–1859* (Waco: Texian Press, 1975).

Robert Simpson Neighbors Papers, Dolph Briscoe Center for

American History, University of Texas at Austin. Rupert N. Richardson, *The Comanche Barrier to South Plains Settlement* (Glendale, California: Clark, 1933; rpt., Millwood, New York:

Kraus, 1973. Dorman H. Winfrey and James M. Day, eds., *Texas Indian Papers* (4 vols., Austin: Texas State Library, 1959–61; rpt., 5 vols., Austin: Pemberton Press, 1966).

PETA NOCONA

Peta Nocona, (?–1860) husband of Cynthia Ann Parker and father of Chief Quanah Parker, was a physically enormous Comanche chief who led a band, the Noconies, in raids on the Texas frontier from the 1830s to December 18, 1860, when he was killed at the Pease River in a battle with Capt. Lawrence Sullivan Ross. Peta Nocona did not know when or where he was born, as Quanah Parker indicated in a letter to Charles Goodnight. He took part in, or perhaps led, the raid on Parker's Fort on May 19, 1836, when the Comanches took Cynthia Ann captive. It is not certain that white settlers knew Peta Nocona's name or distinguished him from other Comanche chiefs until after his death.

Many years later, Quanah raised doubts about the identity of the chief killed at the Pease River, perhaps because of a Comanche belief that ill repute disturbs the peace of the dead. But the preponderance of evidence supports the contention that Peta Nocona was the chief killed at the Pease. Ross's Mexican interpreter, for instance, who said Nocona had taken him as a slave when he was a child, identified the chief. Cynthia Ann Parker wept over the dead man and called him Nocona. And after the battle at the Pease, which was itself big news, no one ever heard anything more about Peta Nocona until Quanah's disclaimer almost four decades later.

Bibliography

Dallas *Morning News*, December 19, 1993. James T. DeShields, *Border Wars of Texas*, ed. Matt Bradley (Tioga, Texas, 1912; rpt., Waco: Texian Press, 1976).

Rupert N. Richardson, *The Comanche Barrier to South Plains Settlement* (Glendale, California: Clark, 1933; rpt., Millwood, New York: Kraus, 1973).

Vertical Files, Dolph Briscoe Center for American History, University of Texas at Austin (Quanah Parker).

Robert H. Williams,, "The Case for Peta Nocona," T*exana* 10 (1972).

SOURCE; Robert H. Williams *Handbook of Texas Online*, s.v.

JOSÉ MARIA (ANDARKO)

JOSÉ MARÍA (ca. 1800–?). José María, Anadarko chief, diplomat, and proponent of peaceful relations with whites, was born sometime around 1800, probably in the region of present-day Nacogdoches. The residents of that area, a small Caddoan-speaking group known as the Anadarkos, were successful agriculturalists, long-range traders, and members of the Hasinai Confederacy, the largest grouping of Caddoan peoples in Texas. By the time of José María's birth, however, centuries of contact with white men and their missions, diseases, and violent rivalries had left the Anadarkos, and most of the Caddoes of East Texas, severely weakened. José María first entered Texas history in 1841 when he was wounded during a skirmish between his group and a company of settlers led by George Bernard Erath. Significantly, this is the only recorded instance where he was involved in hostilities with whites, for over the course of the next two decades José María distinguished himself as a leader favoring peaceful coexistence with the settlers. Accommodation made sense for him and the Anadarkos.

By the mid-nineteenth century, they numbered only about 200 individuals and lived under constant threat of attack from the Wacos and Tawakonis. As a result, José María turned to white authorities for protection nd between 1843 and 1846 signed a series of treaties effectively surrendering Anadarko sovereignty in return for protection and trade goods. During these important councils, José María emerged as a spokesman ot only for the Anadarkos but for other small Caddoan groups, and he often served as an intermediary between agents and other tribes. Yet despite his relatively high standing among government officials and his efforts to encourage the Anadarkos to pursue the path of peace and adopt some aspects of white culture, José María was unable to prevent frontier whites from encroaching on the lands reserved for his people under the various treaties.

In 1848 he expressed concern that his people were in danger of being driven from their farms, a fear that was heightened when, in the following year, tribal lands were surveyed. The situation was so unstable that when, in 1854, the Texas legislature voted to establish reservations for the state's remaining Indians, the Anadarkos accepted the offer readily. In 1855, joined by a number of Indians from destitute tribes in East Texas, they moved to the Brazos Indian Reservation on the Brazos River near the mouth of the Clear Fork. While there, the Anadarkos established farms and began serving as scouts for army expeditions into Comanche country.

But the Texas Indian reservations lasted just four years. According to the familiar frontier pattern, the lands set aside for the Anadarkos and their compatriots on the Brazos became a focal point for frontier violence. In February 1859, after the murder of several Indians, José María was indicted for horse stealing by the Palo Pinto County grand jury. Evidently the charges had no basis in fact, but given the tense situation, innuendo was enough to stimulate the Texas legislature into taking action. Legislation passed on July 11 abolished the Texas Indian reservations and placed the Indians under the charge of federal officials whose plan was to remove the Indians from the state. The removal process was set in motion on July 31, 1859, when 1,112 Indians, including 235 Anadarkos, were escorted by a military regiment to the Wichita reservation in western Oklahoma. There, gradually losing their distinctive identity, they remained. José María, it is said, died on the reservation during the Civil War years.

Bibliography

Grant Foreman, *The Last Trek of the Indians* (University of Chicago Press, 1946; 2d ed., New York: Russell, 1972).

William W. Newcomb, *The Indians of Texas* (Austin: University of Texas Press, 1961).

Rupert N. Richardson, *The Comanche Barrier to South Plains Settlement* (Glendale, California: Clark, 1933; rpt., Millwood, New York: Kraus, 1973).

Brian C. Hosmer - *Handbook of Texas Online*, s.v. http://www.tshaonline.org/handbook/online/articles/JJ/fjo74.html.

Bowl

BOWL (ca. 1756–1839). Chief Bowl (also known as Duwali, Diwal'li, Chief Bowles, Colonel Bowles, Bold Hunter, and the Bowl), the principal chief of the Cherokees in Texas, was born in North Carolina around 1756. He was the son of a Scottish father and a full-blooded Cherokee mother. Duwali was leader of a village at Little Hiwassee (in western North Carolina). In 1791 he signed the Treaty of Holston, and in 1805 he signed an unauthorized cession treaty, a move that proved unpopular with the majority of Cherokees.

In early 1810, to access better hunting ground and to escape growing pressures of settlement in the southern states, he and his band moved across the Mississippi River and settled in the St. Francis River valley, near New Madrid, Missouri. In 1812–13 his people moved into northwestern Arkansas, south of the Arkansas River, and in 1819 they once more moved on, stopping briefly in southwestern Arkansas and at the three forks of the Trinity River before settling north of Nacogdoches. In Texas Chief Bowl became the primary "civil" chief or "peace chief" of a council that united several Cherokee villages. In 1822 he sent diplomatic chief Richard Fields to Mexico to negotiate with the Spanish government for a land grant or title to land occupied by Cherokees in East Texas. In 1827 he cooperated with the Mexican government in putting down the Fredonian Rebellion.

In 1833 he made another attempt to secure from the Mexican government land on the Angelina, Neches, and Trinity rivers, but negotiations were interrupted by political unrest in Texas. In February of 1836 Sam Houston negotiated a treaty with Bowl's council, guaranteeing the tribe possession of lands occupied in East Texas. After the Texas Revolution, however, the treaty was invalidated by the Senate of the Republic of Texas. In desperation, Bowl briefly allied with agents soliciting allies for a Mexican reinvasion of Texas. Shortly thereafter, President Mirabeau B. Lamar ordered him and his people to leave Texas. After negotiations failed, Bowl mobilized his warriors to resist expulsion. On July 16, 1839, Chief Bowl was killed in the battle of the Neches.

Bibliography

Mary Whatley Clarke, *Chief Bowles and the Texas Cherokees* (Norman: University of Oklahoma Press, 1971).

Dianna Everett, *The Texas Cherokees: A People between Two Fires, 1819–1840* (Norman: University of Oklahoma Press, 1990).

Dorman Winfrey, "Chief Bowles and the Texas Cherokees," *Chronicles of Oklahoma* 32 (Spring 1954).

E. W. Winkler, "The Cherokee Indians in Texas," *Quarterly of the Texas State Historical Association* 7 (October 1903).

Albert Woldert, "The Last of the Cherokees in Texas and the Life and Death of Chief Bowles," *Chronicles of Oklahoma* 1 (June 1923).

SOURCE; Dianna Everett - *Handbook of Texas Online*, s.v.

TEXAS JACK (JOHN BURWELL OMONUNDRO, JR)

OMOHUNDRO, JOHN BURWELL, JR. (1846–1880). John Burwell (Texas Jack) Omohundro, Jr., frontier scout, actor, and cowboy, was born on July 26, 1846, to J. B. and Catherine (Baker) Omohundro at Pleasure Hill, near Palmyra, Virginia, the fourth of twelve children. He received some formal elementary education. In his early teens he left home, made his way alone to Texas, and became a skilled cowboy. Unable to join the Confederate Army in 1861 because of his youth, he nevertheless entered Confederate service as a courier and scout.

In 1864 he enlisted in Gen. J. E. B. Stuart's command as a runner, scout, and spy. After the Civil War he spent several months in Florida and worked briefly as a schoolteacher. In 1866 Omohundro resumed his life as a *Texas cowboy*. He participated in *several early cattle drives*, including a drive across Arkansas to meat-short Tennessee. Grateful citizens of Tennessee nicknamed him "*Texas Jack*." On one of his drives he rescued a small boy whose parents had been killed in an Indian raid. The boy took the name Texas Jack, Jr., and went on to lead Texas Jack's Wild West Show and Circus in 1903.

Omohundro may also have served for a time as a Texas Ranger. In 1869 he moved to Cottonwood Springs, Nebraska, near Fort McPherson and became a scout and buffalo hunter. Here he met William F. (Buffalo Bill) Cody. Together they participated in Indian skirmishes and buffalo hunts, including acting as guides in the royal hunt of 1872, a highly publicized excursion with Russian grand duke Alexis and a group of prominent American military figures. Omohundro and Cody traveled to Chicago in December 1872 to debut in *The Scouts of the Prairie,* one of the original Wild West shows. Critics described him as physically impressive and magnetic in personality.

He was credited with introducing roping acts to the American stage. During the 1870s Texas Jack divided his time between the Eastern stage circuit and the hunting ranges of the Great Plains. He guided hunting parties that included European nobility. He headed his own acting troupe in St. Louis in 1877. That year he also wrote articles about his hunting experiences, published in the *Spirit of the Times*, a popular magazine. The Texas Jack legend grew in many "dime novels," particularly those written by Col. Prentiss Ingraham.

In 1900 Joel Chandler Harris featured Texas Jack in a series of fictional accounts of the Confederacy for the *Saturday Evening Post*. Thereafter, the legend faded. On August 31, 1873, Omohundro married Giuseppina Morlacchi, a dancer-actress from Milan, Italy, who starred with him in the *Scouts of the Prairie* and other shows. Texas Jack died on June 28, 1880, of pneumonia, in Leadville, Colorado, and was buried in Evergreen Cemetery in that town.

Bibliography

J. C. Dykes, "Dime Novel Texas; or the Sub-Literature of the Lone Star State," *Southwestern Historical Quarterly* 49 (January 1946).

Joel Chandler Harris, O*n the Wing of Occasions* (New York Doubleday, Page, 1900). Prentiss Ingraham, T*exas Jack, The Mustang King* (New York: Ivers, 1899).

Herschel C. Logan, *Buckskin and Satin: The Life of Texas Jack and His Wife* (Harrisburg, Pennsylvania: Stackpole, 1954). *Southwestern Historical Quarterly*, Texas Collection, October 1958.

Windham Thomas Wyndham-Quinn, Earl of Dunraven, *Hunting in the Yellowstone* (New York: Macmillan, 1925).

JOSEFA RODRIGUEZ (CHIPITA)

Josefa (Chipita) Rodríguez (? – 1863) was for many years considered to be the *only woman legally hanged*[1] *in Texas.*

Most of her story verges on legend; facts surrounding her arrest, trial, and execution are scant, and many aspects of her story, including the name Josefa, cannot be verified. She is believed to have been the daughter of Pedro Rodríguez, who is said to have fled from Antonio.

López de Santa Anna. Chipita moved with her father to San Patricio de Hibernia, Texas, while quite young, and for many years after Rodríguez's death furnished travelers with meals and a cot on the porch of her lean-to shack on the Aransas River. When Cotton Road traveler John Savage was murdered with an ax, presumably for the $600 in gold which he had been carrying, Chipita was accused of robbery and murder. Recovery of the gold from the Aransas River north of San Patricio, where Savage's body was found in a burlap bag, raised substantial doubt about the motive for the crime, but Josefa Rodríguez and Juan Silvera (who sources suggest may have been her illegitimate son) were indicted on circumstantial evidence and tried before Fourteenth District Court judge Benjamin F. Neal at San Patricio.

After Chipita pleaded *not guilty, the jury recommended mercy*, but *Neal ordered her executed* on November 13, 1863. For some time she was held at sheriff William Means's home in Meansville, where two attempts by a lynching mob were thwarted. According to legend, Chipita was kept in leg irons and chained to a wall in the courthouse. There, local children brought her candy and shucks to make cigarettes. At the time, she was described as "very old" or "about ninety," but was probably in her sixties.

The court records, except for a week of transcripts, were burned in a courthouse fire or lost in a flood, and many discrepancies exist in trial accounts. From these it has been determined that no list of qualified jurors

---

[1] Jane Elkins, a slave convicted of murder, was hanged on May 27, 1853, in Dallas. She was the first woman legally hanged in the state.

existed, but the sheriff, instructed as jury foreman to produce "at least twenty qualified men," produced closer to thirty; at least three members of the grand jury also served on the trial jury; the foreman of the grand jury was the sheriff who arrested her; members of both juries had been indicted on felony charges; Chipita had little in the way of defense counsel, and her sole defense was the words "not guilty." There was no appeal or motion in arrest of judgment, and though some talk of a retrial may have occurred, none took place.

Lore says that resident Kate McCumber drove off hangman John Gilpin when he came for her wagon to transport Chipita to the hanging tree. At least one witness to the hanging claimed he later *heard a moan from the coffin,* which was placed in an unmarked grave. Many tales have arisen as a result of the trial and the hanging, one of which claims that Chipita *was protecting her illegitimate son.*

Lore says that resident Kate McCumber drove off hangman John Gilpin when he came for her wagon to transport Chipita to the hanging tree. At least one witness to the hanging claimed he later *heard a moan from the coffin,* which was placed in an unmarked grave. Many tales have arisen as a result of the trial and the hanging, one of which claims that Chipita *was protecting her illegitimate son.*

Other sources indicate she may have been involved in gathering information to influence the state's decision about which side to take in the Civil War and was framed as a political act. Her ghost is said to haunt the area, especially when a woman is sentenced to be executed. She is pictured as a specter with a noose around her neck, wailing from the riverbottoms. She has been the subject of two operas, numerous books, newspaper articles, and magazine accounts.

In 1985 state senator Carlos Truan of Corpus Christi asked the Texas legislature to absolve Chipita Rodríguez of murder. The Sixty-ninth Legislature passed the resolution, and it was signed by Governor Mark White on June 13, 1985.Jane Elkins, a slave convicted of murder, was hanged on May 27, 1853, in Dallas. She was the first woman legally hanged in the state.

Bibliography

Francis Edward Abernethy, ed., *Legendary Ladies of Texas*, Publications of the Texas Folklore Society 43 (Dallas: E-Heart, 1981). Dallas *Morning News*, November 13, 1994.

Rachel Bluntzer Hebert, *Shadows on the Nueces* (Atlanta: Banner, 1942).

Ruel McDaniel, "The Day They Hanged Chipita," *Texas Parade*, September 1962. *San Patricio County in 1976: A Bicentennial Perspective* (Sinton, Texas: Sinton Bicentennial Celebrations, 1976).

Vernon Smylie, *A Noose for Chipita* (Corpus Christi: Texas News Syndicate Press, 1970).

Ruthe Winegarten, *Finder's Guide to the 'Texas Women: A Celebration of History' Exhibit Archives* (Denton: Texas Woman's University Library, 1984).

Marylyn Underwood - *Handbook of Texas Online*, s.v.

FAMOUS TEXANS

Augustus Chapman Allen

(1806-1864), along with his younger brother, John Kirby Allen, founded the City of Houston in the U.S. state of Texas. He was born on July 4, 1806, in Canasareaugh, New York, to Sarah (Chapman) and Roland Allen.

Early years
Not long after turning seventeen, Augustus graduated from the Polytechnic Institute at Chittenango, New York and started teaching mathematics there. In 1827, he changed careers and resigned his professorship. He then went to work as bookkeeper for the H. and H. Canfield Company, New York. After two years, he and his brother John bought an interest in the business. In the summer of 1832, the Allen brothers left Canfield to move to Texas, where they settled in San Augustine. By June 1833, the brothers had moved to Nacogdoches.

In Texas
The Allen brother arrived first in Galveston, Texas and then moved to the small town of Saint Augustine. In 1833, Augustus Allen and his brother associated with a group of entrepreneurs in Nacogdoches and started operating a business as land speculators.

During the Texas Revolution
Instead of joining the army when the Texas War of Independence started, Augustus and his brother engaged in the business of keeping supply channels open. At their own expense they outfitted a ship, the *Brutus*, for the purpose of protecting the Texas coast and assisting troops and supplies from the United States to arrive safely in Texas.Nevertheless, some members of the Texas provisional government objected to the Allen brothers' activities, and there were rumors that they were engaged in privateering. In January 1836, they sold the *Brutus* to the Texas Navy, and it became only the second ship in the fledgling Texas navy. Augustus and John Allen continued to raise money and operate as receivers and dispensers of supplies and funds for the war effort without charge. In spite

of the brothers' services, gossip and censure were aimed at the Allens because they were not in the armed services.

After the Revolution

Financed by an inheritance received by Augustus's wife, Charlotte, on August 26, 1836, the Allen brothers purchased 6,600 acres (27 km²) along the Buffalo Bayou for $5,000, for the purpose of establishing a new city. At the suggestion of Charlotte, they named their townsite for the hero of the time, General Sam Houston.

In the years after his brother John's death in 1838, the Allen family grew apart and began to squabble about the family's various businesses and finances, which had been somewhat commingled. These family disputes eventually led to Allen's separation with his wife in 1850.

Later life and legacy

Soon, Allen's health began to fail, and he decided to leave Houston, signing over to his wife, whom he had never divorced, most of what remained of his many enterprises. Augustus Allen relocated to Mexico to tend to his health and a new start in life.

In 1852, Augustus was appointed United States consul for the port of Tehuantepec on the Pacific Ocean, and in 1858 he was given the same position for the port of Minotitlán. These offices gave him control of the consular affairs of the United States for the entire Isthmus of Tehuantepec, a commercially important position.

Allen, in partnership with an Englishman named Welsh, developed an extensive private shipping business. He was never able to recover his health, however, and realized, in 1864, that he was critically ill. Augustus then closed his business and went to Washington, D.C. to resign his consulships. Soon after arriving there he contracted pneumonia, from which he would not recover.

Augustus Allen died on June 11, 1864 in Washington, D.C. and was buried in the Green-Wood Cemetery in Brooklyn, New York.

John Wesley Hardin

"The fact that he had more than thirty notches on his gun is evidence that no more dangerous gunman ever operated in Texas."

John Wesley (Wes) Hardin, outlaw, son of James G. and Elizabeth Hardin, was born in Bonham, Texas, on May 26, 1853. His father was a Methodist preacher, circuit rider, schoolteacher, and lawyer.

Hardin's violent career started in 1867 with a schoolyard squabble in which he stabbed another youth. At fifteen, in Polk County, he shot and killed a black man as a result of a chance meeting and an argument. With the Reconstruction government looking for him, he fled to his brother's house, twenty-five miles north of Sumpter, Texas, where in the fall of 1868 he claimed to have killed three Union soldiers who sought to arrest him. Within a year, he killed another soldier at Richard Bottom.

In 1871 Hardin went as a cowboy up the Chisholm Trail. He killed seven people en route and three in Abilene, Kansas.

After allegedly backing down city marshal Wild Bill Hickok, who may have dubbed him "Little Arkansas," Hardin returned to Gonzales County, Texas, where he got into difficulty with Governor Edmund J. Davis's State Police. Hardin then settled down long enough to marry Jane Bowen. Out of that marriage came a son and two daughters.

Hardin added at least four names to his death list before surrendering to the sheriff of Cherokee County in September 1872. He broke jail in October and began stock raising but was drawn into the Sutton-Taylor Feud in 1873-74. He aligned himself with Jim Taylor of the anti-Reconstruction forces and killed the opposition leader, Jack Helm, a former State Police captain. In May 1874 he started two herds of cattle up the trail; while visiting in Comanche he killed Charles Webb, deputy sheriff of Brown County.

From that time, Hardin was constantly pursued in Texas. He went with his wife and children to Florida and Alabama, adding one certain and five possible names to his death list before the Texas Rangers captured him in Pensacola, Florida, on July 23, 1877. He was tried at Comanche for the murder of Charles Webb and sentenced, on September 28, 1878, to twenty-five years in prison. During his prison term he made repeated efforts to escape, read theological books, was superintendent of the prison Sunday school, and studied law.

He was pardoned on March 16, 1894, and admitted to the bar.

In 1895 he went to El Paso to appear for the defense in a murder trial and to establish a law practice. Despite efforts to lead a decent life, he was soon in trouble. He took as his lover the wife of one of his clients, Martin Morose, and when Morose found out about the affair, Hardin hired a number of law officials to assassinate him. On August 19, 1895, Constable John Selman, one of the hired killers, shot Hardin in the Acme Saloon, possibly because he was not paid for the murder of Morose. Hardin died instantly and was buried in Concordia Cemetery, El Paso.

His autobiography, completed to the beginning of his law studies in prison, was the subject of some litigation and was published in 1896. Hardin was an unusual type of killer, a handsome, gentlemanly man who considered himself a pillar of society, always maintaining that he never killed anyone who did not need killing and that he always shot to save his own life. Many people who knew him or his family regarded him as a man more sinned against than sinning. The fact that he had more than thirty notches on his gun, however, is evidence that no more dangerous gunman ever operated in Texas.

Source; Leon Metz

Sam Bass - Outlaw

Sam Bass, outlaw, was born on a farm near Mitchell, Indiana, on July 21, 1851, a son of Daniel and Elizabeth Jane (Sheeks) Bass. He was orphaned before he was thirteen and spent five years at the home of an uncle. He ran away in 1869 and worked most of a year in a Sawmill at Rosedale, Mississippi. Bass left Rosedale on horseback for the cattle country in the late summer of 1870 and arrived in Denton, Texas, in early fall. For the winter he worked on Bob Carruth's ranch southwest of town. But, finding cowboy life not up to his boyhood dreams, he went back to Denton and handled horses in the stables of the Lacy House, a hotel. Later he worked for Sheriff William F. Egan, caring for livestock, cutting firewood, building fences, and spending much of his time as a freighter between Denton and the railroad towns of Dallas and Sherman.

Before long Bass became interested in horse racing, and in 1874, after acquiring a fleet mount that became known as the Denton Mare, he left Egan's employ to exploit this horse. He won most of his races in North Texas and later took his mare to the San Antonio area.

When his racing played out in 1876, he and Joel Collins gathered a small herd of longhorn cattle to take up the trail for their several owners. When the drovers reached Dodge City they decided to trail the cattle farther north, where prices were higher. After selling the herd and paying the hands, they had $8,000 in their pockets, but instead of returning to Texas, where they owed for the cattle, they squandered the money in gambling in Ogallala, Nebraska, and in the Black Hills town of Deadwood, South Dakota, which was then enjoying a boom in gold mining.

In 1877 Bass and Collins tried freighting, without success, then recruited several hard characters to rob stagecoaches. On stolen horses they held up seven coaches without recouping their fortunes.

Next, in search of bigger loot, a band of six, led by Collins and including Bass, rode south to Big Springs, Nebraska, where, in the evening of September 18, they held up an eastbound Union Pacific passenger train. They took $60,000 in newly minted twenty-dollar gold pieces from the express car and $1,300 plus four gold watches from the passengers. After dividing the loot the bandits decided to go in pairs in different directions. Within a few weeks Collins and two others were killed while resisting arrest. But Bass, disguised as a farmer, made it back to Texas, where he formed a new outlaw band.

He and his brigands held up two stagecoaches and, in the spring of 1878, robbed four trains within twenty-five miles of Dallas. They did not get much money, but the robberies aroused citizens, and the bandits were the object of a spirited chase across North Texas by posses and a special company of Texas Rangers headed by Junius Peak. Bass eluded his pursuers until one of his party, Jim Murphy, turned informer. As Bass's band rode south intending to rob a small bank in Round Rock, Murphy wrote to Maj. John B. Jones, commander of the Frontier Battalion of Texas–the rangers. In Round Rock on July 19 Bass and his men became engaged in a gun battle, in which he was wounded. The next morning he was found lying helpless in a pasture north of town and was brought back to Round Rock. He died there on July 21, his twenty-seventh birthday. He was buried in Round Rock and soon became the subject of cowboy song and story.

Source; Leon Metz

John Henry Selman - Outlaw

John Henry (Old John, Uncle John) Selman, outlaw come lawman, was born in Madison County, Arkansas, on November 16, 1839, the son of Jeremiah Selman, an Englishman. In 1858 the Selmans moved to Grayson County, Texas, where Jeremiah died, and on December 15, 1861, John joined the Twenty-second Texas Cavalry as a private. He deserted from Fort Washita, Choctaw Nation (Indian Territory), in April 1863 and joined his family at Fort Davis, a Stephens County settlement at the fort on the Clear Fork of the Brazos. On August 17, 1865, John married Edna DeGraffenreid. They had four children. In the mid-1870s the Selmans moved to Fort Griffin, Texas, where John became a deputy for Shackelford County sheriff John M. Larn. The two controlled the vigilantes, rustled cattle, and at times terrorized the county, until the vigilantes locked Larn in his own jail and shot him to death. After Selman's wife died in 1879, he fled to Lincoln County, New Mexico, and organized the "Selman Scouts," a band of desperados accused of murder and rape during the late 1870s. In 1880 strong law enforcement drove Selman to Fort Davis in Jeff Davis County, Texas, where the TexasRangers captured him; they took him to Shackelford County for trial. He escaped, however, and fled to Chihuahua, where he lived until 1888, when Texas charges were dropped. That year Selman moved to El Paso, where he married Romula Granadine on August 23, 1893. He lived mostly as a gambler and city constable. On April 5, 1894, he killed former Texas Ranger Baz (Bass) Outlaw during a wild brawl in Tillie Howard's brothel.

Selman killed the famous gunman John Wesley Hardin on August 19, 1895, by putting three bullets in him as he rolled dice in the Acme Saloon. Selman went on trial for murder, but because of a hung jury he was released on bond. After leaving the Wigwam Saloon late one night, he met United States deputy marshal George Scarborough, and the two fought. Scarborough shot Selman four times, and he died on the operating table on April 6, 1896. Scarborough was acquitted of murder.Selman was buried in El Paso's Concordia Cemetery in the Catholic section, but his grave was unmarked, and all attempts to locate it have been unsuccessful.

Source; FrontierTimes

APPENDIX

Background to the early period leading up to the Mexican Revolution subsequently the Battle for Texas prior to the establishment of the Republic of Texas.

The Mission '*Presidio La Bahia*' in *Goliad, Texas was the scene of the martyrdom of Fannin's Company*

Diarist Sanchez with the Terán inspection expedition 1828

> "All Texas is greatly in debt to Bexar for the remonstrance of 19th Dec. last—That paper was reprinted here [in Mexico City] and has had more weight in favor of Texas than all that has been done or said. I doubt much whether the memorial of the Convention [of 1833] would have been even looked at had not the minds of Govt. been prepared by the Bexar representation. It came from natives and is believed."
> —Stephen F. Austin

> "...our own countrymen, who had joined the enemy's cause, assailed us in our own language with such a volley of threats, insults, and abuse that the tongue of that vile and recreant Mexican seemed to have been wrought in the very caves of hell and set in motion by Lucifer himself. "Now you shall see," he said, "contemptible and faithless assassins, if you do not pay with your vile blood for your murders at the Alamo and La Bahia. The time has come when the just cause that we defend triumphs over you; you shall pay with your heads for the arson, robberies, and depredations that you have committed in our country, etc., etc."
> —Centralista Col. Pedro Delgado describing the Battle of San Jacinto

> One of them [a captured Mexican female camp follower at San Jacinto] throwing herself into the posture of a suppliant addressed him [a Hispanic Tejano patriot in the battle] as follows; "Señor God demme no me mata por el amor de Dios y por la villa cle su

madrecita! " On being answered in her own language she turned to her sister & said, "Hermanita! mira! ese señor God demme habla la lingua cristiana como nosotros!" [Mr. God damn don't kill me, for the love of God and for the life of your dear mother! Look here sister this Mr. God damn speaks the Christian Language like us!]
—Anecdote from The Lamar Papers.

...Captain (now Colonel) Juan Nepemucene Seguin, a native of Bexar and whom I have known from a boy, commanded 25 men, all natives of the same place, and performed wonders; every man signalized himself in the most distinguished manner. One of them, with a Bowie knife, killed 25 of his countrymen, and one of them, Colonel Batres, whom I knew well.
—29 May 1836 letter published in The Mobile Morning Chronicle.

By kind permission of; Presidio La Bahia, Goliad, Texas.

HISPANIC TEXIAN PATRIOTS

The Struggle for Independence

As a proportion of the population, the active participation of Hispanic native and immigrant residents in the struggle for independence of Texas from Spain and Mexico was equal to or greater in specific battles than that of resident immigrants from the United States of the North--Don Guillermo...with the North Americans, they [Mexicans] have adopted their customs and habits, and one may truly say that they are not Mexicans except by birth, for they even speak Spanish with marked incorrectness.
—Diarist Sanchez with the Terán inspection expedition 1828

All Texas is greatly in debt to Bexar for the remonstrance of 19th Dec. last—That paper was reprinted here [in Mexico City] and has had more weight in favor of Texas than all that has been done or said. I doubt much whether the memorial of the Convention [of 1833] would have been even looked at had not the minds of Govt. been prepared by the Bexar representation. It came from natives and is believed.
—Stephen F. Austin

...our own countrymen, who had joined the enemy's cause, assailed us in our own language with such a volley of threats, insults, and abuse that the tongue of that vile and recreant Mexican seemed to have been wrought in the very caves of hell and set in motion by Lucifer himself. "Now you shall see," he said, "contemptible and faithless assassins, if you do not pay with your vile blood for your murders at the Alamo and La Bahia. The time has come when the just cause that we defend triumphs over you.

As a proportion of the population, the active participation of Hispanic native and immigrant residents in the struggle for independence of Texas from Spain and Mexico was equal to or greater in specific battles than that of resident immigrants from the United States of the North—Don Guillermo...with the North Americans, they [Mexicans] have adopted their customs and habits, and one may truly say that they are not Mexicans except by birth, for they even speak Spanish with marked incorrectness.
—Diarist Sanchez with the Terán inspection expedition 1828

All Texas is greatly in debt to Bexar for the remonstrance of 19th Dec. last—That paper was reprinted here [in Mexico City] and has had more weight in favor of Texas than all that has been done or said. I doubt much whether the memorial of the Convention [of 1833] would have been even

> looked at had not the minds of Govt. been prepared by the Bexar representation. It came from natives and is believed.
> —Stephen F. Austin

> ...our own countrymen, who had joined the enemy's cause, assailed us in our own language with such a volley of threats, insults, and abuse that the tongue of that vile and recreant Mexican seemed to have been wrought in the very caves of hell and set in motion by Lucifer himself. "Now you shall see," he said, "contemptible and faithless assassins, if you do not pay with your vile blood for your murders at the Alamo and La Bahia. The time has come when the just cause that we defend triumphs over you; you shall pay with your heads for the arson, robberies, and depredations that you have committed in our country," etc., etc.

By kind permission of; Presidio La Bahia, Goliad, Texas.

PUBLIC OPINION IN TEXAS PRECEDING THE REVOLUTION

By Eugene Barker (From Annual Report of the American Historical Association for the Year 1911).

Earnest patriots like Benjamin Lundy, William Ellery Channing, and John Quincy Adams saw in the Texas revolution a disgraceful affair promoted by sordid slaveholders and land speculators. Even to the critical ear of the modern historian their arguments sound plausible, and it is not strange that in a period distinguished by sectionalism they were accepted by partisans at full value. The fundamental defect of these arguments lay in the fact that their authors knew too little of contemporary opinion in Texas. The truth is, so far as one may judge from the absence of discussion of the subject in Texas, that slavery played no part in precipitating the revolution; while it is certain that land speculation, of which there was unquestionably a great deal, tended rather to retard than to hasten the outbreak.

[Author Barker footnotes that: I have found but three contemporary references which might indicate a potential connection between the slavery question and the revolution: (1) In a Fourth of July address intended to stir the colonists to resistance R. M. Williamson, a prominent radical, declared that the *Mexicans were coming to Texas to compel the Texans, among other things, to give up their slaves* (a broadside in the Bexar archives; "Publications" of So. Hist. Assn., VIII, 7-18). (2) In a letter of August 21,1835, Stephen F. Austin said "Texas must be a slave country. It is no longer a matter of doubt" (Quarterly of Tex. State Hist. Assn., XIII, 271). (3) On August 28 the radicals issued a circular in which they quoted H. A. Alsberry, who had recently returned from Mexico, as saying that the *Mexicans boasted that they would free the slaves of the Texans and set them against their masters* (Broadside in the Austin Papers).

Von Holst in Constitutional History of the United States 11, 553:

"Settlers came with their slaves from the slave States [to Texas]. In this the heads of individual persons may have been haunted by far-reaching projects; but I can find no support for the assertion that back of it there was a definite plan of the South." Stephen F. Austin to William H. Ashby, Oct. 10, 1832, MS. Austin Papers: "It is not our interest to separate [from Mexico) if such a thing can be avoided, unless indeed we should float into the Northern Republic with the consent of all parties, ourselves included."

During 1834 and the spring of 1835 the legislature of Coahuila and Texas made a number of large grants of land in Texas under circumstances that pointed to bribery. Some Texans shared in these questionable transactions, and it was from them that the first rumors of disturbances in the interior reached Texas. The people suspected that for purposes of their own the speculators were trying to raise a revolt against the Federal Government, which had annulled some of the objectionable grants, and for that reason their eyes remained closed to Santa Anna's designs longer than would have otherwise happened. Their attitude toward the speculators is indicated by the fact that the constitution of 1836 expressly annulled the large grants made in 1834 and 1835. (See "Land Speculation as a Cause of the Texas Revolution," by the writer, in Quarterly of Tex. State Hist. Assn., XI, 76-95)]

A study from within reveals some *interesting parallels between the Texan revolt and the American Revolution.* In each the general cause of revolt was the same—a sudden effort to extend imperial authority at the expense of local privilege. In each the method of molding public opinion by town meetings and committees of correspondence was the same, but with the fundamental difference in aim that, whereas in the American colonies it was the object of such agencies to organize resistance, in Texas their purpose was at first to quiet opposition. In each the divided state of opinion which preceded and accompanied the resort to arms was similar.

One cause of exasperation existed in Texas, however, which the American colonists never felt. At the close of summer in 1835 the Texans saw themselves in danger of becoming the alien subjects of a people to whom they deliberately believed themselves morally, intellectually, and politically superior. This racial feeling, indeed, underlay and colored Texan-Mexican relations from the establishment of the first Anglo-American colony in 1821. It was a fertile soil in which throve mutual suspicion and distrust. And on the Mexican side it was powerfully stimulated by the efforts of Adams and Jackson to purchase Texas. With the ultimate aim of the Texan immigrants this inquiry is not concerned. One may find it difficult to believe that they considered their relation to Mexico as permanent, but there is little evidence before 1835 that they thought of the matter at all. And the minutes of public meetings, town councils, and committees of correspondence, as well as private letters, furnish a mass of testimony from which it is hard to resist the conclusion that at that time the people were almost unanimous in wishing to avoid a breach.

This first becomes apparent in the spring of 1835, through a conflict between State and Federal authority. Toward the end of April Governor Viesca called upon the local departments of Coahuila and Texas for militia to guard the capital from Federal troops who were known to have been ordered against it *[the three departments of Texas at the time were Bexar with capital San Antonio, Brazos with capital San Felipe and Nacogdoches with capital Nacogdoches. Each department had a political chief]*. The contingent desired from Texas was a hundred men from each of the three departments, and it was only by a show of military force that the Mexican inhabitants of San Antonio were prevented from responding at once; but the political chief of the Anglo-American department of the Brazos frankly declined to muster the militia, and wrote the governor that his people had no sympathy with the State government. He was commended for this by an editorial in The Texas Republican, the only paper then published in Texas. On May 21 the legislature adjourned to avoid capture, and a few days later the governor fled. He was arrested on June 6.

While this unfortunate episode was being smoothed over, a serious clash developed between certain individuals and the customs collectors. The innate aversion of the average American to paying duties had been strengthened in the Texans by local circumstances. To encourage immigration the Government had practically suspended the tariff until 1831. Collectors appointed in 1832 had left the Province the same year, when the colonists expelled the garrisons from several towns. And between 1832 and 1834 political convulsions had prevented the Government from giving the matter attention.

Mexicans as far away as Santa Fe and Chihuahua improved the opportunity to import goods through Texas, and for 1834 Col. Almonte estimated this contraband trade at $270,000. Friction was not long delayed, therefore, when the opening of the new year brought a collector and several deputies to Texas with a small detachment of soldiers. The collector stopped to reconnoiter on the Brazos River, while the deputies and the soldiers went on to Anahuac, which was regarded as the port of Galveston. The collector later established himself at Velasco, near the mouth of the Brazos, but made no effort to collect anything but tonnage duties, while the deputies at Anahuac, backed by the garrison began a rigid enforcement of the tariff.

The first and chief sufferer from this or at least the most outspoken was Andrew Briscoe, a merchant of Anahuac. Since the collector at Velasco was demanding only the tonnage duties, he felt that he was the victim of unjust discrimination, and his indignation waxed accordingly. His own statement of his grievance is an important testimonial to the attitude of the people: "The people would calmly stand by and see me lose all, damn them. * * * My business has been delayed, our provisions and groceries seized as contraband, and the whole duties claimed on the balance * * * and all this by a deputy collector and forty soldiers. And when I talk of resistance the people blame me and talk of a jury trial, and the devil knows what all."

This testimony is supported by the action of the town council (ayuntamiento) of Liberty, the capital of the municipality in which Anahuac lay. On April 17 this body declared that it was the undoubted right of every nation "to establish its own system of revenue," and that "obedience to the laws" was "the first duty of a good citizen." This duty extended farther than mere personal obedience; it entailed the obligation to make others obey. The present tariff laws might be, and probably were, unwise in some of their restrictions on Texas, but to resist them by force "would be more unwise and ill timed than the laws themselves." The way to obtain relief was to memorialize Congress for modification of the laws. In the meantime, could the people not abstain from using imported goods? Turning directly to Briscoe's case, the council suggested that a prudent merchant should inform himself of the revenue laws of the country in which he carried on business, and that he had no right to complain if his failure to do so involved him in trouble. In conclusion it cautioned all persons against interfering with the customs officers, and enjoined the civil and military officers of the municipality to assist, if necessary, in the maintenance of the revenue, laws."

Briscoe was willing to try the effect of a memorial, and on May 4 he and some twenty others petitioned the governor to intercede with Congress for a suspension or amendment of the law. But instead of adopting the suggestion for a non-importation agreement, they declared that, until the result of their petition was learned, they would pay no more duties, unless collections were equally enforced throughout Texas. The petition never reached its destination, but a few weeks later the collectors abandoned Texas, and on June 30 a small force commanded by William B. Travis captured the garrison at Anahuac. This action, as we have seen, had been secretly authorized by the radicals at San Felipe on June 22. When news arrived of Travis's attack on the garrison, several communities had already expressed disapproval of Briscoe's petition.

By kind permission of; Presidio La Bahia, Goliad, Texas

Author Barker footnotes:

Municipality of Columbia, June 28:
"Your committee, particularly protest against the proceedings of those persons at Anahuac who gave the collector of customs, Don Jose Gonzales, a series of resolutions declaring that they would not obey the revenue laws of Mexico. They denounce such persons as foreigners, and disclaim all participation in the act whatever"

The Texas Republican, July 4, 1835;
Publications 11 of So. Hist. Assn., VIT, 86) Lavaca and Mina on July 4 indorsed these Columbia resolutions (The Texas Republican, July 18, 1835;) Publications" of So. Hist. Assn., VII, 88-90, VIII, 2-6)
Municipality of Gonzales ,July 7:

We protest against those acts which tend to a resistance to the revenue laws of the Government and sincerely invite the Supreme Executive to carry them into effect, suggesting at the same time a modification of those laws, in order that the duties shall all be collected.

We believe reasonable duties, received by collectors understanding both languages, without favor or collusive arrangement would be cheerfully submitted to by the merchants, but in contrary case we pledge ourselves to aid the Federal Government in their collection."
Colonial Archives (MS., Texas State Library.

Nearly every municipality in the department of the Brazos now adopted resolutions condemning Travis's act or professing loyalty to Mexico. The Texas Republican, July 18, 1835; "Publications" of the So. Hist. Assn., Vill, 18 Tenorio to Ugartechea, July 15, 1835.(Bexar Archives)

Tenorio, the captain of the dislodged garrison was invited to attend a meeting at San Felipe and see for himself that "it was not by the vote of the majority nor by the will of the inhabitants that those persons were authorized to commit that outrage against the supreme Government."

Travis himself was so affected by the general disavowal that he expressed to Col. Ugartechea a strong desire to make amends: "I am extremely anxious to bring all our difficulties to a happy and peaceable termination, and to see the Government firmly established on a permanent basis. It matters not to me what form of government Mexico adopts, so that we are guaranteed in the security of person and property.

Therefore, I am disposed to lend my feeble aid to any plan which may be fallen on to bring about these things. There only wants a good understanding between the Government and the people of Texas to set all things right. If you will condescend to open a correspondence with me on this subject, I think that this good understanding may be brought about, as I assure you the inhabitants are in the best disposition for it."

Travis to Ugartechea, July 31,1 &15. Mexican Archives, Archivo de Guerra y Marina, Frac. 1, Op. Mil. Texas, 1835, Legajo 1, Exp. 9.

For several weeks he printed a card in The Texas Republican, asking the people to suspend judgment against him until he had time to make an explanation. This was somewhat tardily prepared on September 1, but it seems never to have been offered to the public. Down to the middle of August, with three exceptions, every popular meeting whose proceedings are available manifested a conciliatory disposition. [The exceptions were the secret meeting of the radicals at San Felipe on June 22, which, under the excitement produced by the arrest of the governor and news of the contemplated military occupation, voted the attack on the garrison at Anahuac; a meeting at Harrisburg on July 14

The Texas Republican, Aug. 22, 1835;
Publications of So. Hist. Assn., VIII, 107-109); and a meeting of the district of the Navidad and Lavaca on July 17

(Brown, History of Texas, 1, 299).

The people were soothed by the assurance that the Government meant them no harm, and an evident effort was made to convince the Government that the people were loyal.

By kind permission of; Presidio La Bahia, Goliad, Texas.

A meeting at Gonzales on July 7 resolved, 113rd.

* * * every act and deed, tending to interrupt the harmony and good understanding existing between Texas and the Federal Government deserves the marked disapprobation and contempt of every friend of constitutional order in the country. 4th. * * * resolved, that we have full confidence in the favorable disposition of His Excellency the President and the General Congress toward Texas, and we believe that when the wants of Texas are fully made known to them they will be provided for. 5th.

* * * resolved, that the course pursued by the citizens of Texas when called on by the governor of the State to move against the Federal troops, with offers of reward to those who should obey the order, in refusing to leave Texas to interfere in the quarrels of the Republic, if duly considered, furnishes conclusive proof of the loyalty of the inhabitants of Texas toward the Nation and their unwillingness to become embroiled with them."---Colonial Archives (MS.), Texas State Library.

The ayuntamiento of Columbia "would represent to you that the citizens of this jurisdiction hold themselves to be true, faithful, loyal, and unoffending Mexican citizens; that they do not violate the laws and constitution of the land, nor will they countenance others in doing it."

—To chairman of public meeting at San Felipe, July 11, 1&35, in The Texas Republican, July 18, 1835.Public meeting at San Jacinto on August 8: "We have always considered and do still consider the aggregate Mexican Nation the rightful sovereign of the territory we occupy. We consider names as the mere signification of things; and we are not so obstinately prejudiced in favor of the term 'federal republic' as peremptorily and without inquiry to reject another Government purely because it has assumed a different external sign or denomination. There are certain essential, sacred, and imprescriptible rights which must be guaranteed to every citizen we believe those rights may be as well secured under a consolidated as under a federative government, provided that government be wisely and liberally organized. * * * Resolved, that the dissolution of a government does not of necessity requisite that the constituent parts of the nation should separate finally; that the abstract right to do a thing does not always render the doing of it wise or commendable; that although the citizens of Texas may have the political right to reject the new Government of Mexico and to adopt one more consonant to their habits and feelings, we do very seriously question the policy of doing so, unless constrained by imperious circumstances, such as, we trust, do not and will not exist; that as adopted citizens we ought to exercise even our absolute rights with some diffidence and with a peculiar regard to the moral obligations that may rest upon us."

The Texas Republican, September 19, 1835; Publications of So. Hist. Assn., VIII, 110-116]

Two of the three meetings whose resolutions were less moderate, while expressing a lack of confidence in President Santa Anna, declared that they were ready to support the Federal constitution of 1824. These meetings represented pretty fully the population of the department of the Brazos. The other department, inhabited chiefly by Anglo-Americans, was that of Nacogdoches, in the eastern part of the State, and its political chief wrote on July 28 that his people were tranquil. But pacific efforts did not stop with holding meetings and adopting resolutions.

A joint committee representing several municipalities sent to Col. Ugartechea at San Antonio an exposition of public opinion, and asked him to forward it to his superior, Gen. Cos. The committee thought that recent acts, if left unexplained, might justify the belief that the colonists were hostile to the Government, but—the people at large we know have not participated either in the feelings which prompted the aggressions or any acts opposed to the legal authorities of the Mexican Republic, and do and ever will disavow the course pursued by a few impetuous and misguided citizens whose conduct *might implicate the whole community.*

The citizens of Texas generally have become adopted citizens of the Mexican Republic from choice, after a full knowledge of the constitution and laws. *They entertain a grateful sense of the liberality of the Government toward her colonies in the distribution of lands to settlers, and other advantages tending to their convenience and prosperity in agriculture and manufactures. They will be prepared on every constitutional call to do their duty as Mexican citizens in the enforcement of the laws and promotion of order.* They will cherish those principles which most clearly demonstrate their love of peace, respect for their Mexican fellow citizens, and attachment to the free and liberal institutions of their adopted country.The same committee appointed commissioners to visit Cos and make a personal explanation of the situation, but they were halted at San Antonio by Ugartechea and never reached him.

Ugartechea said that it would be useless to make explanations, unless the colonists would prove their loyalty by surrendering the radical leaders to the military authorities. This was later confirmed by Cos. The radicals were not surrendered and the commissioners went no further

Barrett and Gritten to Cos ., Aug. 9, 1835, Bexar Archives; Publications of So. Hist. Assn., VIII, 343-344]

At the same time, however, the colonists were organizing the militia, and they went so far as to propose a committee to raise subscriptions for the purchase of arms and munitions. The menacing activity of the Indians was sufficient to justify this ;*See Yoakum, "History of Texas," 1, 336; Brown, "History of Texas," 1, 290; Navarro to Political Chief of Nacogdoches, June 4, 1835, Bexar Archives; Lynch to Brown, July 20,1835, Colonial Archives, Tex. State Library],* and it was used as a pretext; but the real motive seems to have been fear that the Government would push them to extremes. *Even the most earnest advocates of peace approved the organization of the militia.(Columbia advised it on June 28, Lavaca on July 4, Mina on the 5th, Matagorda on the 13th, and Gonzales on the 19th The Texas Republican, July 4 and 18,1835, for Columbia and Lavaca; Colonial Archives, Tex. State Library, for other references). The double purpose of the organization is shown by the Proclamation in which the political chief ordered it: "The critical and peculiar situation" of the country, he said, demanded "speedy and decisive measures for defense against military usurpation on the one hand and the depredations of the Indians on the other."*

It is further illustrated by a writer who thought that every exertion should be made to avoid a conflict, but, lest this should fail, advised the organization of the militia. The Mexican view of the situation is shown by a letter from Ugartechea. Cos wrote him Sept. 28, to find out what the four tribes were of which the political chief of the Brazos was complaining and against which he was keeping 300 men in arms. Ugartechea replied that he had no information that the tribes of that department were hostile, and "as for placing 300 carbineros under arms in San Felipe, there is no other motive than to contribute to the revolution which is fomenting in the colonies." BexarArchives

Students familiar with the development of the American Revolution will not consider this necessarily inconsistent with a desire for peace. Nevertheless, it raises a question of the sincerity of the Texans, which demands some attention. The best evidence that the professions of loyalty were made in good faith by the public meetings is the fact that they convinced certain contemporaries. Travis, a prominent radical, leader of the attack on the garrison at Anahuac, and an intelligent lawyer who would have been likely to detect subterfuge, wrote on July 30, "The peace party, as they style themselves, I believe are the strongest and make much the most noise. Unless we could be united, had we not better be quiet and settle down for awhile?" Three weeks later he wrote, "I found the tories and cowards making a strong effort, and for a time they were but too successful. I was, therefore, disgusted and wrote you but little, as I had nothing to communicate but what I was ashamed of as a free man and a friend of my country."

By kind permission of ; Presidio La Bahia, Goliad, Texas.

On July 25 Dr. James H. C. Miller, an intense loyalist, who viewed these proceedings with feelings directly the opposite of Travis's, wrote from San Felipe, "All here is in a train for peace; the war and speculating parties are entirely put down." So convinced was he that the people would make almost any sacrifice for peace that he advised Col. Ugartechea to demand the surrender of the radical leaders to the military authorities. Finally we have the testimony of a neutral.

Edward Gritten, an Englishman who was making at this time a semi-official inspection of the colonies for Ugartechea, wrote on July 5, "From what I have observed and the conclusions that I have drawn, the greater part of the colonists desire to avoid any breach with the Government. If the executive could adopt a conciliatory policy it would meet the support of the sane portion of Texas, which is truly numerous." And again on the 17th he said, "All, even to the Sabine, unanimously desire to preserve the peace, and they manifest without exception their disapprobation of the offenses committed against the nation. In my opinion no more is needed to consummate the work of pacifying this country than to abstain from bringing troops into it for offenders."

It seems clear that until well into August the war spirit was confined to a few individuals *[the prominent ones were W. B. Travis, John A. and William H Wharton, R. M. Williamson, Henry Smith, Mosely Baker, and James Bowie. But until the end of July we have no record of any active agitation from any of them except Travis, Williamson, and Bowie].*

Public meetings, which may be said to have expressed organized opinion, were almost unanimously opposed to anything that might precipitate a breach. Conservative leaders from the beginning of the colonies had deliberately advocated the policy of standing aloof from Mexican party dissentions; and they hoped that a continuance of the policy might avert the present threatening storm from Texas. There were sacrifices, however, which they would not make, and these were now unfortunately demanded. Col. Ugartechea made a requisition on the political chief for the surrender of the radical leaders to the military authorities *[Ugartechea to Tenorio, July 31, and to Cos, August 1, 1835, Bexar Archives*]; and it became apparent that Santa Anna was fixed in the determination to establish a number of strong garrisons in Texas.

Had there been no atmosphere of racial distrust enveloping the relations of Mexico and the colonists, a crisis might not have followed. Mexico might not have thought it necessary to insist so drastically on unequivocal submission; or the colonists might not have believed so firmly that submission would endanger their liberty. As it was, the Texans at first evaded and then categorically declined to make the arrests; while the same municipalities that had professed unswerving loyalty to Mexico were entirely outspoken in opposing the military occupation of the province. If their action needs palliation, it may be somewhat to the point to remember two things. In the first place, these men who refused to deliver their neighbors to military authority for trial outside the colony were sons of the generation that indignantly protested against Great Britain's revival of Henry VIII's law of treason and against the extraordinary powers of the Gaspee commission; and in the second place, they had already experienced some inconvenience from Mexican garrisons in 1832.

It remains to say a word about the motives of those who at first opposed a breach with Mexico. Most of the rank and file, perhaps, was merely indifferent to the alleged abuses of Santa Anna. They recognized certain rights of the central Government in Texas, and, desiring peace, were slow to believe that these rights would be exceeded. More thoughtful conservatives who may have seen the danger from Santa Anna's plans probably feared that Texas could not sustain a struggle with Mexico, and advocated non-resistance as a policy of expediency. But beyond indifference, a readiness to recognize constitutional rights of the Federal Government, and resignation to present ills lest worse should be provoked by resistance, the loyalty of the colonists did not perhaps extend.

By kind permission of; Presidio La Bahia, Goliad, Texas.

Correspondence Related to Public Opinion in Texas, 1835

Miller to Public 21 Jun 1835
Chieftaincy of' the Department of Brazos Fellow Citizens

It has become my duty as chief of the Department to advise you of the critical situation in which you and your rights as freemen & as citizens of the state of Coahuila & Texas are placed by the usurpation of the Military that Garrison the towns of the three Eastern States. So far back as September of last year, they in conjunction with a faction at Saltillo overthrew the constitutional authorities of the State and placed a Military officer in the supreme executive, power of the state as its Governor.

In December last you were told by that officer that certain individuals had incurred the displeasure of the President General of the Republic and you were instructed not to vote them as officers of the state. As freemen & lovers of liberty and the rights guaranteed by a federal republican Government, you chose to exercise your suffrage freely & openly, you voted for Augustine Viesca as your Governor, your votes thus openly and manfully given called loudly upon that individual to yield to your wishes although he knew he would be driven from his office by the military force of the nations: he took the oath of office and entered upon its duties. He saw too soon the clouds gathering around film and applied to the legislature for permission to establish his government in some place of safety ere the storm burst in its violence upon him.

Texas was hailed by him as that favored spot, its inhabitants had been born and nursed in the cradle of liberty, their bosoms expanded with the breath of freeman & their proud hearts spurned the yoke of military power. Permission was granted to the Governor to come to Texas; he gathered together the documents of his office important to the country and started, he was compelled to return to Monclova because the military had declared he should not pass to Texas. Persuaded that could he reach Texas military rule and misrule would fall paralyzed, he, made an attempt to *[illegible*] the vigilance of the military in and reach Texas, but who can escape the bayonets of a sanguinary soldierasy he has been taken and is thrown into a dungeon without a hope, without a prospect, except such as your proud spirits may determine on.

Let me ask you as one in common with yourselves, are you prepared to receive such a government as it may please the commandant General Cos and his masters to give you, and again receive as military officer as your Governor? or will you support and sustain the officer your own voluntary vote place din office, and who now lies in prison on account of that vote made in his favor. I think by the feelings which I have that I can answer, you will never submit tamely to such a course--The object is to establish the supreme executive authority of the State of Texas, this is highly important and it behooves everyman to strain every nerve to accomplish so desired an object--Then let me call upon you in furtherance of your interest and in obedience to the orders we have received to turn out immediately, organize and march to his relief and bring him to a place of safety, in this our favored Texas You will march to this place as soon as possible and wait for further orders.

Given at office in the town of San Felipe de Austin this 21st day of June 1835 J B Miller

By kind permission of; Presidio La Bahia, Goliad, Texas.

COLUMBIA RESOLUTIONS 23RD JUNE 1835

At a meeting of the citizens held in the town of Columbia, on Tuesday the 23d of June, 1835, Silas Dinsmore, Jr., was called to the chair. On motion of John A. Wharton, seconded by Win. H. Jack, Esq. (the letter of the Political Chief being under consideration) the following resolutions were unanimously adopted:

1st. Resolved. That in the sense of the present meeting it is inexpedient to adopt any measures of committal, until all the citizens of this Jurisdiction can be consulted in general meeting.

2nd. Resolved. That the Political Chief be requested to take the sense of the citizens of his Department, in regard to the most proper political measures to be adopted on the present occasion.

3rd. Resolved. That we recommend to the citizens of Texas, union, concert, and moderation in the adoption of measures to meet the present crisis; and that we pledge our fortunes, lives and honors In support of such measures as the majority may adopt.

4th. Resolved. That a general meeting of the citizens of this Jurisdiction be called to take place in the town of Columbia, at 12 o'clock on Sunday, the 28th inst.

5th Resolved. That the Chairman be requested to address a letter to the Political Chief, enclosing him a copy of these resolutions, and assuring him that he will find us at all times ready and prompt to discharge our duty as good citizens.

6th. Resolved. That the proceeding together with the Political Chief's letter be published in handbill form, and in the Texas Republic. On motion of Wm. J. Russell, Henry Smith, Branch T. Archer, Silas Dinsmore, Jr., Robert H. Williams, W. H. Sledge, William H. Jack and John A. Wharton, were appointed to draft a report for the next meeting. On motion of Branch T. Archer, the thanks of the meeting were presented to the Chairman.

On motion the meeting adjourned until Sunday next, at 12 o'clock. SILAS DINSMORE, JR, Chairman.

Columbia Meeting 28th June 1835

At a very large and respectable meeting of the citizens of the jurisdictions of Columbia on the 28th day of June, 1835, held in pursuance of a previous call, for the purpose of considering the present situation of the country, and determining upon the course of conduct for the people of this jurisdiction to pursue, in the present and approaching crisis---Col. W. D. C. Hall was called to the chair and Byrd B. Waller elected secretary. The letter of the Political Chief of this department together with a number of documents throwing light upon the real situation of the country, having been read to the meeting, Messrs. John A. Wharton, W. D. C. Hall, H. Smith. J. F. Perry, J. W. Bell, S. Whiting, G. B. McKinstry, W. C. White, P. B. McNeel, F. Dingham, J. E. Phelps, Edwin Waller, E. Andrews, J. P. Caldwell and E. G. Head, were unanimously chosen to prepare a report and resolution to be submitted for the consideration of this meeting. The committee retired, but shortly returned with the following report and resolutions which were unanimously adopted:

To the citizens of the jurisdiction of Columbia:---Your committee having given the subject on which they were to report, as thorough an investigation as time and circumstances will permit, beg leave to present the following report: Your committee view with the deepest regret, and greatest alarm *the present political situation of Texas,* and recommend to this meeting, and their fellow citizens generally, union, moderation, organization and a strict adherence to the laws and constitution of the land. Your committee protest against the conduct and acts of any act of individuals (less than a majority) *calculated to involve the citizens of Texas in a conflict with the Federal Government of Mexico,* and Particularly Protest against the proceedings of those persons at Anahuac who gave the collector of customs, Don Jose Gonzales, a series of resolutions declaring that they would not obey the revenue laws of Mexico. The denounce said persons as foreigners, and disclaim all participation in the act whatsoever.

Your committee further declare that they are the faithful and loyal citizens of Mexico, and that they are disposed and desirous to discharge their duty as such, and that it Is their wish and interest to remain attached to the Federal Government of Mexico. Your committee recommend to the

Political Chief. The adoption of the most prompt & energetic measures to chastise the savages that have lately committed depredations upon our frontier citizens: and beg leave to present the following resolutions:

Resolved, That it is the duty of the citizens of Texas to write in the support of the constitution and laws of their adopted Country.

Resolved. That inasmuch as Texas is left in a state of anarchy, and without governor, vice-governor of council, that we recognize the Political Chief as the highest executive officer, and that we earnestly recommend an immediate organization of the militia for the protection of the frontier, and that we suspend further orders until the whole people are consulted, and also that we recommend a similar course to the Chief's of the other departments of Texas.

Resolved, That the Political Chief be requested to correspond with the other chiefs of department in Texas and request them to cooperate with him in electing three deputies from each jurisdiction of their several departments, to meet the chiefs of departments in public council, with full powers to form for Texas a Precisimial Government, on the principles of the constitution, during the reign of anarchy in the state, and that they meet as soon as circumstances will possibly permit.

Resolved, That a committee of five be chosen to wait on the Political Chief with the view of this meeting, and that they remain a permanent committee of vigilance, correspondence and safety.

Resolved, That the Political chief be requested [illegible] the Executive of the Federal Government of Mexico, representing to him the peaceable and loyal disposition of the citizens of Texas, and their great desire to remain attached to the Federal government.

Resolved, That the Political Chief be requested to address the citizens of his department, commanding thern to adhere strictly to the laws and constitution of the land.

Resolved, That we will support the Political Chief in the discharge of all constitutional duties.

Resolved, that the chairman of this meeting be requested to address a letter to the Political Chief, enclosing him a copy of the proceedings of this meeting. Messrs. W.D.C. Hall, I.A. Wharton, W. H. Jack. J. G. McNeel, and and G. B. McKinstry were chosen by the meeting, the committee of vigilance, correspondence and safety, and to wait on the political Chief with the views of this meeting. On motion, it was resolved: that the thanks of this meeting to the Chairman and Secretary. And then the meeting adjourned. W. D. C. Hall. Ch`n, Byrd C. Waller, Secretary.

By kind permission of; Presidio La Bahia, Goliad, Texas.

H. Austin to James Perry 24 Jun 1835

Columbia 24 June 1835 Mr J F Perry

An attempt has been made here to day to involve us in an immediate Revolution, by sending troops forthwith in obedience to a call by the chief of Police to fight the federal force a rport and resolution were produced Cut and dried in Caucus last night, Compromiting us at once- I moved as an amendment- That the further consideration of the subject matter before the meeting should be postponed until the great body of the people of this Municipality could be convened to express their sentiments as to the expediency of a measure involving the Security of the rights and property and the safety and lives of the families of the people this was not admitted by the agitators as an amendment, when it was determined to put the Report and Resolutions to Vote first and then take the Vote upon my motion, on division 2/3 were against their report- They then without taking a vote upon my motion, so modified their resolutions as to effect the same purpose which being agreed to

They appointed a committee to draft a report and resolution to be proposed to the meeting on Sunday- Archer Wharton Jack and Henry Smith It was proposed to add me and R. Williams I declined to aid in forestaling the sentiments of the people. Thinking the meeting on Sunday ought to be left free to appoint their own committee and the people will reject their report on that ground if it be put to them You and Pleasant McNeel must be here all the McNeels Grev, Crosby-, every one, who can give a vote, for the cast is to be which will lose or win all our hopes in Texas The court is adjourned until Monday I am still quite unwell with the cold I took below--12 oclock is the hour---best remembrances to Emily and truly your friend H. Austin [Adressed:] Mr. Jas F Perry Peach Point

By kind permission of; Presidio La Bahia, Goliad, Texas.

Rueg to Political Chief 28 Jul 1835
Chieftaincy of the Department of Nacogdoches I have received your address to the people of Bexar and about the minutes of the Commission of public security. I agree with your honor on every point of his content. The excitment which has existed in this dept. has much ceased since we heard of the measures which have been taken in yours. I do most regret that the time does not allow us the chance of sending on a committee of a couple of Citizens of this department to join your depaticher to the comand. Genl. for the inform the jenl. Gov by this channel of our govd dispositions for peace and order, but I do cooperate with your views by adrissing by todys's mail, Genl. Cos with an official Staterant of our Situation and our good dispositions to obey the laws of our adopted Country and to respect the Constitutional authorities, with request to be forwarded to the Supreme powers of the Union. I take this opportunity to present to Your honor my best respects God and Liberty Nacogdoches July 28 of 1835 E. Henry Brigs [Addressed:] to Political Chief of the depart.of Brassos.

Centralista Col. Pedro Delgado describing the Battle of San Jacinto

....one of them [a captured Mexican female camp follower at San Jacinto] throwing herself into the posture of a suppliant addressed him [a Hispanic Tejano patriot in the battle] as follows; "Señor God demme no me mata por el amor de Dios y por la villa cle su madrecita! " On being answered in her own language she turned to her sister & said, "Hermanita! mira! ese senor God demme habla la lingua cristiana como nosotros!" [Mr. God damn don't kill me, for the love of God and for the life of your dear mother!---Look here sister this Mr. God damn speaks the Christian Language like us!]---Anecdote from The Lamar Papers.

"....Captain (now Colonel) Juan Nepemucene Seguin, a native of Bexar and whom I have known from a boy, commanded 25 men, all natives of the same place, and performed wonders; every man signalized himself in the most distinguished manner. One of them, with a Bowie knife, killed 25 of his countrymen, and one of them, Colonel Batres, whom I knew well."--29 May 1836 letter published in *The Mobile Morning Chronicle*

As a proportion of the population, the active participation of Hispanic native and immigrant residents in the struggle for independence of Texas from Spain and Mexico was equal to or greater in specific battles than that of resident immigrants from the United States of the North--Don Guillermo...with the North Americans, they [Mexicans] have adopted their customs and habits, and one may truly say that they are not Mexicans except by birth, for they even speak Spanish with marked incorrectness.

In the years 1835 and 1836, the period of the Texas Revolution, La Bahia was in the hands of the Texas-Americans. After the San Jacinto victory, Vicente Filisola and of the Mexican Army, and under orders from the captured General Santa Anna lead the retreating Mexican Army back to Mexico. When he neared Goliad he asked to be allowed come through the town, but was refused permission as the authorities feared that the Texas garrison and returning Texian soldiers there would object and probably would attack the Mexicans. Filisola bypassed the town.

By kind permission of; Presidio La Bahia, Goliad, Texas.

Now the Anglo-Americans took over. After San Jacinto, the volunteers from the United States, having joined the fighting here for the purpose of securing lands and homes in Texas in payment for their armed services, were quick to file on the most desirable tracts, therefore many stopped in the Goliad, Victoria, and Refugio region. Almost overnight the country changed from the Spanish culture to that of the Anglo-Americans.

Local families that supported the Texas Revolution, such as the De Leon Family, (founders of a), bought $35,000 worth of supplies and ammunition to fight the war. They also supplied many of the men to fight in the war, including all the De León men and all the husbands of the De León women. After the war for Texas Independence the De León family was subjected to horrible injustices. As one historian put it "They became the victims of the most unjust discrimination ever known in Texas". The family had been robbed of their dignity and all of their lands. They did not even have monies to pay for tombstones for their dead.

The municipalities became counties and the ayuntamiento (City Council) became towns and were reorganized and established under the Republic of Texas.The Texas Revolution had concluded, and all Mexican Government holdings were taken over for the republic, and so the Presidio and Mission of La Bahia del Espiritu Santo de Zunega became public property, and as such the state and the town of Goliad assumed ownership of both.

Within eleven months of the victory at San Jacinto and the gaining of the freedom of Texas, a small group of Texas-Irish veterans were in New Orleans loading a vessel with building materials and merchandise with which to rebuild their homes, which had been destroyed by the advance of Santa Anna's army. Before leaving, they named one of their own to John J.Linn, as the spokesman to present a memorial (request) to Bishop Blanc of New Orleans, requesting him to present it to the Council of Bishops of the United States, then assembled in Baltimore, Maryland. The memorial requested priests for Texas and they asked that the memorial be passed on to Rome. The memorial was signed on March 20, 1837 in New Orleans. [1]

The petition was signed by: John J. Linn, of Victoria, Robert Hearn from Mission Refugio; William R. C. Hayes, Robert O'Boyle, and Andrew A. Boyle from San Patricio; and John McMullen for the Catholics in San Antonio de Bexar.[2]

First Texas Declaration of Independence

On December 20, 1835 ninety two men gathered in the Our Lady of Loreto Chapel and signed the First ce from Mexico. At the time they flew the severed arm and bloody sword flag which is now called the Goliad flag. The statement that they were making with the flag was that they would rather sever an arm than to continue

November 4, 1835: occurred on the east bank of the Nueces River two miles up stream ofSin today's San Patricio County, directly across from Fort Lipantitlán. A Texas force of around seventy men under Adjutant engaged a Mexican force of about ninety men under Capt. Nicolás Rodríguez. The Texans scored an important victory. to live under the dictatorship of Santa Anna.

Gen. Cos, the commander of northeastern Mexico, appointed a military governor, pending a new election. This news reached San Felipe by special courier from Cos himself on June 21. Besides being the capital of the department of the Brazos, San Felipe was also headquarters for a small party of radicals who probably desired separation from Mexico. Some members of this party seized the courier, and rifling his dispatches to Capt. Tenorio, who was commanding a garrison at Anahuac, near the head of Galveston Bay, learned that the Government was contemplating a military occupation of Texas.

This seemed a favorable occasion for alarming the people and hurrying them into a position from which it would be difficult to withdraw. The leaders induced the political chief to issue a proclamation calling on the men of the department to march to the governor's rescue, and in a public meeting adopted resolutions enumerating abuses which Texas had suffered through Santa Anna's enlargement of the central power and urging both Texan and Coahuilans to defend the constitution. At the same time, in secret meeting, they authorized an expedition to expel the troops from Galveston Bay. But they soon found that they had reckoned without the host.

A meeting of the neighboring municipality of Columbia on June 23 refused by a large majority to indorse the political chief's proclamation (see H. Austin letter to Perry); while a second and larger meeting on the 28th went much farther. It protested against "the acts and conduct of any set of individuals (less than a majority) calculated to involve the citizens of Texas in a conflict with the Federal Government of Mexico;" declared that the participants in this meeting were "faithful and loyal citizens of Mexico," whose "wish and interest" it was "to remain attached to the Federal Government;" and requested the political chief to inform the president of these sentiments, while "commanding" the inhabitants of his department to adhere strictly to the laws and the constitution." It recommended the organization of the militia for protection from the Indians, and advised the establishment of a local provisional government.

Within two weeks a number of municipalities adopted the Columbia resolutions on bloc, or passed others of like tenor; while the political chief wrote a public letter apologizing for his proclamation. The news of the governor's arrest and of the plans for occupying Texas had overwhelmed him with surprise, he explained, and he had yielded to "the earnest protestations of a number of influential citizens" in calling out the people.

By kind permission of ; Presidio La Bahia, Goliad, Texas.

The Unanimous Declaration of Independence made by the Delegates of the People of Texas in General Convention at the town of Washington on the 2nd day of March 1836

When a government has ceased to protect the lives, liberty and property of the people, from whom its legitimate powers are derived, and for the advancement of whose happiness it was instituted, and so far from being a guarantee for the enjoyment of those inestimable and inalienable rights, becomes an instrument in the hands of evil rulers for their oppression.

When the Federal Republican Constitution of their country, which they have sworn to support, no longer has a substantial existence, and the whole nature of their government has been forcibly changed, without their consent, from a restricted federative republic, composed of sovereign states, to a consolidated central military despotism, in which every interest is disregarded but that of the army and the priesthood, both the eternal enemies of civil liberty, the everready minions of power, and the usual instruments of tyrants.

When, long after the spirit of the constitution has departed, moderation is at length so far lost by those in power, that even the semblance of freedom is removed, and the forms themselves of the constitution discontinued, and so far from their petitions and remonstrances being regarded, the agents who bear them are thrown into dungeons, and mercenary armies sent forth to force a new government upon them at the point of the bayonet. When, in consequence of such acts of malfeasance and abdication on the part of the government, anarchy prevails, and civil society is dissolved into its original elements.

In such a crisis, the first law of nature, the right of self-preservation, the inherent and inalienable rights of the people to appeal to first principles, and take their political affairs into their own hands in extreme cases, enjoins it as a right towards themselves, and a sacred obligation to their posterity, to abolish such government, and create another in its stead, calculated to rescue them from impending dangers, and to secure their future welfare and happiness. Nations, as well as individuals, are amenable for their acts to the public opinion of mankind.

A statement of a part of our grievances is therefore submitted to an impartial world, in justification of the hazardous but unavoidable step now taken, of severing our political connection with the Mexican people, and assuming an independent attitude among the nations of the earth. The Mexican government, by its colonization laws, invited and induced the Anglo-American population of Texas to colonize its wilderness under the pledged faith of a written constitution, that they should continue to enjoy that constitutional liberty and republican government to which they had been habituated in the land of their birth, the United States of America. In this expectation they have been cruelly disappointed, inasmuch as the

Mexican nation has acquiesced in the late changes made in the government by General Antonio Lopez de Santa Anna, who having overturned the constitution of his country, now offers us the cruel alternative, either to abandon our homes, acquired by so many privations, or submit to the most intolerable of all tyranny, the combined despotism of the sword and the priesthood.

It has sacrificed our welfare to the state of Coahuila, by which our interests have been continually depressed through a jealous and partial course of legislation, carried on at a far distant seat of government, by a hostile majority, in an unknown tongue, and this too, notwithstanding we have petitioned in the humblest terms for the establishment of a separate state government, and have, in accordance with the provisions of the national constitution, presented to the general Congress a republican constitution, which was, without just cause, contemptuously rejected. It incarcerated in a dungeon, for a long time, one of our citizens, for no other cause but a zealous endeavor to procure the acceptance of our constitution, and the establishment of a state government.

By kind permission of; Presidio La Bahia, Goliad, Texas.

It has failed and refused to secure, on a firm basis, the right of trial by jury, that palladium of civil liberty, and only safe guarantee for the life, liberty, and property of the citizen. It has failed to establish any public system of education, although possessed of almost boundless resources, (the public domain,) and although it is an axiom in political science, that unless a people are educated and enlightened, it is idle to expect the continuance of civil liberty, or the capacity for self government.

It has suffered the military commandants, stationed among us, to exercise arbitrary acts of oppression and tyrrany, thus trampling upon the most sacred rights of the citizens, and rendering the military superior to the civil power. It has dissolved, by force of arms, the state Congress of Coahuila and Texas, and obliged our representatives to fly for their lives from the seat of government, thus depriving us of the fundamental political right of representation.

It has demanded the surrender of a number of our citizens, and ordered military detachments to seize and carry them into the Interior for trial, in contempt of the civil authorities, and in defiance of the laws and the constitution. It has made piratical attacks upon our commerce, by commissioning foreign desperadoes, and authorizing them to seize our vessels, and convey the property of our citizens to far distant ports for confiscation.

It denies us the right of worshipping the Almighty according to the dictates of our own conscience, by the support of a national religion, calculated to promote the temporal interest of its human functionaries, rather than the glory of the true and living God.

It has demanded us to deliver up our arms, which are essential to our defence, the rightful property of freemen, and formidable only to tyrannical governments. It has invaded our country both by sea and by land, with intent to lay waste our territory, and drive us from our homes; and has now a large mercenary army advancing, to carry on against us a war of extermination.

It has, through its emissaries, incited the merciless savage, with the tomahawk and scalping knife, to massacre the inhabitants of our defenseless frontiers. It hath been, during the whole time of our connection with it, the contemptible sport and victim of successive military revolutions, and hath continually exhibited every characteristic of a weak, corrupt, and tyrranical government. These, and other grievances, were patiently borne by the people of Texas, untill they reached that point at which forbearance ceases to be a virtue.

We then took up arms in defence of the national constitution. We appealed to our Mexican brethren for assistance. Our appeal has been made in vain. Though months have elapsed, no sympathetic response has yet been heard from the Interior. We are, therefore, forced to the melancholy conclusion, that the Mexican people have acquiesced in the destruction of their liberty, and the substitution therfore of a military government; that they are unfit to be free, and incapable of self government.

The necessity of self-preservation, therefore, now decrees our eternal political separation. We, therefore, the delegates with plenary powers of the people of Texas, in solemn convention assembled, appealing to a candid world for the necessities of our condition, do hereby resolve and declare, that our political connection with the Mexican nation has forever ended, and that the people of Texas do now constitute a free, Sovereign, and independent republic, and are fully invested with all the rights and attributes which properly belong to independent nations; and, conscious of the rectitude of our intentions, we fearlessly and confidently commit the issue to the decision of the Supreme arbiter of the destinies of nations.

(signed)

Richard Ellis - *President of the Convention and Delegate from Red River.*

By kind permission of; Presidio La Bahia, Goliad, Texas.

Charles B. Stewart
Tho. Barnett
John S. D. Byrom
Francisco Ruis
Jose. Antonio Navarro
Jesse B. Badgett
Wm D. Lacy
William Menifee
Jn. Fisher
Matthew Caldwell
William Motley
Lorenzo de Zavala
Stephen H. Everett
George W. Smyth
Elijah Stapp
Claiborne West
Wm. B. Scates
M. B. Menard
A. B. Hardin
J. W. Burton
Thos. J. Gazley
R. M. Coleman
Sterling C. Robertson
James Collingsworth
Edwin Waller
Asa Brigham
Geo. C. Childress
Bailey Hardeman
Rob. Potter
Thomas Jefferson Rusk
Chas. S. Taylor
John S. Roberts
Robert Hamilton
Collin McKinney
Albert H. Latimer
James Power
Sam Houston
David Thomas
Edwd. Conrad
Martin Palmer
Edwin O. Legrand
Stephen W. Blount
Jms. Gaines
Wm. Clark, Jr.
Sydney O. Pennington
Wm. Carrol Crawford
Jno. Turner
Benj. Briggs Goodrich
G. W. Barnett
James G. Swisher
Jesse Grimes
S. Rhoads Fisher
John W. Moore
John W. Bower
Saml. A. Maverick (from Bejar)
Sam P. Carson
A. Briscoe
J. B. Woods
H. S. Kimble, Secretary

Provided courtesy of Randall Tarin from the files of Alamo de Parras
By kind permission of; Presidio La Bahia, Goliad, Texas

LETTER OF GRIEVANCE ADDRESSED TO STEPHEN H. DARDEN, STATE COMPTROLLER, FROM HISPANIC TEXIAN HEROES, DATED JANUARY 12, 1875

"Sir: We the undersigned citizens of this country respectfully address you this communication to remove from your mind what seems to us an unjust impression as regards the application of certain Mexicans for pensions who participated in the Revolution which separated Texas from Mexico. We assert that the following named persons commanded companies at the taking of San Antonio in 1835. On the 20th of October 1835, Juan N. Seguin followed by thirty seven men of Mexican birth, joined on the Salado Creek according to previous appointment, the first Texan forces that gathered in order to oppose the Central Government proclaimed by Santa Anna in violation of the Federal government constitutionally existing. Placido Benavides of (La Bahia).

Goliad joined on the same creek with the revolutionary troops with 26 or 28 men, so that at the Battle of Concepcion, the Mexicans who took part in that fight numbered some seventy men if we add some isolated soldiers. Directly after the Concepcion fight it was agreed between the Texian Leaders to put the siege to the city of San Antonio and to remove the camp to the Northern part of the city. But before the removal, Salvador Flores was detailed to the Mexican ranches on the San Antonio River, and Manuel Leal to the Mexican with the object of raising new forces that were very much needed; these two patriots returned soon after, Flores with 15 new men, and Manuel Leal with 26.

A conflict of authority took place at that moment between Juan N. Seguin and Placido Benavides both claiming to be Captain; it was amicably settled in favor of Seguin for the reason that he had raised more men than Benavides, but with the understanding that although Seguin was to be the Superior officer, Benavides would preserve the direct and immediate Command over the men he had brought from Goliad, and that agreement was intended to Manuel Leal and Salvador Flores; as soon as the troops reach their new camp, on the old mill, they were joined by fourteen privates of the old Company of the Alamo for the most part sons of San Antonio who deserted from Mexican forces of Gen. Cos and joined Seguin's Command with arms and baggage.

There was not at that time any thing like a muster roll, or a regular register of enlistment; every volunteer who offered his services was readily accepted, and the men joined the party that suit them best, they acted with a liberty that had nothing in common with the disciplining of a regular army: generally the private followed the order of the officer who had brought them to Camp. During the stay of the troops, before San Antonio, several parties of Mexicans joined the patriots; namely, Miguel Aldrete who in company with Mayor Collensworth came from Goliad with twenty odd men; Col. J. C. Neil and Philip Dimmit who arrived also with a Mexican Company raised in Victoria and in the lower country, without counting isolated enlistment that took place every day. In fact, the company of Seguin alone amounted to over one hundred and sixty men on the day of the Storming of Bexar.

After the taking of the place, that company was sent out to protect the people of the Ranches, against the devastation made by the retiring Mexican troops. On their return they found that the Mexican volunteers of Benavides and Dimmitt had left for home, as well as the American patriots. There were not fourteen Americans in San Antonio, after the taking of the place. Col. J. C. Neil had received, first the military Command, but he was soon after superceded by B. Travis who had under his former company and that of Seguin. They continued in active service, for several months and relying on the false report that all was quiet on the Mexican borders, a large quantity of the Mexicans were authorized to retire in order to protect their families against Indian depredations

At the coming of Santa Anna, the company of Seguin had been reduced, and the arrival of the enemy being entirely unsuspected; the most part of the men received the authorization to secure the safety of their families and to join the Texians at the Alamo; it is due to that circumstance that fifteen Mexicans only entered the Alamo with Travis.

At the gathering of the Texian Army at Gonzalez, Seguin had a large Company, in fact the largest of the Army, but it was a new Company quite different from the one he had commanded at the taking of Bexar. He had above one hundred men; out of whom 25 were detached to protect the invaded population. From 15 to 20 were at the order of Deaf Smith; thirty odd were sent Eastward to escort and protect American families, three men were sick at San Felipe, about ten at least were with the baggage at Harrisburg, four or five remained behind in charge of the horses at the moment of the battle of San Jacinto., so that he mustered only twenty two men, when he was ordered to give the names of those who had actually fought.

We would respectfully remind you that we and our comrades took up arms against our own kindred and country, believing we were right, and now we feel humiliated to find that when we have testified on oath to the services rendered by us and our (own) old companions, many of whom are not only suffering from the infirmities of age but also from extreme poverty that their claims should be disregarded and forced to wait for weeks and months for their pensions, when Americans have been promptly paid upon what we consider no better evidence than our friends have furnished. We feel assured, Honored Sir, that you must have been misled or misinformed as to the parties who have applied for pensions as well as their witnesses, and we address you this communication to disabuse your mind of any prejudice you may entertained, and to assure you that we entertained for you personally the kindest feelings and only ask for our old companions simply justice and nothing more."

Signed: Juan Jimenes, Ygnacio Espinaso, Martine Maldonado, Ignacio Arocha, Tomas Martines, Narciso Leal, Juan Martines, Antonio Oliva, Estevan Uron, Manuel Montalvo, Crescencio Montes, Pablo Salinas, Quirino Garza, Nepomuceno Flores, Juan N. Seguin, Antonio Menchaca, Jose Antonio Rodriguez, Antonio Vasquez, Damaso de Los Reyes

By kind permission of ; Presidio La Bahia, Goliad, Texas.

THE REPUBLIC OF THE RIO GRANDE IN 1840

The Republic Of The Rio Grande e was an effort on the part of Federalist leaders in Tamaulipas, Nuevo Leon, and Coahuila to break away from the centralistic government of Mexico in 1840 and to form a new confederation. After much Federalistic flurry in the northern frontier Mexican states, leaders of the party met at Laredo, Texas, in convention on January 17, 1840. The convention declared independence from Mexico and claimed for its territory the areas of Tamaulipas and Coahuila north to the Nueces and Medina rivers, respectively, and Nuevo León, Zacatecas, Durango, Chihuahua, and New Mexico. The red, black and white of the Republic of the Rio Grande contained three stars, representing the Mexican states of Tamaulipas, Nuevo Leon, and Coahuila.

Officers and a general council were elected as follows: Jesús de Cárdenas, president; Antonio Carlos Rivera?, commander-in-chief of the army; Juan Nepomuceno Molano, delegate and member of the council for Tamaulipas; Francisco Vidaurri y Villaseñor, for Coahuila; Manuel María de Llano, for Nuevo León; and Jose Maria Jesus Carbajal, , secretary to the council. The government was moved to Guerro, Tamaulipas, where it was to have remained temporarily. Canales with his force took the field against the Centralist army under General Mariano Arista, and on March 24-25, 1840, met Arista in battle at Morales, Coahuila, and was disastrously defeated. Colonel Antonio Zapata, cavalry commander of Canales, was captured and executed.

An army of approximately 500 - 600 men consisting of Mexicans, Americans, and Indians was organized in San Patricio. The principal leader of the Americans was Colonel Samuel W. Jordan and ninety men were ordered to the Rio Grande as the vanguard of the army late in June. They proceeded into the interior of Tamaulipas and captured Ciudad Victoria without a battle. From there treacherous subordinate officers led them toward San Luis Potosí, but, suspecting the treachery, Jordan changed direction and marched toward Saltillo. There, on October 25, 1840, he was attacked by Gen. Rafael Vásquez, the Centralist commander at Saltillo, but in spite of the desertion of part of his command, managed to defend himself and return to Texas. Early in November commissioners of Canales and Arista met, and Canales capitulated at Camargo on November

6, 1840. He was taken into the Centralist army as an officer, and Federalism was dead for the time being.

After the failure of The Republic of the Rio Grande movement, in December of 1840, Colonel Jordan would attempt to kill Sam Houston with an ax in Austin, Texas, but it was prevented by Adolphus Sterne.

MIER AND SOMERVELL EXPEDITIONS IN 1842

President Santa Anna and the central government of Mexico did not recognize the Republic of Texas. In 1842, Mexican troops crossed the Rio Grande and fought several Mexican Invastions in 1842 with the Texan army. This lesson follows these military encounters, including the sacking of Laredo (November, 1842) and the Mexican victory at Mier Expedition (December 26, 1842), which led to the infamous Black Bean episode (March 25.1843)

The Somervell expedition was a punitive expedition against Mexico in retaliation for three predatory raids made by Mexican armies upon Texas in 1842: Rafael Vasquez, Antonio Canales Rosillo's descent upon Fort Lipantitlan and the captures of San Antonio by Rafael Vasquez and Wolls

By kind permission of; Presidio La Bahia, Goliad, Texas.

PRESIDIO LA BAHIA (SITE OF THE GOLIAD MASSACRE)

A History of the Mission

Bishop Blanc was instructed by Rome in March of 1838 to prepare a report as to the conditions in vast regions of Texas. He requested that Father Timon lead the expedition into Texas and prepare the report for Rome.

Father Timon and his expedition landed at Galveston on December 27, 1838. From there they proceeded to Houston where the congress was in session and where they met the president and many prominent members of the congress. After consulting with these men, Catholic and Protestant alike, Father Timon was told that his plan of visiting San Antonio, East and south Texas was impractical because of the winter rains; the peril of Indian attack, and the Mexican guerillas were known to be roaming the country along the Guadeloupe and San Antonio Rivers.

Among those he consulted were Judge John Dunn Senator from Goliad, Refugio, and San Patricio; Senator John J.Linn from Victoria; Senator Juan N.Seguin of San Antonio and other Representatives. On their recommendations and information of state of religion in Texas, Timon returned to New Orleans and made his report.

In his recommendation, Timon stated: ".and "Old La Bahia District." He reported that in the region of La Bahia, now better known as Goliad, there were about twenty Catholic families, mostly Mexicans, that "The beautiful old church of Our Lady of Loreto still bore ugly scars of war, but was not beyond repair, standing abandoned, without sacred vessels, vestments, or furnishings."

Presidio La Bahia

Presidio La Bahia sold in 1848

The Goliad city council on December 19, 1848, leased to Reverend John F. Hillyer r for a term of five years, "The lot of land of twenty acres, including the old Aranama Mission [8] (Mission Espiritu Santo, across the San Antonio River from Presidio La Bahia was known to locals as Aranama Mission) with its appurtenances, for the purpose of having a Female School intuited and maintained at that place." The city council authorized and required the mayor to offer for sale on the next sale day the old Fort near, its walls and fixtures adjoining the Church in the Old Town of Goliad on a credit of one and two years in equal installments.", etc., provided as much as one thousand dollars shall be bid for same. It must have been under this authorization that Pryor Lea had a secured possession of the whole Presidio. A map in the Goliad County Archives shows that the Presidio was within the large acreage allotted to Lea.

By kind permission of; Presidio La Bahia, Goliad, Texas.

Presidio La Bahia in 1850

Judge Pryor Lea, as he was usually addressed, was said to have been related to Sam Houston's last wife. He came to Texas some time after annexation. He was a railroad promoter, and was in Goliad in 1847, surveying for the railroad he proposed to connect Goliad and Lamar and to make the latter place a deep-water port and railroad terminal. Goliad was his headquarters for his various promotional schemes, and he brought his family here and domiciled them in the old chapel of Our Lady of Loreto, to the horror and disgust of the few old Mexican families remaining at La Bahia and Carlos Ranch. He was in the Presidio and Church in 1850 if the dates of the following account of a trip John Russell Bartlett made through this country in 1850 is correct. His account of his travels state that in September, 1850 on his route over alternate prairies and woodland, between Victoria and Goliad, he reached the latter place where he had some business to transact.

In his later published narrative he leaves the following graphic account of the sad state of the old fort at that time.

In the various domestic wars of Mexico *this was an important place, and frequently changed hands*; nor was its importance lost during the struggle for Texas independence, when it was occupied by the Mexicans as well as the Texan forces. Its original name was La Bahia del Espiritu Santo, the Bay Town of Espiritu Santo, because it was originally the place for collecting the revenue of the small ports upon the bay. Hence, all persons arriving on the bay with merchandise were obliged to go forty miles into the interior duties... This name of La Bahia was changed by the Spaniards about thirty years since, when it began to decay as a religious establishment, to that Goliad, on account of its great strength." At this time, the facts of the name change from La Bahia to Goliad was being confused, or lost.

"Around the church are some twenty or more ruined buildings of stone, with nothing but their walls standing. One of these extends about 150 feet southward, and appears, from its small apartments, to have been constructed for barracks; its wall, like those of the church, are very massive. A height wall seems once to have surrounded the church, but much of it lies now prostrate. The other buildings, which are detached and of various dimensions, were chiefly used as dwellings.

The whole town is in ruins, and presents a scene of desolation, which to an American is at once novel and interesting. Each succeeding capture, of course, impaired the buildings; and after the decisive battle of San Jacinto, the Mexicans evacuated it and destroyed it as far as they were able. The material of these buildings is a soft white sandstone, which underlay the town, and which appears to become hardened when exposed to the air."

"The church is especially *notorious as having been the place where Fannin and his men were confined and massacred.* We were fortunate enough to meet with a gentleman, Judge H. W.L .Hunter) *who was one of the prisoners..."*

By kind permission of; Presidio La Bahia, Goliad, Texas.

Presidio La Bahia returned to the Church in 1853

The few people left in La Bahia were Catholics, and were in distress at the uses to which the chapel had been put, notably in the usurpation of it by Pryor Lea , and his actual occupancy of the old chapel of Loreto, and logged for its restoration to its rightful use. They humbly appealed to the Corporation of the new town of Goliad to restore it to its Sacred use. Happily the authorities of the town seeing their distress, and probably ashamed of having allowed its desecration, moved to restore La Bahia to the Church, which they now claimed to belong to the Corporation of Goliad, in spite of the legal transfer of the property to the Church by the Texas Congress, January 13, 1841.

Moved by the desire to clear up the situation, the Mayor of Goliad, A.H. Briscoe, wrote to Bishop Odin as follows:

Rt. Revd. Bishop Odin:
Goliad, Texas, Sept. 19 1853

Revd. And Dear Sir:

By an ordinance passed by the "Common Council of the town of Goliad", I (the Mayor) was empowered to open a correspondence with you with a view of adjusting by sale or otherwise the title to the old church and fort situated in the old town of Goliad, most usually called Labiahia. "The Common Council" are aware that the property is claimed by you as "Chief Pastor" of the Catholic Church in Texas. They however have no doubt as to the rights of the town of Goliad to said property and will claim and defend it to the last her right should they be disputed.

Knowing that it is the wish of the citizens of old Goliad, that the said church should belong to the Catholic Church, and that it only be used for the purpose for which it was set apart and consecrated, the Common

Council in view of this and prompted by feelings of kindness and liberality, offer to sell to you as Chief Pastor of the Catholic Church, said property, provided terms can be agreed upon. You are aware, I presume, that the said property, with other property was granted to the town for a specific purpose, and that the same *cannot be alienated or disposed of in any other manner* - than according to the conditions set forth in said grant - although the "Common Council" entertain liberal feelings, yet their duty as public officers require that they be just and true to their trust, through activated and governed by genuine motives.

By kind permission of; Presidio La Bahia, Goliad, Texas.

SOURCES

The Lure of Texas – Pre-Columbian Era

Lawrence E. Aten, *Indians of the Upper Texas Coast*, (New York: Academic Press, 1983).

Thomas N. Campbell, *The Indians of Southern Texas and Northeastern Mexico: Selected Writings of Thomas Nolan Campbell* (Austin: Texas Archeological Research Laboratory, 1988).

David S. Dibble and Dessamae Lorrain, *Bonfire Shelter: A Stratified Bison Kill Site, Val Verde County, Texas* ,Texas Memorial Museum Miscellaneous Papers [Austin: University of Texas, 1968]).

Thomas Hester, *Digging into South Texas Prehistory: A Guide for Amateur Archaeologists* (San Antonio: Corona Press, 1980).

T. R. Hester et al., *From the Gulf to the Rio Grande: Human Adaptation in Central, South and Lower Pecos Texas* (Arkansas Archeological Survey Research Series 33, Fayetteville: University of Arkansas, 1989]).

T. R. Hester et al., *Field Methods in Archaeology* (Palo Alto, California: Mayfield, 1975).

J. L. Hofman et al., *From Clovis to Comanchero: Archeological Overview of the Southern Great Plains* (Arkansas Archeological Survey Research Series 35(Fayetteville: University of Arkansas, 1989).

Jack T. Hughes, "Prehistoric Cultural Developments on the Texas High Plains," *Bulletin of the Texas Archeological Society* 60 (1989)

. Alex D.Krieger, *Culture Complexes and Chronology in Northern Texas with Extension of Puebloan Datings to the Mississippi Valle y* (University of Texas Publication 4640, Austin, 1946).

H. P. Newell and A. D. Krieger, *The George C. Davis Site, Cherokee County, Texas* (Society for American Archaeology Memoir 5, Menasha, Wisconsin, 1949).

Parker Nunley, *A Field Guide to Archeological Sites in Texas* (Austin: Texas Monthly Press, 1989).

R. A. Ricklis, *Aboriginal Life and Culture on the Upper Texas Coast: Archaeology at the Mitchell Ridge Site, 41GV66, Galveston Island* (Corpus Christi: Coastal Archaeological Research, 1994).

Dee Ann Suhm, "An Introductory Handbook of Texas Archeology," *Bulletin of the Texas Archeological Society* 25 (1954). A. J. Taylor and

C. L. Highley, *Archeological Investigations at the Loma Sandia Site (41LK28*

Thomas R. Hester and Ellen Sue Turner, Handbook of Texas Online, s.v.

Spanish Texas

From Handbook of Texas on line- Handbook of Texas Online, s.v. "," http://www.tshaonline.org/handbook/online/articles

Fray Marcos of Niza

'Ancient Peoples of the American Southwest, Stephen Plog -2nd.Edition-1997, Thames & Hudson

Quivera

William B.Workman; Prehistory of the Aishihik-Kluane Area, Southwest Yukon Territory, Mercury Series, Archaeological Survey Paper no.74, Ottawa; National Museums of Canada, 1976

Missions Attacked by Comanches and Allies

Great Indian Tribes, Daniel Jacobson, Hamond Incorporated. 1970

Mexican War of Independence

Rupert N. Richardson, *Texas: The Lone Star State* (New York: Prentice-Hall, 1943; 4th ed., with Ernest Wallace and Adrian N. Anderson, Englewood Cliffs, New Jersey: Prentice-Hall, 1981).

Robert S. Weddle, *San Juan Bautista: Gateway to Spanish Texas* (Austin: University of Texas Press, 1968).

Donald E. Chipman, *Handbook of Texas Online, s.v.*

Fort Anahuac

Margaret S. Henson, *Anahuac in 1832: The Cradle of the Texas Revolution* (Anahuac, Texas: Fort Anahuac Committee of the Chambers County Historical Commission, 1982).

Margaret S. Henson, *Juan Davis Bradburn: A Reappraisal of the Mexican Commander of Anahuac* (College Station: Texas A&M University Press, 1982).
James Wright Steely, comp., *A Catalog of Texas Properties in the National Register of Historic Places (Austin: Texas Historical Commission, 1984).*
Margaret Swett Henson; *Handbook of Texas On Line, s.v.*

Fort Lipantitlán

Eugene C. Barker, ed., *The Austin Papers* (3 vols., Washington: GPO, 1924–28). William Campbell Binkley, ed., *Official Correspondence of the Texan Revolution, 1835–1836* (2 vols., New York: Appleton-Century, 1936).
Bethel Coopwood, "The Route of Cabeza de Vaca,"
Quarterly of the Texas State Historical Association 3 (October 1899, January, April 1900).
Charles Adams Gulick, Jr., Harriet Smither, et al., eds., *The Papers of Mirabeau Buonaparte Lamar* (6 vols., Austin: Texas State Library, 1920–27; rpt., Austin: Pemberton Press, 1968).
Hobart Huson, *Refugio: A Comprehensive History of Refugio County from Aboriginal Times to 1953* (2 vols., Woodsboro, Texas: Rooke Foundation, 1953, 1955). Hobart Huson, *Souvenir Program, Refugio County Centennial Celebration* (Refugio: Refugio *Timely Remarks*, 1936?).
John J. Linn, *Reminiscences of Fifty Years in Texas* (New York: Sadlier, 1883; 2d ed., Austin: Steck, 1935; rpt., Austin: State House, 1986). Joseph Milton Nance, *Attack and Counterattack: The Texas-Mexican Frontier, 1842* (Austin: University of Texas Press, 1964).

Source; Keith Guthrie; Handbook of Texas Online, s.v.

Fort Tenoxtitlán

Source; Burleson County Historical Society, *Astride the Old San Antonio Road: A History of Burleson County, Texas* (Dallas: Taylor, 1980). Malcolm D. McLean, "Tenoxtitlán, Dream Capitol of Texas," *Southwestern Historical Quarterly* 70 (April 1967).
Source; *Charles Christopher Jackson, Handbook of Texas Online*, s.v.",

Fort Teran

Juan N. Almonte, "Statistical Report of Texas," *Southwestern Historical Quarterly* 28 (January 1925). Wallace W. Atwood, *The Physiographic Provinces of North America* (Boston: Ginn, 1940). Eugene C. Barker, *Readings in Texas History* (Dallas: Southwest Press, 1929). David J. Weber, *The Mexican Frontier, 1821–1846* (Albuquerque: University of New Mexico Press, 1982). James E. Wheat, "The Story of Fort Teran on the Neches," *It's Dogwood Time in Tyler County*, March 1951. James E. and Josiah Wheat, "Tyler County under Mexico," *It's Dogwood Time in Tyler County*, March 1966. Amelia W. Williams and Eugene C. Barker, eds., *The Writings of Sam Houston, 1813–1863* (8 vols, Austin: University of Texas Press, 1938–43; rpt., Austin and New York: Pemberton Press, 1970).

Source; Howard N. Martin; *Handbook of Texas Online*, s.v.

Old Stone Fort

Archie P. McDonald- Handbook of Texas Online, s.v.

The Alamo

Frederick Charles Chabot, *The Alamo: Mission, Fortress and Shrine* (San Antonio: Lenke, 1935). Anne A. Fox,. Anna B. Story, The Alamo from Its Founding to 1937 (M.A. thesis, University of Texas, 1938). Henry Ryder Taylor, *History of the Alamo and of Local Franciscan Missions* (San Antonio: N. Tengg, 1908). SOURCE; *Amelia W. Williams - Handbook of Texas Online*, s.v.

Henry Perry

Henry Stuart Foote, *Texas and the Texans* (2 vols., Philadelphia: Cowperthwait, 1841; rpt., Austin: Steck, 1935). Julia Kathryn Garrett, *Green Flag Over Texas: A Story of the Last Years of Spain in Texas* (Austin: Pemberton Press, 1939). Margaret S. Henson, *Juan Davis Bradburn: A Reappraisal of the Mexican Commander of Anahuac* (College Station: Texas A&M University Press, 1982). Harris Gaylord Warren, *The Sword Was Their Passport: A History of American Filibustering in the Mexican Revolution* (Baton Rouge: Louisiana State University Press, 1943).

Goliad - The Massacre

Source; Presidio La Bahia, Goliad, Texas

The Angel of Goliad

Source; Presidio La Bahia, Goliad, Texas

Reminiscences of Andrew Boyle

Source; *From The Forgotten Colony: San Patricio de Hibernia by Rachel Bluntzer Hébert* also from; Presidio La Bahia, Goliad, Texas

Dillard Cooper's Remembrances of the Fannin Massacre

From Rangers and Pioneers of Texas by A.J. Sowell 1884 as reprinted from the American Sketch Book 1881.

Nicholas W. Eastland and Capt. Horton's Troop

From Recollections of Early Texas: Memoirs of John Holland Jenkins also; Presidio La Bahia, Goliad, Texas

William L. Hunter

Source; No. 1650 Historical Notes. Lamar Papers also Presidio La Bahia, Goliad, Texas.

The Escape of Kentuck

Presidio La Bahia, Goliad, Texas

An Account of the Goliad Massacre by Major Miller's Secretary

Barber and Howes History of the United States 1861
Source; by kind permission of Presidio La Bahia, Goliad, Texas

Fannin's Retreat and the Battle of Coleto / An account by Ms. Cash

Source; The Rise, Progress and Prospects of the Republic of Texas' William Kennedy, 1841.
Source; by kind permission of Presidio La Bahia, Goliad, Texas

Fannin and his Command

[Lewis M. H. Washington in the Georgia Citizen also; Presidio La Bahia, Goliad, Texas

The Alamo

Presidio La Bahia, Goliad, Texas

My Early Times in Texas by J. C. Duval

H. P. N. GAMMEL & CO., PUBLISHERS, AUSTIN, 1892

Edward Burleson

Eugene C. Barker, ed., *The Austin Papers* (3 vols., Washington: GPO, 1924–28).

Mary Starr Barkley, *History of Travis County and Austin, 1839–1899* (Waco: Texian Press, 1963).

Edward Burleson, Jr.,Papers, Dolph Briscoe Center for American History, University of Texas at Austin.

John H. Jenkins and Kenneth Kesselus, *Edward Burleson: Texas Frontier Leader* (Austin: Jenkins, 1990).

Frank W. Johnson, *A History of Texas and Texans* (5 vols., ed. E. C. Barker and E. W. Winkler [Chicago and New York: American Historical Society, 1914; rpt. 1916]).

Kenneth Kesselus, *History of Bastrop County, Texas, Before Statehood* (Austin: Jenkins, 1986).

Frances Stovall et al., *Clear Springs and Limestone Ledges: A History of San Marcos and Hays County* (San Marcos: Hays County Historical Commission, 1986).

Dudley Goodall Wooten, ed., *A Comprehensive History of Texas* (2 vols., Dallas: Scarff, 1898; rpt., Austin: Texas State Historical Association, 1986).

Source; Helen Burleson Kelso - *Handbook of Texas Online*, s.v.

David (Davy) Crockett

Michael A. Lofaro, "Crockett, David," *Handbook of Texas Online*, s.v." by kind permission of "Handbook of Texas Online."

'Big Foot' Wallace

John C. Duval, Macon, Georgia - THE J. W. BURKE COMPANY -1870

Cynthia Ann Parker

Source; *Margaret Schmidt Hacker- Handbook of Texas Online*, s.v.

Rebecca Jane Gilliland Fisher

Elizabeth Brooks, Prominent Women of Texas (Akron, Ohio: Werner, 1896)

Rebecca J. Gilleland Fisher, "Capture and Rescue of Mrs. Rebecca J. Fisher, née Gilleland,"

Quarterly of the Texas State Historical Association 3 (January 1900)

Annie Doom Pickrell, Pioneer Women in Texas (Austin: Steck, 1929).

Handbook of Texas Online, s.v.

Matilda Lockhart

T. R. Fehrenbach, *Lone Star: A History of Texas and the Texans* (New York: Macmillan, 1968).

Rena Maverick Green, ed. *Memoirs of Mary A. Maverick* (San Antonio: Alamo Printing, 1921 rpt., Lincoln: University of Nebraska Press, 1989)

Andrew Jackson Sowell; *Early Settlers and Indian Fighters of Southwest Texas*

(Austin: Ben C. Jones, 1900; rpt., Austin: State House Press, 1986).

J. W. Wilbarger, *Indian Depredations in Texas* (Austin: Hutchings, 1889;

Buffalo Hump

E. W. Henderson, "Buffalo Hump, a Comanche Diplomat," West Texas Historical Association *Yearbook* 35 (1959).

John Holland Jenkins , *Recollections of Early Texas*, ed. John H. Jenkins III (Austin: University of Texas Press, 1958; rpt. 1973). Kenneth F. Neighbours,

Dorman H. Winfrey and James M. Day, eds., *Texas Indian Papers* (4 vols., Austin: Texas State Library, 1959–61; rpt., 5 vols., Austin: Pemberton Press, 1966). *Jodye Lynn Dickson Schilz -Handbook of Texas Online*, s.v.

Peta Nocona

Dallas *Morning News*, December 19, 1993. James T. DeShields, *Border Wars of Texas*, ed. Matt Bradley (Tioga, Texas, 1912; rpt., Waco: Texian Press, 1976).

Rupert N. Richardson, *The Comanche Barrier to South Plains Settlement* (Glendale, California: Clark, 1933; rpt., Millwood, New York: Kraus, 1973).

Vertical Files, Dolph Briscoe Center for American History, University of Texas at Austin (Quanah Parker).

Robert H. Williams, "The Case for Peta Nocona," T*exana* 10 (1972).

Robert H. Williams - *Handbook of Texas Online*, s.v.

José Maria (Andarko)

Grant Foreman, *The Last Trek of the Indians* (University of Chicago Press, 1946; 2d ed., New York: Russell, 1972).

William W. Newcomb, *The Indians of Texas* (Austin: University of Texas Press, 1961).

Rupert N. Richardson, *The Comanche Barrier to South Plains Settlement* (Glendale, California: Clark, 1933; rpt., Millwood, New York: Kraus, 1973).

Brian C. Hosmer- Handbook of Texas Online, s.v.,"

Bowl

Mary Whatley Clarke, *Chief Bowles and the Texas Cherokees* (Norman: University of Oklahoma Press, 1971).

Dianna Everett, *The Texas Cherokees: A People between Two Fires, 1819–1840* (Norman: University of Oklahoma Press, 1990).

Dorman Winfrey, "Chief Bowles and the Texas Cherokees," *Chronicles of Oklahoma* 32 (Spring 1954). E. W. Winkler, "The Cherokee Indians in Texas,"

Quarterly of the Texas State Historical Association 7 (October 1903).

Albert Woldert, "The Last of the Cherokees in Texas and the Life and Death of Chief Bowles,"

Chronicles of Oklahoma 1 (June 1923). ; *Dianna Everett - Handbook of Texas Online*, s.v.

Texas Jack (John Burwell Omonundro, Jr)

J. C. Dykes, "Dime Novel Texas; or the Sub-Literature of the Lone Star State," *Southwestern Historical Quarterly* 49 (January 1946).

Joel Chandler Harris, *On the Wing of Occasions* (New York Doubleday, Page, 1900).

Prentiss Ingraham, T*exas Jack, The Mustang King* (New York: Ivers, 1899). Herschel C. Logan,

Buckskin and Satin: The Life of Texas Jack and His Wife (Harrisburg, Pennsylvania: Stackpole, 1954). *Southwestern Historical Quarterly*, Texas Collection, October 1958.

Windham Thomas Wyndham-Quinn, Earl of Dunraven, *Hunting in the Yellowstone* (New York: Macmillan, 1925). *Edgar P. Sneed - Handbook of Texas Online*, s.v.

Josefa Rodriguez (Chipita)

Francis Edward Abernethy, ed., *Legendary Ladies of Texas*, Publications of the Texas Folklore Society 43 (Dallas: E-Heart, 1981).

Rachel Bluntzer Hebert,*Shadows on the Nueces* (Atlanta: Banner, 1942).

Ruel McDaniel, "The Day They Hanged Chipita," *Texas Parade*, September 1962. *San Patricio County in 1976: A Bicentennial Perspective* (Sinton, Texas: Sinton Bicentennial Celebrations, 1976). Vernon Smylie, *A Noose for Chipita* (Corpus Christi: Texas News Syndicate Press, 1970).

Ruthe Winegarten, *Finder's Guide to the 'Texas Women: A Celebration of History' Exhibit Archives* (Denton: Texas Woman's University Library, 1984). *Marylyn Underwood - Handbook of Texas Online*, s.v.

John Wesley Hardin

Source; Leon Metz

Sam Bass

Source; Leon Metz

Diarist Sanchez with the Terán inspection expedition 1828

Presidio La Bahia, Goliad, Texas
Hispanic Texian Patriots *in the Struggle for Independence*
Presidio La Bahia, Goliad, Texas
Public Opinion in Texas Preceding the Revolution
By Eugene Barker (From Annual Report of the American Historical Association for the Year 1911)
Publications" of So. Hist. Assn., VIII, 7-18, Quarterly of Tex. State Hist. Assn., XIII, 271
"Presidio La Bahia, Goliad, Texas - *Municipality of Columbia, Barker – 1835 -*
The Texas Republican, July 4, 1835; "Publications 11 of So. Hist. Assn., VIT, 86): Lavaca and Mina on July 4 indorsed these Columbia resolutions
(The Texas Republican, July 18, 1835; "Publications" of So. Hist. Assn., VII, 88-90, VIII, 2-6). Municipality of Gonzales,July 7, "Colonial Archives (MS.), Texas State Library.]
The Texas Republican, July 18, 1835; "Publications" of the So. Hist. Assn., Vill, 18 Tenorio to Ugartechea, July 15, 1835. Bexar Archives]
Publications" of the So. Hist. Assn., Vill, 18 Tenorio to Ugartechea, July 15, 1835. Bexar Archives
(Travis to Ugartechea, July 31, 1 &15. Mexican Archives, Archivo de Guerra y Marina, Frac. 1, Op. Mil. Texas, 1835, Legajo 1, Exp. 9)
The Texas Republican, Aug. 22, 1835; Publications of So. Hist. Assn., VIII, 107-109
The Texas Republican, Aug. 22, 1835; Publications of So. Hist. Assn., VIII, 107-109); and a meeting of the district of the Navidad and Lavaca on July 17 (Brown, History of Texas, 1, 299).
The Texas Republican, September 19, 1835; Publications of So. Hist. Assn., VIII, 110-116]
Barrett and Gritten to Cos .,Aug. 9, 1835, Bexar Archives; Publications of So. Hist. Assn., VIII, 343-344]

Yoakum, "History of Texas," 1, 336; Brown, "History of Texas," 1, 290; Navarro to Political Chief of Nacogdoches, June 4, 1835, Bexar Archives; Lynch to Brown, July 20,1835, Colonial Archives, Tex. State Library

The Texas Republican, July 4 and 18,1835, for Columbia and Lavaca; Colonial Archives, Tex. State Library,Ugartechea to Tenorio, July 31, and to Cos, August 1, 1835, Bexar Archives

Presidio La Bahia, Goliad, Texas

Correspondence related to Public Opinion in Texas 1835

Miller to Public 21 Jun 1835- Cheftaincy of' the Department of Brazos Fellow Citizens

Presidio La Bahia, Goliad, Texas - town of San Felipe de Austin this 21st day of June 1835 J B Miller

Columbia Resolutions 23 Jun 1835

Presidio La Bahia, Goliad, Texas

H. Austin to James Perry 24 Jun 1835 Columbia 24 June 1835 Mr J F Perry – Source; *Presidio La Bahia, Goliad, Texas*

Rueg to Political Chief 28 Jul 1835

Source; *Presidio La Bahia, Goliad, Texas*

Centralista Col. Pedro Delgado describing the Battle of San Jacinto - Source; *Presidio La Bahia, Goliad, Texas*

First Texas Declaration of Independence

Source; *Presidio La Bahia, Goliad, Texas*

The Unanimous Declaration of Independence made by the Delegates of the People of Texas in General Convention at the town of Washington on the 2nd day of March 1836. Presidio La Bahia, Goliad, Texas

LETTER OF GRIEVANCE ADDRESSED TO STEPHEN H. DARDEN, STATE COMPTROLLER, FROM HISPANIC TEXIAN HEROES, DATED JANUARY 12, 1875 Presidio La Bahia, Goliad, Texas

The Republic if The Rio Grande in 1840
Presidio La Bahia, Goliad, Texas

Mier and Somervell Expeditions In 1842
Presidio La Bahia, Goliad, Texas

Presidio La Bahia - GOLIAD, TEXAS
Presidio La Bahia, Goliad, Texas

Presidio La Bahia Returned To The Church In 1853 Source; *Presidio La Bahia, Goliad, Texas*